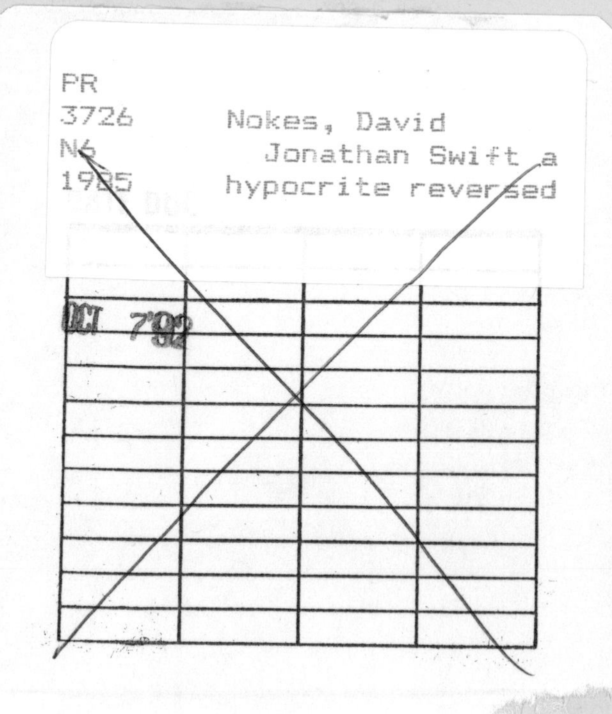

JONATHAN SWIFT,
A HYPOCRITE REVERSED

DAVID NOKES

Jonathan Swift, A Hypocrite Reversed

A Critical Biography

He had, early in life, imbibed such a strong hatred to hypocrisy, that he fell into the opposite extreme; and no mortal ever took more pains to display his good qualities, and appear in the best light to the world, than he did to conceal his, or even to put on the semblance of their contraries . . .

Lord Bolingbroke, who knew him well, in two words summed up his character in this respect, by saying that Swift was a hypocrite reversed.

Sheridan, *Life of Jonathan Swift* (1784)

OXFORD UNIVERSITY PRESS
1985

Oxford University Press, Walton Street, Oxford OX2 6DP
London New York Toronto
Delhi Bombay Calcutta Madras Karachi
Kuala Lumpur Singapore Hong Kong Tokyo
Nairobi Dar-es-Salaam Cape Town
Melbourne Auckland
and associated companies in
Beirut Berlin Ibadan Mexico City Nicosia

Oxford is a trade mark of Oxford University Press

Published in the United States
by Oxford University Press, New York

British Library Cataloguing in Publication Data
Nokes, David
Jonathan Swift, a hypocrite reversed: a critical biography.
1. Swift, Jonathan–Biography 2. Authors, Irish–18th century–Biography
I. Title
828'.509 PR3726
ISBN 0-19-812834-7

Library of Congress Cataloging in Publication Data
Nokes, David.
Jonathan Swift, a hypocrite reversed.
Includes index.
1. Swift, Jonathan, 1667-1745. 2. Authors, Irish–18th century–Biography. I. Title.
PR3726.N6 1985 828'.509 [B] 85-5657
ISBN 0-19-812834-7

Set by Cotswold Typesetting Ltd, Cheltenham
Printed in Great Britain by The Alden Press, Oxford

For My Parents

When I am reading a book, whether wise or silly, it seemeth to me to be alive and talking to me.

Swift, *Thoughts on Various Subjects*

PREFACE

A TRUE critic, we are told in *A Tale of a Tub,* is 'like a dog at a feast . . . apt to snarl most when there are the fewest bones'. Swift has been most obliging to his critics and biographers, leaving a goodly scatter of bones buried just beneath the surface of his works. The more enthusiastic burrowers have dragged whole skeletons into the light of day: his uncertain parentage; his secret marriage to Stella; his enigmatic relationship with Vanessa; his alleged misanthropy and madness. Generations of critics have picked over the bones without yet gnawing them clean.

Few authors' reputations have fluctuated as violently as Swift's. From the beginning, commentators arranged themselves in rival camps. For some he was a champion of liberty, the hero of an oppressed nation; for others, a sadistic misanthrope, the mysterious tormentor of mankind in general and of two unfortunate young women in particular. Lord Orrery's *Remarks,* published seven years after Swift's death, presented the picture of a man whose disappointments 'rendered him splenetic, and angry with the world'. Patrick Delany, Deane Swift, and Thomas Sheridan rallied to Swift's defence, emphasizing his sense of fun, his private acts of charity and his public commitment to principles of liberty and justice. Yet to most nineteenth-century critics Swift appeared as quite simply a monster. Writing in the *Edinburgh Review* (1816) Francis Jeffrey observed that Swift used 'at one and the same moment, his sword and his poisoned dagger–his hands and his teeth and his envenomed breath'. But it was Thackeray who most famously expressed the full force of Victorian horror at Swift's life and works.

> As is the case with madmen, certain subjects provoke him, and awaken his fits of wrath. Marriage is one of these; in a hundred passages in his writings he rages against it; rages against children . . . What had this man done? What secret remorse was rankling at his heart? What fever was boiling in him that he should see all the world bloodshot? (*The English Humourists of the Eighteenth Century,* p. 33.)

Contemporary criticism has removed the bones from the charnel house to the laboratory. The modern true critic is a specialist, trained to analyse one aspect, one phase, even one work from among Swift's total output. Over the past forty years dozens of books and hundreds of articles have appeared examining different elements in Swift's life and works and greatly expanding our understanding of both. For years, biographies of Swift were bedevilled by legends of his madness and by romantic mysteries surrounding his relationships with Stella and Vanessa. By a familiar process of transposition the more grotesque and misanthropic images from Swift's works were used as metaphors for his life; and macabre anecdotes accumulated around the morose and lonely Dean in his latter years which lost nothing in the telling and re-telling. It is only in the present century that any serious attempts have been made to disentangle facts from fiction and to study Swift's works in the context of the intellectual and political developments of his age. We now know a considerable amount about his career in the Church, about his contribution to politics in both England and Ireland, about his attitude to the 'new science', about his literary mentors and models. Freudian critics have psychoanalysed Swift and, most recently a group of persuasive studies have insisted on Swift's significance as a poet.

It is the aim of this biography to offer a new, comprehensive view of the man and his works, based on the accumulated wealth of evidence which is now available. In the Preface to his monumental study *Swift, the Man, his Works and the Age,* Irvin Ehrenpreis declared, 'I have been less concerned to add than to eliminate fables'. In consequence he excluded a number of traditional anecdotes from his biography and avoided all discussion of such doubtful incidents in Swift's life as the possible secret marriage to Stella. At the time when Ehrenpreis began his biography in the late 1950s this was both a valuable and a courageous decision, which enabled students of Swift to see him for the first time free of the dubious accretions of legend. It may seem somewhat perverse, therefore, that I have chosen to reinstate some of these anecdotes in the present biography. I have done so not because I necessarily believe them to be true–in most cases there can be no way of deciding that; nor simply to lend colour to a life which has suffered too much from overcolourful interpretations. They have been restored because they have played such an important part in the transmission of Swift's reputation through the ages that to have omitted them would have been to ignore an important element in the enigmatic record of a man who deliberately cultivated false images of himself.

Reading through the ever-lengthening bibliography of Swift studies, I have become aware of two main tendencies in critical approaches to Swift. Specialist works often produce partial or lop-sided views of his writings by giving undue prominence to a particular theme or genre, while the broad sweeps of the generalizing critic run the risk of missing the teasing particularity of Swift's ironies altogether. In this critical biography I have attempted to walk a middle path by offering some fresh points of interpretation of Swift's life and works without losing the general reader in academic groves that have sprouted into thickets. Swift's works are full of warnings for those who attempt such a middle course. Yet, unless I have completely misread his irony, he also insists that it is the only way to avoid shipwreck on the opposing rocks of pedantry or populism. Throughout his life Swift adhered to an ideal of conservative humanism which saw specialization itself as a first dangerous step towards that distorted simplification of complex human phenomena which characterized the views of all factions and fanatics. I have therefore attempted to present a portrait of the whole man in his multifarious roles as satirist, politician, churchman and friend and, in particular, have sought to re-establish the balance between his public and private lives which has been missing from some other recent biographies.

I have included discussions of all Swift's major works and of many of his minor ones. To have examined all his writings would have meant expanding what is already a substantial volume to Brobdingnagian proportions with little compensating gain in our understanding of his significance. To those who may have anticipated a fresh analysis of *Pethox the Great,* or a new insight into *The Dying Words of Ebenezor Elliston,* I apologize. I leave those as bones for others to chew upon. I have modernized spellings and regularized punctuation in accordance with normal modern practice.

I should like to acknowledge with gratitude the help, inspiration, and guidance that I have received from the following people in the preparation of this book and in the years of study which preceded it: Barbara Ansell, Richard Axton, Gerald Baker, Tom Deveson, Alan Downie, Howard Erskine-Hill, Hanif Kureishi, Frank Miles, Mike Neve, Tristram Powell, Clive Probyn, David Profumo, John Rathmell, Claude Rawson, Christine Rees, Pat Rogers.

David Nokes
London, 1984

CONTENTS

ILLUSTRATIONS

ABBREVIATIONS

The place of publication is London, unless otherwise indicated.

Corr.	*The Correspondence of Jonathan Swift,* ed. Harold Williams, 5 vols., Oxford, 1963-5.
Deane Swift	*An Essay upon the Life, Writings and Character of Dr Jonathan Swift,* by Deane Swift, 1755.
Delany	*Observations upon Lord Orrery's Remarks on the Life and Writings of Dr Jonathan Swift,* by Patrick Delany, 1754.
Ehrenpreis	*Swift, the Man, his Works, and the Age* (3 vols.), by Irvin Ehrenpreis; i. *Mr Swift and his Contemporaries,* 1962; ii. *Dr Swift,* 1967; iii. *Dean Swift,* 1983.
Elias	*Swift at Moor Park,* by A. C. Elias, Philadelphia, 1982.
Forster	The Forster Collection of the Victoria and Albert Museum, including Swift's manuscript account books.
Johnson	*Lives of the English Poets,* by Samuel Johnson, ed. G. B. Hill, 3 vols., Oxford, 1905.
Journal	*Journal to Stella,* ed. Harold Williams, 2 vols., Oxford, 1948.
Landa	*Swift and the Church of Ireland,* by Louis A. Landa, Oxford, 1954.
Longe	*Martha, Lady Giffard, Her Life and Correspondence,* by Julia Longe, 1911.
Murry	*Jonathan Swift: A Critical Biography,* by John Middleton Murry, 1954.
Poems	*The Poems of Jonathan Swift,* ed. Harold Williams, 2nd edn., 3 vols., Oxford, 1958.
PW	*The Prose Works of Jonathan Swift,* ed. Herbert Davis *et al.,* 14 vols., Oxford, 1939-68.
Sherburn	*The Correspondence of Alexander Pope,* ed. George Sherburn, 5 vols., Oxford, 1956.

Sheridan	*The Life of the Reverend Dr Jonathan Swift,* by Thomas Sheridan, 1784.
Spence	*Observations, Anecdotes and Characters of Books and Men,* by Joseph Spence, ed. J. M. Osborn, 2 vols., Oxford, 1966.
Thackeray	*The English Humourists of the Eighteenth Century,* by W. M. Thackeray, 1853.
Woodbridge	*Sir William Temple,* by H. E. Woodbridge, New York, 1940.

CHRONOLOGICAL TABLE

1667	Swift was born in Dublin on 30 November.
1669	Swift was taken by his nurse to Whitehaven in Cumberland where he remained for almost three years.
1673	Began his education at Kilkenny School.
1682	Entered Trinity College, Dublin on 24 April.
1686	Graduated BA *speciali gratia* in February.
1689	Swift left Ireland as a consequence of the troubles following the fall of James II. He spent some time with his mother in Leicester before entering the household of Sir William Temple where he made the acquaintance of the eight-year-old Esther Johnson (Stella).
1690	Returned to Ireland for the benefit of his health.
1691	Returned to England where he joined Sir William Temple's household at Moor Park.
1692	Took his MA from Hart Hall, Oxford in July; began the composition of his early poems.
1694	Left Moor Park and returned to Ireland to take holy orders; ordained deacon on 25 October.
1695	Ordained priest and presented to the prebend of Kilroot in Ulster, where he became acquainted with Jane Waring.
1696	Proposed marriage to Jane Waring in April. Subsequently left Ulster and rejoined Temple at Moor Park.
1699	Death of Temple on 27 January; Swift returned to Dublin in August as chaplain to the Earl of Berkeley (Lord Justice of Ireland).
1700	Instituted Vicar of Laracor in March and installed as prebendary of St Patrick's Cathedral in September.
1701	Travelled to England with Lord Berkeley; published his first political pamphlet, the *Discourse of the Contests and Dissensions.* Stella and Rebecca Dingley left England to settle in Ireland.
1702	Awarded degree of Doctor of Divinity from Trinity College, Dublin in February.
1704	Publication of *A Tale of a Tub.*

1707 Returned to England as emissary of the Church of Ireland to seek remission of the First Fruits from Queen Anne.

1708 Met Addison and Steele; published *Bickerstaff Papers*; wrote tracts on Church and State, including *An Argument against Abolishing Christianity*.

1709 Failed to gain Whig support for his mission and returned to Ireland in June.

1710 In September Swift returned to England, still seeking remission of the First Fruits; began his *Journal to Stella* and his visits to Hester Vanhomrigh (Vanessa). In October he met Harley, the head of the new Tory administration, who agreed to support his petition. In return Swift began writing *The Examiner* on Harley's behalf.

1711 Swift wrote several poems and tracts in support of the Tory ministry, including *The Conduct of the Allies* (November). As his friendship with ministers grew, so his acquaintance with former Whig friends like Addison and Steele cooled.

1712 Continuing his propaganda work, he published *Remarks on the Barrier Treaty* (February); new friendships with Pope, Gay, and Arbuthnot developed.

1713 Installed as Dean of St Patrick's Cathedral, Dublin in June; returned to London in September.

1714 Published *The Public Spirit of the Whigs* (February) for which he was threatened with prosecution. Attended meetings of the Scriblerus Club in the spring before retiring to Letcombe Bassett in June. After the death of Queen Anne on 1 August, he returned to Dublin to take up his duties as Dean. Vanessa soon followed him to Ireland.

1715 Swift's letters intercepted on suspicion of Jacobite sympathies.

1716 Swift's possible secret marriage to Stella.

1718 Swift began friendships with Thomas Sheridan and Patrick Delany.

1720 Published *A Proposal for the Universal Use of Irish Manufacture,* his first political pamphlet since 1714.

1723 Death of Hester Vanhomrigh.

1724	Published *The Drapier's Letters*; a reward offered for the discovery of the Drapier.
1725	Wood's halfpence defeated and Swift celebrated as an Irish patriot.
1726	Swift returned to England in March and discussed Irish problems with Walpole. Meanwhile his poem *Cadenus and Vanessa* was published and Stella became very ill. He returned to Ireland in August, having arranged for the publication of *Gulliver's Travels* in London, in October.
1727	Swift's last visit to England (April to September); death of George II, but Tory hopes of power again frustrated.
1728	Death of Stella on 28 January; Swift visited Market Hill and began writing *The Intelligencer* with Sheridan.
1729	Returned to Market Hill; published *A Modest Proposal* in October.
1730	Last visit to Market Hill; published *A Libel on D-- D--*.
1731	Wrote several poems, including the scatalogical verses and *Verses on the Death of Dr Swift*.
1733	Published *An Epistle to a Lady* and *On Poetry: A Rhapsody*.
1736	Wrote *The Legion Club*, an attack on the Irish House of Commons.
1738	Published *Polite Conversation*.
1742	A commission of lunacy found Swift to be 'of unsound mind and memory'.
1745	Death of Swift on 19 October.

Part One
THE CONJURED SPIRIT

I

Lo here I sit at holy head
With muddy ale and mouldy bread . . .
I never was in haste before
To reach that slavish hateful shore:
Before, I always found the wind
To me was most malicious kind,
But now the danger of a friend
On whom my fears and hopes depend,
Absent from whom all climes are cursed,
With whom I'm happy in the worst,
With rage impatient makes me wait
A passage to the land I hate.[1]

ON SUNDAY 24 September 1727 Jonathan Swift, accompanied by his servant Wat, arrived at Mrs Welch's inn at Holyhead. The wind had changed the previous night and was now unfavourable for crossing to Ireland. Only the day before a number of people had sailed on the last of the wind. Had not Swift's horse cast a shoe on the rocky ways of Llangueveny he might have been in time to join them, but to spare the creature he completed his land journey on foot. As a result he found not only an absence of wind but also of wine. The departing travellers had drunk the cellar dry, save for some stale beer, 'the worst ale in the world'.

Swift was now within three months of his sixtieth birthday, a famous, even a notorious figure–but not at Holyhead. He was still physically robust and not much altered in appearance from the portrait of him done by Jervas fifteen years earlier. That shows a determined face, heavy jowls and piercing eyes which Pope described as being as azure as the heavens. He was a witty, charming man who could be amusing or tyrannical according to mood, but whose pleasure in social life had been greatly diminished by increasingly severe bouts of deafness. During such

[1] *Poems,* ii. 420.

attacks even the company of close friends was irksome to him. He had suffered badly from deafness while in London that summer, which had hastened his decision to leave. The previous eighteen months had brought him considerable triumphs. His *Drapier's Letters* had inflicted an embarrassing defeat on the English government, and *Gulliver's Travels* had met with immediate and universal acclaim. Yet as he left England for what he knew was the last time, his pleasure at these successes was overshadowed by the knowledge that his dearest friend, Esther Johnson–Stella–was dangerously ill in Dublin. With every post from Ireland he expected to hear the worst, and could scarcely bear to return while her condition remained in doubt.

> I am determined not to go to Ireland to find her just dead or dying–nothing but extremity could make me so familiar with those terrible words applied to such a dear friend.[2]

Throughout his life Swift sought to distance himself from emotional demands. Now, as Stella was dying, he sent instructions that she should be removed from her lodgings in the Deanery. He could not endure even so distant a contact with her suffering as that she should die under his roof.

While becalmed in this rocky corner of Anglesey Swift kept a journal in which he acted out his isolation like a shipwrecked mariner. From infancy he had divided his life between England and Ireland without ever feeling truly at home in either country. He relished the irony of his new-found reputation as an Irish patriot leader yet made no secret of his contempt for the Irish themselves. 'I reckon no man thoroughly miserable', he declared, 'unless he be condemned to live in Ireland.' Significantly, he arranged that his body should be sent to Holyhead for burial. Though condemned to Ireland for the remainder of his life, he was determined to spend eternity if not in English, at least not in Irish soil. He chose as a final resting place this Welsh haven, midway between the extremes of London and Dublin. In this, as in so many things, he was disappointed, and Swift's body lies in St Patrick's Cathedral, Dublin.

Holyhead was no conventional haven, and Swift's *Holyhead Journal* is a catalogue of complaints about uneatable food, unmannerly servants, unwashed clothes, and unventilated

[2] *Corr.* iii. 141.

rooms. Yet there is a tone of self-mockery behind the censure and a feeling of psychic release. 'I come from being used like an Emperor, to be used worse than a dog at Holyhead.' He found something salutary about affronts applied openly and universally, without respect for reputations. 'By my conscience, I believe Caesar would be the same without his army at his back.' He concluded his litany of complaints on a very different note.

Yet here I could live with two or three friends, in a warm house, and good wine–much better than being a slave in Ireland.[3]

There is a deliberate comic exaggeration in Swift's deadpan self-portrait as the victim of idle servants, rascally captains, and ignorant Welshmen. The blunders of his servant Wat are like a running joke throughout the journal. One day Swift climbed the mountain of Holy Head to look across at the Wicklow hills but he was caught in a rainstorm and forced to shelter in a peasant cabin.

There was only an old Welshwoman sifting flour, who understood no English, and a boy who fell a roaring for fear of me. Wat (otherwise called unfortunate Jack) ran home for my coat, but stayed so long that I came home in worse rain without him, and he was so lucky to miss me, but took care to carry the key of my room where a fire was ready for me. So I cooled my heels in the parlour till he came.*

The *Holyhead Journal* places Swift in the tradition of stoic comedians and his grumbling is a kind of surface noise, designed to divert more unpleasant thoughts. Underlying the journal with its domestic anecdotes, its jottings of rhymes, riddles, and dreams, is the memory of the daily *Journal to Stella* that he wrote between 1710 and 1713. Swift strives hard to recapture the intimate playful idiom of that earlier work.

Is this strange stuff? Why, what would you have me do? I have writ verses and put down hints till I am weary . . . so adieu till I see you at the Deanery.

Yet the irony is that Stella, to whom he addresses these musings, was, he believed, probably dead; or, if alive, certainly not at the Deanery from which he had ordered her removal. It is as if, by cataloguing his penances in the sanctuary of this Welsh no man's land, Swift sought absolution from the spirit of the one person

[3] 'Holyhead Journal', *PW* v. 201-8.

who had been able to see through his customary guise of misanthropy to the humanity beneath.

In fact Stella did not die for another four months. Even then Swift could not bring himself to attend her funeral, but moved to another apartment 'that I may not see the light in the church, which is just over against the window of my bed-chamber'.[4] Though long anticipated, her death was still a shock to him and led to a period of anxious self-examination. As part of that process he began an autobiography, but did not get very far.[5] Ten quarto sheets preserve a few anecdotes of his own early life and of his ancestors but come to an abrupt halt in 1704. A later hand has commented in the margin of the manuscript: 'either indolence, sickness, old age or carelessness hindered the Dean from proceeding further in these *Memoirs*.' These may have been the reasons. Or it may be that after the initial autobiographical urge declined, deeper and more characteristic instincts took over: a despair of being understood; a disinclination to explain himself to the vulgar Prince Posterity; a desire to leave mysteries still mysterious. For the autobiography introduces as many enigmas as it solves. Not only is it brief; it is also demonstrably inaccurate at times and teasingly imprecise at others. It is written in the third person, maintaining a typical pose of impersonality. Swift wishes to gain a reader's sympathy, but cannot ask for it directly and so writes of himself as of someone else. The evident unreliability of the autobiography has led subsequent biographers to question not only Swift's facts but also his motives in supplying them. Here are a few deceptively simple statements.

> J.S.D.D., and D of St P was the only son of Jonathan Swift, who was the seventh or eighth son of (that eminent person) Mr Thomas Swift. . . . He was born in Dublin on St Andrews day. . . .

Not one of these facts can be proved and each has been contested at some time, occasionally by Swift himself. He was a posthumous child born, probably, some seven months after Jonathan Swift senior's death, but just in time, as he told Laetitia

[4] 'On the Death of Mrs Johnson', *PW* v. 229.

[5] 'A Fragment of Autobiography' *PW* v. 187-95. Ehrenpreis (iii, 879) suggests that the fragment of autobiography may have been composed in 1738. While there is some circumstantial evidence to recommend this view, I tend to agree with Davis's opinion–based on the authority of Deane Swift–that the fragment was written between 1727 and 1729. (*PW* v, p. xxii.)

Pilkington 'to save his mother's credit'.[6] Several early biographers wondered whether he was in fact the son of his father and even suggested that Sir William Temple, who was later Swift's patron, may have been Swift's real father. The place of his birth has also been debated. In later years, when expressing his vehement hatred of all things Irish, Swift would sometimes claim to be a native Englishman, though usually he admitted to the misfortune of having been 'dropped' in Dublin. Even the year of his birth has been questioned and ingenious arguments in support of 1669 or 1670 have been offered in place of the generally accepted 1667.

The uncertainty about such matters is taken by some biographers as implicit confirmation that a certain judicious confusion was part of Abigail Swift's method of saving her credit. For Herbert Davis the inaccuracies of the autobiography are the result of simple carelessness. 'He seems to have written entirely from memory and has not even taken the trouble to verify all his dates'.[7] While for Denis Johnston they are part of a deliberate cover-up, as Swift prepares an 'account of what he wished to have said of himself'.[8] For most readers of Swift these are academic quibbles, yet it seems ironically appropriate that a writer whose works specialize in the creation of false identities should, from the moment of conception, have no settled identity of his own. By accident or design, the 'real' Jonathan Swift of the autobiographical fragment is as much a creature of the imagination as the memory.

II

'THE FAMILY of the Swifts was ancient in Yorkshire' is the opening statement of the autobiography.[9] Swift makes two immediate and important assertions: that the family is of some antiquity, and that it is *English*. The first ancestor mentioned in the synoptical summary of this ancient family is one 'Cavaliero Swift, a man of wit and humour' who 'was made an Irish peer by King James or K. Charles I . . . but never was in that kingdom'. How

[6] *Memoirs of Mrs Laetitia Pilkington,* 3 vols., 1758, i. 68.
[7] *PW* v, p. xxii.
[8] Denis Johnston, *In Search of Swift,* 1959, p. 14.
[9] *PW* v. 187.

Swift envied him that. The rest of the family tree is sketched in briefly, although for two of his ancestors he does have more to say. The first was the 'heiress of Philpot' who married William Swift in the reign of James I, 'a capricious, ill natured and passionate woman' who 'absolutely disinherited her only son, Thomas, for no greater crime than that of robbing an orchard when he was a boy'. That disinherited son was Swift's grandfather and first hero, Thomas Swift, vicar of Goodrich, whose fierce loyalty to Charles I caused him to be 'plundered by the roundheads six and thirty times'. Swift proudly relates how once his grandfather sewed all his money up into a quilted waistcoat, rode to a town that was held for the king, and presented the waistcoat to the governor,

> who, ordering it to be unripped found it lined with three hundred broad pieces of gold, which as it proved a seasonable relief must be allowed an extraordinary supply from a private clergyman with ten children of a small estate, so often plundered and soon after turned out of his livings in the church.[10]

Three hundred broad pieces of gold must be reckoned very extraordinary indeed for a man who had been completely disinherited, whose revenues brought in only £100 per annum, and who had been plundered thirty-six times. Swift was always liable to exaggerate when recalling the sufferings of this revered ancestor. Ten years later he informed Pope that his grandfather had been 'persecuted and plundered two and fifty times by the barbarity of Cromwell's hellish crew'.[11]

Thomas Swift married Elizabeth Dryden of Northampton, thereby establishing the link with Dryden of which Swift at first boasted, and subsequently complained. They had, according to Swift 'ten sons and three or four daughters' (it seems churlish to correct such an imperiously casual estimate, but the actual figures are six sons and five daughters). It was this generation of Swifts who made the move to Ireland, 'driven thither by their sufferings, and by the death of their father'. Note the word *driven.* It was Swift's settled conviction that no one would ever choose Ireland who had any reasonable alternative.

At this time the pickings in Ireland, after the Cromwellian Settlement, were very good. The Swift sons, Godwin, William,

[10] Ibid., p. 189. [11] *Corr.* v. 150.

Adam, and Jonathan, were part of the army of occupation of landless gentry who took advantage of that act of arbitrary power by which the whole territory of Ireland was declared confiscated.[12] After the Restoration, attempts were made to regularize and legitimize the chaotic state of Irish property rights. Very roughly, the Act of Settlement (1662) divided the land so that a third reverted to the native Catholic landlords, a third went to the older established Protestant families, and a third was granted to the newcomers, including the Swifts. In those days, Deane Swift later observed, 'Ireland was very moderately supplied with lawyers.'[13] Nothing makes more work for lawyers than a contentious system of land tenure, and the Swifts soon set about supplying this shortage. Of the brothers, Godwin was the most successful, Jonathan the least. Consequently after the death of Jonathan in early 1667, it was upon the charity of uncle Godwin that his yet unborn son would have to rely. Swift's lifelong resentment of any kind of dependence dates from these earliest years. However, in Godwin's defence it must be said that Swift could not have been the easiest or most grateful of nephews to support. Godwin's four marriages had resulted in thirteen children of his own, all probably more agreeable than his sulky nephew, and he may be forgiven for having found the quality of his charity somewhat strained towards this cuckoo in the nest. In later years Swift remarked that his family were 'of all mortals what I despise and hate' and evidently a sense of neglect of his special qualities burned into him deeply as a child. He explained his poor performance at university as a result of the 'ill treatment of his nearest relations' which made him 'so discouraged and sunk in his spirits, that he too much neglected his academical studies'.[14] Swift's description of his uncle is grudging but shrewd. He was, he reports 'a little too dextrous in the subtle parts of the law'. It was a recipe for success and Godwin soon rose to be Attorney General to the County of Tipperary. It was he who brought the Swift family into contact with the Ormondes and the Temples, two of Ireland's most influential families who represented the two traditions of Anglo-Irish set-

[12] See O. W. Ferguson, *Jonathan Swift and Ireland,* University of Illinois Press, 1962.

[13] Deane Swift, p. 15.

[14] *PW* v. 192.

tlers. The Ormondes were well established grandees, who identified their own interests with those of the Irish; whereas the Temples, more recently arrived, saw themselves as colonial administrators of a dependent territory. Both families were to be of major importance throughout Swift's career.

The family of Swift's mother, Abigail Erick, derived its lineage, he claimed, from Erick the Forester, 'a great commander, who raised an army to oppose the invasion of William the Conqueror'. From these heroic origins it had gradually declined until the family was now 'in the condition of very private gentlemen'.[15] Swift was in no doubt that his parents' marriage was a mistake. This was not a matter of love or temperament–merely of money. His lifelong aversion to impetuous matches derived from the neglect he had suffered in infancy.

This marriage was on both sides very indiscreet, for his wife brought her husband little or no fortune, and his death happening so suddenly before he could make a sufficient establishment for his family: And his son (not then born) hath often been heard to say that he felt the consequences of that marriage not only through the whole course of his education, but during the greater part of his life.[16]

Disapproval of his parents' marriage carried with it an implied desire not to have been born. This was a characteristic of adolescence which Swift never lost and which, in fact, he later developed into a careful ritual, repeating upon his birthday these grim verses from the Book of Job (3:3-5).

Let the day perish wherein I was born, and the night in which it was said, there is a man child conceived. Let that day be darkness; let not God regard it from above, neither let the light shine upon it. Let darkness and the shadow of death stain it; let a cloud dwell upon it; let the blackness of the day terrify it.

In all probability, Jonathan Swift was born on 30 November 1667, at 7 Hoey's Court in the parish of St Werburgh's in Dublin. The house belonged to Godwin, but the infant Swift was not, initially at least, a burden to his uncle.

When he was a year old, an event happened to him that seems very unusual; for his nurse, who was a woman of Whitehaven, being under an absolute necessity of seeing one of her relations, who was then

[15] Ibid., p. 191. [16] Ibid., p. 192.

extremely sick, and from whom she expected a legacy; and being at the same time extremely fond of the infant, she stole him on shipboard unknown to his mother and uncle, and carried him with her to Whitehaven, where he continued for almost three years. For when the matter was discovered, his mother sent orders by all means not to hazard a second voyage, till he could be better able to bear it.[17]

To lose one parent, as Lady Bracknell observes, is a misfortune; to lose both seems a little like carelessness. There is a feeling of something akin to carelessness about that rather absent-minded remark 'when the matter was discovered' and about Swift's mother's apparent lack of concern. His 'loss' of both parents at this formative stage may perhaps have caused problems of identity that might help to account for his later predilection for ironic masks. Swift's relationship with his mother was the first of his several enigmatic relationships with women. The evidence suggests that she neglected him in an age when neglect of children was not the invariable rule that has sometimes been claimed. After his return to Ireland he was packed off to school at Kilkenny, seventy miles south-west of Dublin, while his mother appears to have joined the rest of her family in Leicester. Nevertheless, Swift always excepted his mother from the general contempt which he expressed towards his family. As a young man he wrote to her and visited her often, and at her death, wrote movingly of her virtues. It may have been Abigail Swift who instilled in him an instinctive association of love and distance. One senses that the infant Swift was anxious to earn the love of this remote mother, and was encouraged by his nurse to believe that good works might gain him the affection he craved. 'The nurse was so careful of him that before he returned he had learnt to spell, and by the time that he was three years old he could read any chapter in the bible.'

From the ages of six to fourteen Swift was a pupil at Kilkenny College, the foremost Anglican school in Ireland. School statutes prescribed a rigorous schedule of Greek, Latin, morality, and prayers. His cousin, Thomas Swift, was a contemporary there, and Congreve enrolled in Swift's final year. Tom Sheridan tells a story that shows Swift's developing character in childhood as moody, self-dramatizing, and subject to the extremes of exultation and despair.[18] One day he spent all the money that he had in

[17] Ibid., p. 192. [18] Sheridan, p. 468.

the world on a horse that was on its way to the slaughterhouse, for the momentary glory of riding his own horse through Kilkenny. Many years later Swift reflected on the unreliability of most childhood recollections.

I have observed from myself and others . . . that men are never more mistaken, than when they reflect upon past things, and from what they retain in their memory, compare them with the present. Because when we reflect on what is past, our memories lead us only to the pleasant side, but in present things our minds are chiefly taken up with reflecting on what we dislike in our condition. So I formerly used to envy my own happiness when I was a schoolboy, the delicious holidays, the Saturday afternoon, and the charming custards in a blind alley; I never considered the confinement ten hours a day to nouns and verbs, the terror of the rod, the bloody noses and broken shins.[19]

However this wise old saw about the unreliability of our memories is itself unreliable. Swift's memory is not stored in this conventional way, with nostalgic recollections of golden days. On the contrary he harbours the memories of snubs, slights, and disappointments, which he tells over to himself obsessively. His main recollection from his schooldays introduces a note that predominates in his presentation of his life.

I remember, when I was a little boy, I felt a great fish at the end of my line which I drew up almost on the ground. But it dropped in and the disappointment vexeth me to this very day and I believe it was the type of all my future disappointments.[20]

It seemed to Swift that the pattern of his life was already determined on those sunny Saturday afternoons in childhood.

Feelings of waste and neglect afflicted him strongly during adolescence, and as with Johnson at Oxford, the combination of poverty and pride did not make him an ideal student. Yet possibly Swift exaggerated his own insufficiencies. It is a harmless vice of elderly sages to look back in mock penitence at their youthful misdemeanours. So Swift reports that

he was stopped of his degree for dullness and insufficiency, and at last hardly admitted in a manner little to his credit, which is called in that college *speciali gratia.* And this discreditable mark, as I am told, stands upon record in their college registry.[21]

He blamed his failure on his relations, which was unfair. Godwin

[19] *Corr.* i. 109. [20] Ibid. iii. 329. [21] *PW* v. 192.

lost his fortune at this time in an ill-conceived scheme for an iron-works, but Swift's uncle William stepped in to support him through college. The truth was that Swift had little relish for the course of studies prescribed, and spent much of his time reading poetry and history, instead of getting his syllogisms by rote. Primarily a training college for the Anglican clergy of Ireland, Trinity was expanding both physically and intellectually. The provost during Swift's time, Narcissus Marsh, was noted for his piety and scholarship, though in later years Swift claimed that Marsh's disposition to study was 'the very same with that of an usurer to hoard up money, or of a vicious young fellow to a wench; nothing but avarice and evil concupiscence'.[22] The most enduring influence on Swift at Trinity was his friend and tutor St George Ashe. Like Marsh, Ashe belonged to the Dublin Philosophical Society, that city's centre for experimental natural philosophy and the new science. Ashe was a relentless experimenter. In the harsh winter of 1683 he was busy comparing the effects of freezing on eggs and urine. In the summer he turned his attention to a solar eclipse. It is one of the ironies of Swift's life that he, so often regarded as a critic of the new scientific methods, should have had as two of his closest friends men who were enthusiastically committed to them, St George Ashe and John Arbuthnot.

The daily routine at Trinity College was similar to that at Kilkenny. Undergraduates were required to attend three services daily, with a penny fine for each absence. Lectures were in Latin, and students were expected to produce a weekly commentary, also in Latin, on some moral or political subject. The central method of instruction was disputation: a formal system of debates, conducted according to strict syllogistic rules. Disputants would be confronted with some contentious metaphysical proposition, such as this:

An praeter esse reale actualis essentiae sit aliud esse necessarium quo res actualiter existat?

Whether besides the real being of actual being, there be any other thing necessary to cause a thing to be?[23]

[22] Ibid., p. 211.

[23] Suarez, *Metaphysicarum Disputationem,* Dis. xxxi, sec. 5. (1605). This particular query is considered by Martin in *The Memoirs of Martinus Scriblerus* by Pope, Swift, Arbuthnot *et al.* (ed. Kerby-Miller, 1966, p. 124).

Each disputant was required to furnish twenty-four arguments in support of the incorrect answer, and twelve arguments in support of the correct (or what was believed to be the correct) one. Though Swift despised the mechanical manipulation of formulas which this method encouraged, it proved a sound training for some of his own favourite satiric techniques. It was at Trinity that he first discovered how to use rules to confute rules, and reason to confound reason. He learnt more than mere techniques, however, for among the tropes and truisms which he was required to debate he encountered formulas which were to stay in his mind for several decades. Many of the themes for disputation were taken from Marsh's own textbook *Institutiones logicae* which provides these classic examples of syllogistic configurations.

homo est animal rationale
nullus equus est rationale
solum animal rationale est disciplinae capax.

Man is a rational animal. No horse is rational. Only rational animals are capable of discipline.[24]

As *Gulliver's Travels* demonstrates, Swift was clearly paying attention when these themes were debated.

In fact the mark sheets which survive show that Swift was a fairly average student. In the Easter term of 1685 he was graded as *bene* for Latin and Greek, *negligenter* for Latin theme, and only *male* for physics. His cousin Thomas was *mediocriter* for everything. Nor was Swift alone in receiving his degree *speciali gratia.*[25] Four of his thirty-seven contemporaries also left Trinity with 'specials'.

Yet he was by no means a model student. In March 1687 he was admonished for 'neglect of duties and frequenting the town'. On his birthday in 1688 he was found guilty of starting tumults in College and insulting the Junior Dean. In the politically sensitive atmosphere of that year the College authorities

[24] *Institutiones logicae* (1681), pp. 116, 185, sig. A5; pp. 175, 42. This point was first made by R. S. Crane in 'The Houyhnhnms, the Yahoos and the History of Ideas' in *Reason and the Imagination: Studies in the History of Ideas,* ed. J. A. Mazzeo, 1962. See also Ehrenpreis, i. 50.

[25] MS of students' marks for the terminal examinations at Trinity College Dublin for Easter 1685, and kept in TCD library. Quoted by Ehrenpreis, i, appendix E, pp. 279-83.

were nervous of all tumults, however innocent the cause, and may well have over-reacted to the high-spirited excesses of Swift and his companions. As a punishment he was suspended from his degree for one month, and required to beg the Junior Dean's pardon publicly, on bended knee. It was exactly the kind of public indignity that Swift most hated, and the humiliation of that punishment rankled with him long afterwards.

Swift later declared that he was 'ashamed to have been more obliged in a few weeks to strangers' (that is to Oxford University for his MA), than he ever was 'in seven years to Dublin College'.[26] This is a typical piece of anti-Irish exaggeration, and Swift's attempts to present himself as an Oxford man are among the more distressing minor snobberies of his latter years. At the very least he made some valuable and enduring friendships at Trinity, and received a sound education in the respective strengths and limitations of rhetoric and logic. It is true, however, that his main education took place elsewhere, not at Oxford, but at Moor Park. His friend Patrick Delany noted that Swift 'had often been heard to say that from the time of taking his degree he studied at least eight hours a day, one with another, for seven years'.[27] It was at the home of Sir William Temple, in Surrey, that Swift accomplished the transition from being a mediocre student to becoming the outstanding writer of his generation.

III

'THE TROUBLES then breaking out, he went to his mother who lived in Leicester.'[28] This latest eruption of troubles in Ireland deprived Swift of his MA. In April 1689 he would have completed the residence requirements for the degree, but in February the College authorities took fright at the political-upheavals and allowed those who wished to leave 'for their better security'.[29] William of Orange, at the head of an army of Dutchmen, English, and Scottish exiles had landed at Torbay on

[26] *Corr.* i. 12.
[27] Delany, p. 50.
[28] *PW* v. 193.
[29] College Register, 19 Feb. 1688/9, quoted in *Trinity College, Dublin,* by W. Macneile Dixon, 1902, p. 63.

5 November, and marched unresisted towards London, where he was crowned in February. James II, having fled from England, returned to Ireland in April in an attempt to rally the Catholics there to his cause. It was a short-lived campaign, and on 1 July 1690, at the battle of the Boyne, William won a victory for Protestantism that has reverberated through Irish history ever since.

Swift took this opportunity to visit his mother, to relax, and to flirt. Indeed his attentions towards Miss Betty Jones became so marked that his mother feared he might be sprung into an imprudent match. She need not have worried, for Swift's impulsiveness was all on the surface. Actually he had his emotions well under control. Miss Jones was the first of a number of girls with whom he flirted in the next two or three years. On another visit to Leicester his behaviour again caused gossip, which he laughed off with the affected swagger of a lady-killer.

> I could remember twenty women in my life to whom I have behaved myself just the same way, and I profess without any other design than that of entertaining myself when I am very idle, or when something goes amiss in my affairs.

Explaining this restlessness he could not resist adding a little mysterious allusion.

> A person of great honour in Ireland (who was pleased to stoop so low as to look into my mind) . . . used to tell me, that my mind was like a conjured spirit, that would do mischief if I would not give it employment.[30]

Whoever this person of great honour was (tradition ascribes the remark to Lord Berkeley), this offers a sharp insight into Swift's character. What more natural outlet for that conjured spirit than to break the hearts of provincial girls, and to confound the tales of local gossips? There was no danger of marriage. Only fools, especially learned fools, were deceived by a pretty face. Swift laughed at them with a callow worldliness.

> Among all the young gentlemen that I have known to have ruined theirselves by marrying (which I assure you is a great number) I have made this general rule, that they are either young, raw and ignorant scholars, who for want of knowing company, believe every silk petticoat includes

[30] *Corr.* i. 4.

an angel, or else they have been a sort of honest young men who perhaps are too literal in rather marrying than burning and entail misery on themselves and posterity by an over-acting modesty.[31]

His father had entailed just such misery on his posterity. Behind these conventional tropes lies a determination much stronger than the usual young man's boast of not getting caught. Throughout his career Swift was fascinated with exposing what exactly was contained within a silk petticoat, if not an angel. Whenever the idea of matrimony entered his head, 'a thousand household thoughts' drove it thence, he claimed.[32] 'Household thoughts' are hardly characteristic of passionate young men, and would seem to indicate the preoccupations of a lonely, defensive child with an overriding need for a secure and comfortable home. Experience, he wrote, had taught him 'not to think of marriage, till I settle my fortune in the world, which I am sure will not be in some years, and even then myself I am so hard to please that I suppose I shall put it off to the other world'.[33] This declaration has a curious blend of tones, from rakishness to a cold determination, and even fear. Swift was clearly attractive to women, and was testing his powers over them. It is interesting that so many of his flirtations took place in Leicester. He may well have liked to demonstrate his powers over women as a method of seeking the affection of the one woman, his mother, who had so hurt him by her neglect. 'I should not have behaved myself after the manner I did in Leicester', he wrote, 'if I had not valued my own entertainment beyond the obloquy of a parcel of very wretched fools.' Arrogance was the cover he assumed for his continuing insecurity.

It was shortly after this that one of Swift's relatives invoked the family tie with the Temples, and in the spring of 1689 Swift entered the service of Sir William Temple firstly at Sheen, and later at Moor Park in Surrey.

Temple, who was the same age as Swift's uncle Godwin, had retired from an active life in politics and diplomacy to enjoy the meditative pleasures of philosophy at Moor Park. Educated at Emmanuel College Cambridge, a centre of intellectual puritanism, he had found gentler pursuits, 'especially tennis', more agreeable to his temperament. His major achievement in

[31] Ibid., p. 4. [32] Ibid, p. 5. [33] Ibid., p. 3.

diplomacy had been to negotiate the Triple Alliance between England, Sweden, and the United Provinces in 1668 to secure 'the general interests of Christendom . . . against the power and attempts of France'.[34] Charles II was never committed to the treaty, however, and within two years abandoned it in favour of an alliance with Louis XIV. Temple's disillusionment with the machinations of power politics dated from this reverse. Charles kept him on as a dishonest broker at The Hague, sweetening his position with promises of peerages and pensions, but the hypocrisy of these years gave Temple 'a distaste to the thoughts of all public employments'.[35] In his essay 'Upon the Gardens of Epicurus' Temple wrote in praise of Horace's choice of the life of retirement and meditation, in preference to that of high public office, and at Moor Park he strove to follow that Horatian example. Yet although the man whom Swift came to serve in 1689 had retired, he was not forgotten. He maintained influential contacts in France, Holland, and Ireland, and frequently entertained members of King William's court, including the King himself, at his new estate in Surrey.

There is some continuing debate about Swift's exact position in Temple's household. Some years after Swift's death Jack Temple, Sir William's nephew, asserted that Swift had been a mere servant in the house, hired 'at the rate of 20L. a year and his board'. He went on to add that 'Sir William never favoured him with his conversation, because of his ill qualities, nor allowed him to sit down at table with him'.[36] This story enjoyed a considerable vogue among Victorian critics such as Macaulay and Thackeray, for whom Swift was a monster of misanthropy 'alone and gnashing in the darkness'.[37] They presented the picture of an embittered young man, smouldering with rage and resentment as he ate his meals at the servants' table. More recently there have been attempts to rehabilitate the story, although the fact is that Jack Temple was not himself present at Moor Park at the time, and had come by the story at second

[34] See H. E. Woodbridge, *Sir William Temple,* New York, 1940, p. 91.

[35] This remark by Lady Giffard is quoted in *Early Essays and Romances of Sir William Temple,* ed. G. C. Moore Smith, Oxford, 1930, p. 20.

[36] The anecdote is contained in a letter from Samuel Richardson to Lady Bradshaigh, 22 April 1752. *Correspondence of Samuel Richardson,* ed. Anna Laetitia Barbauld, 1804, 6 vols., vi. 173-4.

[37] Thackeray, p. 54.

hand.[38] Since Swift quarrelled violently with the Temple family after Sir William's death, little reliance can be placed on an anecdote from so prejudiced a source.

In fact there are sound reasons for believing that Swift's relationship with Temple was far stronger than that between master and servant. Swift entered the Temple household at a peculiarly sensitive time. Only months before, Temple's last surviving son John had committed suicide at the age of twenty-five. On his father's recommendation he had been appointed Secretary for War, and his first act had been to advise the King to free the Irish general Hamilton so that he could persuade the rebel Tyrconnel to surrender. Instead Hamilton joined the rebels. When he heard the news John Temple took a boat out on to the Thames, filled his pockets with stones, and jumped in by London bridge. He left this sad note.

> 'Tis not out of any dissatisfaction with my friends, from whom I have received infinitely more kindness and friendship than I deserve, I say it is not from any such reason that I do myself this violence, but having been long tired with the burden of this life, 'tis now become insupportable. From my father and mother I have had especially of late all the marks of tenderness in the world.[39]

The death was a terrible blow to Temple. 'It brought a cloud upon the remainder of his life,' observed his sister, Lady Giffard, adding that he 'was often heard to say how happy his life had been if it had ended at fifty'.[40] Swift was bound to remind Temple of his loss, and at first a certain awkward reserve was evident on both sides. But the needs of the fatherless son and the sonless father were too great for this initial awkwardness to last. During the next ten years Swift's relationship with Temple went through all the stages of a filial relationship beginning with hero-worship, passing through jealousy and struggles, and concluding with an assertion of independence. It was Temple who moulded Swift's ideas at this formative stage; Temple whose theories and experiences shaped his secretary's views of politics and learning. Swift's early works abound with borrowings from Temple, and although many are ironic, they demonstrate clearly

[38] See A. C. Elias, *Swift at Moor Park,* Philadelphia, 1982, pp. 128-55.

[39] Quoted in *Early Essays and Romances of Sir William Temple,* ed. G. C. Moore Smith, Oxford, 1930, p. 194.

[40] Ibid., p. xii.

that it was through analysing his patron's attitudes that Swift arrived at his own.

The family atmosphere of Moor Park was as important to Swift as the intellectual stimulation that he received there. The love of Sir William and Lady Temple was something of note in a Restoration period not over-sentimental about the joys of matrimony. Their six-year courtship, carried on against parental opposition, was the basis for a marriage which was able to withstand many tribulations, including the deaths of nine children in infancy. Temple's sister, Lady Giffard, also lived at Moor Park. Ten years younger than her brother, she had left home at twenty-three to marry an Irish landowner, who died within a fortnight of the wedding. She returned home, happily independent, and prepared to devote the rest of her life to assisting her brother's career. To Swift she appeared snobbish and stubborn, and her close friend at court, the Duchess of Somerset, was to become one of his greatest enemies. Lady Giffard's waiting-woman was Rebecca Dingley, a spinster, distantly related to the Temple family. The housekeeper was Bridget Johnson, who had three children of whom the eldest was a frail little girl called Esther. Esther Johnson was eight years old when Swift arrived although he, always notoriously imprecise about her age, insisted that she was only six. He also renamed her as Stella. There is even more uncertainty about Stella's status at Moor Park than about Swift's. Writing immediately after her death in 1728 Swift noted only that 'her father was a younger brother of a good family in Nottinghamshire, her mother of a lower degree; and indeed she had little to boast of her birth'.[41] Stella's father, Edward Johnson, may well have been Temple's steward, but it is clear that the little girl was singled out for special treatment from an early age. Swift records that she would often receive presents of gold pieces from 'her mother and other friends' and that 'she grew into such a spirit of thrift, that in about three years, they amounted to above two hundred pounds. She used to show them with boasting.'[42] Bridget Johnson, whose annual salary was £20, could not have been the source of such gifts, which must have come from Temple himself. In his will he left her a lease of lands in Co. Wicklow worth £1,000 and £500 made up of gifts during his

[41] *PW* v. 227. [42] Ibid., p. 231.

lifetime. This is in marked contrast to his bequest of a year and a half's wages to her mother and of nothing at all to her brother and sister. Clearly little Esther Johnson was something of a favourite in the household. She was taught not only French, as most young ladies were, but also physic and anatomy 'in which she was instructed by an eminent physician'. By Swift she was introduced to a wide range of intellectual disciplines generally thought too demanding for softly formed feminine minds.

It is significant that Swift excludes any mention of Temple in his account of her life, saying only that she lived 'generally in the country with a family where she contracted an intimate friendship with a lady of more advanced years'.[43] The favouritism with which Stella was treated inevitably caused gossip and it was often rumoured that she was Temple's natural daughter. This theory has irresistible attractions for those who believe that Swift was, or believed himself to be, Temple's natural son. This could explain the enigma of their relationship in a way that involves romantic heartache, artistic symmetry, a guilty secret, and a vibrant sub-text for a great many of Swift's works. As with most Keys to All Mythologies, there is a strong magnetic charge to the theory, but rather too many facts that it does not fit. Yet it is possible that Swift feared something of the sort. The suppression of all references to Temple in his account of her life may be taken as his reaction to such rumours. He was determined to take her out of Temple's influence, and secure her firmly within his own.

Whether or not she was Temple's child, it was natural enough that he should have virtually adopted her after the death of his last little daughter Diana in 1679. 'My heart is so broken,' he wrote, 'that I have done nothing since as I should do, and fear I never shall again.'[44] Stella was born two years later. After Swift's arrival the little girl who was Temple's favourite soon became his favourite too, and his consistent tendency to make her younger than she was reflects his desire to have known her and formed her from the beginning.

It was at this time that Swift suffered the first attacks of the giddiness and deafness that were to torment him for the rest of his life. They were the result, he believed of 'eating a hundred golden pippins at a time', and so severe that they 'almost

[43] Ibid., p. 227. [44] Woodbridge, p. 207.

brought him to his grave'.[45] The physicians whom he consulted recommended a return to Ireland in the unlikely hope that 'his native air might be of some use to recover his health'. It seems that Temple had only intended to offer him a temporary post until the Irish troubles were over, and he now supplied Swift with a letter of introduction to Sir Robert Southwell, the principal Secretary of State for Ireland. The terms of this testimonial are formal and do not suggest that Temple had yet detected any remarkable qualities in his young secretary.

> He has lived in my house, read to me, writ for me, and kept all accounts as far as my small occasions required. He has latin and greek, some French, writes a very good and current hand, is very honest and diligent . . .[46]

Temple asked Southwell to find a place for Swift either in his own service or as a Fellow of Trinity College, but it was a fruitless expedition. The College was still struggling to regain normality and no new Fellows were appointed till 1692. Even Swift's health grew worse in the atmosphere of his 'native air'. He returned quickly to England, and spent the autumn of 1691 in Leicester, where he flirted once more. In November he visited his cousin Thomas at Oxford, and made arrangements to take his own MA there the following year. By Christmas he was back at Moor Park.

IV

DURING SWIFT's second period of residence at Moor Park, Temple grew increasingly to recognize and rely upon his talents, entrusting to him many of the responsibilities which might naturally have devolved upon a son. 'I never read his writings,' wrote Swift at 25, 'but I prefer him to all others at present in England', adding that the 'likeness of humours' between himself and his patron made this preference for Temple 'all but a piece of self-love'.[47] At this time he was not content with admiring Temple, but strove to imitate him. Eight years later, when his attitude to Temple had changed rather, he still declared that

[45] *Corr.* iii, 232, and *PW* v. 193.
[46] *Corr.* i. 1-2.
[47] Ibid., p. 10.

Temple 'had advanced our English tongue to as great a perfection as it can well bear'.[48] In 1690 he embarked upon his literary career with a series of odes which are, both in form and content, a set of homages to Temple. The odes were a false start to his career, but they give a valuable insight into Swift's ambitions and insecurities at this formative period.

Like most young men embarking upon a literary career, he had a joyous vanity in his own work. 'I am Cowley to myself,' he told his cousin, though confessing that he was hardly an instinctive poet. 'It makes me mad to hear you talk of making a copy of verses next morning, which . . . are what I could not do under 2 or 3 days . . . I seldom write above 2 stanzas in a week . . . and when all's done, I alter them a hundred times, and yet I do not believe myself to be a laborious dry writer'.[49] Most of those who have set themselves to read these early odes have found them both laborious and dry. Even Deane Swift was unable to 'drudge through' more than fifty or sixty lines. The ode 'To the Athenian Society', a miracle of dispatch which was 'all rough drawn in a week' is not conspicuously better for this greater speed of composition though recently some ingenious arguments have been advanced to reveal unsuspected ironies beneath the dry conceits.[50]

The opening line of Swift's literary *œuvre* could hardly be more uncharacteristic of his subsequent works and reputation.

> Sure there's some wondrous Joy in doing Good.

This philanthropic exclamation begins Swift's ode 'To the King, on his Irish Expedition', a shameless piece of place-seeking in which Swift celebrates King William as the embodiment of goodness and greatness combined. The conceits and allusions of the poem, though not indecipherable are certainly opaque. The circuitous syntax and hyperbolic imagery strain after an inspired effect.[51] Each of these early odes is couched in the form of a dialogue between Swift and an inspirational muse such as 'fame', 'philosophy', or 'poetry', usually clouded in veils of mystic glory. King William is presented as a 'bold romantic knight'

[48] Preface to Temple's *Letters,* vol. i, 1699.

[49] *Corr.* i. 8.

[50] See Elias, pp. 79-80.

[51] *Poems,* i. 4-10.

rescuing the 'airy goddess' Fame from 'the giant's fort'. Against him, Louis XIV is shown to be not the Sun King, but merely a gilded meteor, that has already overshot its zenith.

> Giddy he grows, and down is hurled
> And as a mortal to his vile disease,
> Falls sick in the posteriors of the world.

This 'vile disease' we are informed in a footnote, was a *fistula in ano,* a royal ailment which so beautifully typified Swift's view of the vanity of human wishes that he returned to it in *A Tale of a Tub*; 'The same spirits which, in their superior progress would conquer a kingdom, descending upon the anus, conclude in a fistula'.[52] If this early appearance of the posteriors of the world strikes us as typically Swiftian, it is worth noting that the axiom illustrated here can be traced directly back to Temple who, as a retired diplomat, knew that the fates of nations often depended upon the whims of princes. In his *Introduction to the History of England* Temple declared that whoever examined the causes of political actions would 'often be forced to derive them from the same passions and personal dispositions which govern the affairs of private lives'.[53]

Swift's next ode, 'To the Athenian Society' was written in a week after Temple had spoken to him 'so much in their praise' that Swift was 'zealous for their cause'.[54] Temple's enthusiasm for this 'society', which was in reality no more than a group of enterprising hacks, is an indication of the uncertain state of learning at the time. According to Johnson this was the poem which Swift showed to Dryden, only to receive the rebuff 'Cousin Swift, you will never be a poet.' One can sympathize with Dryden. Seldom can enlightenment have been celebrated with such obscurity. The syntax is tortuous, the imagery hideously contrived. Borrowing from Cowley's 'Ode to the Royal Society', Swift invokes Philosophy to his aid. But whereas Cowley's Philosophy was male, Swift's is female and he quickly loses himself in her wardrobe.

[52] *PW* i. 104.
[53] Sir William Temple, *Introduction to the History of England,* 1695, p. 284.
[54] *Corr.* i. 8.

> With a huge fardingal to swell her fustian stuff,
> A new commode, a top-knot and a ruff,
> Her face patched o'er with modern pedantry.[55]
> (ll. 161-4)

One would like to think that some of Swift's confusion comes from a sense of the absurdity of celebrating John Dunton and the other 'Athenians' as 'ye great, unknown and far-exalted men', but we have no reason to think so. No less exalted but far better known, Temple himself was the subject of Swift's third and least successful ode. The tone of this poem is sometimes hectoring and sometimes shrill as Swift takes it upon himself to resent the slights that Temple has suffered from an ignorant court. Not only good and great, like his friend and monarch, Temple is learned, good, and great, uniting the virtues of Virgil, Epicurus, and Caesar.

> Your happy frame at once controls
> This great triumvirate of souls.[56]
> (ll. 68-9)

Swift concludes the poem prostrate with humility, marvelling that he should walk so close a path with greatness.

> Shall I believe a spirit so divine
> Was cast in the same mould with mine?
> (ll. 178-9)

Such intense hero-worship was bound to lead to disillusionment, and even here one may sometimes detect a tell-tale hint of irony lurking within the hyperbole.

As Swift began to copy out the essays and letters that flowed from Temple's pen, an intimacy developed between the two men stronger even than that between blood relatives. For, as Temple wrote, 'we are constrained . . . in our demeanour towards our parents by our respect and an awful sense of their arbitrary power over us'.[57] There were, however, certain topics upon which they always disagreed, most notably religion. Temple's education at Cambridge, and his long years in the Netherlands

[55] *Poems,* i. 14-25.
[56] Ibid., pp. 26-33.
[57] Woodbridge, p. 21.

had led him to recommend a greater degree of toleration than Swift could ever countenance. Temple's plan to encourage Protestant immigration into Ireland 'by some large degree of liberty in matters of religion' came very close to the policy which Swift vehemently condemned ten years later, when he wrote:

> These men take it into their imagination that trade can never flourish unless the country becomes a common receptacle for all nations, religions and languages.[58]

Swift found his religious hero in the person of Archbishop Sancroft, the subject of his fourth ode. Sancroft was the leader of the seven bishops condemned to the Tower by James II in 1688 for refusing to read his Declaration of Indulgence from their pulpits. He subsequently confirmed his commitment to principle–or obstinacy–by refusing to acknowledge the authority of William III either. Sancroft became a symbol of an uncompromising spiritual independence that sought no accommodation with the temporal powers. He was, said Swift, 'a gentleman I admire at a degree more than I can express'.[59] He was precisely the kind of austere moral figure to appeal to an idealistic young man who was cynical about the 'petty engines' that drove state affairs. Swift began his Ode to Sancroft in January 1692, but five months later was forced to abandon it. 'I have done nine stanzas and do not like half of them', he confessed.[60] His difficulty, an interesting prefigurement of many subsequent dilemmas in his career, was to find a way of praising Sancroft without implicitly criticizing the king. His abandonment of this poem was an unhappy omen for his many later struggles to reconcile the conflicting claims of Church and State.

For his next ode, 'To Mr Congreve', Swift finally discarded the clumsy Pindaric form and turned instead to couplets.[61] Still only twenty, Congreve had already gained a considerable reputation as a writer, and there is an interesting ambiguity of tone in this poem as Swift compares his own situation with that of his younger schoolfriend. While applauding Congreve's success, Swift affects to sympathize with him for the exposure to fools which is the inevitable price of fame. But an element of envy is

[58] *PW* iii. 48 (*Examiner*, no. 21).
[59] *Corr.* i. 8.
[60] Ibid., p. 9.
[61] *Poems*, i. 43-50.

unmistakable in these lines, though it strives hard to masquerade as pity.

> Truth I could pity you; but this is it
> You find, to be the fashionable wit;
> These are the slaves whom reputation chains
> Whose maintenance requires no help from brains.
> (ll. 161-4)

By contrast, Swift presents himself as happily obscure in his rural seclusion 'by a mountain's side' at Moor Park. But these are remarks which cry out to be read by contraries. By November 1693 Swift too wanted the excitement and acclaim of literary success and was tired of being a captive audience for a superannuated diplomat's reminiscences in the Horatian calm of Moor Park.

There was now a clash of interests between Swift and Temple. Swift looked upon his period at Moor Park as an education in philosophy and statecraft from a patron who would then help to launch him upon a career. In his poems we can hear him attempting to tune his youthful treble to his master's sombre tones, attacking society fops and political jugglers, singing the praises of retirement. But the resulting tone is hollow and unconvincing. Yet as Temple came to rely more heavily on Swift's services he was correspondingly reluctant to part with him for a career in the metropolitan world that he had grown to despise. The most he would do was to send Swift on little diplomatic errands, 'treats' that would stem his complaints. In 1693 the Whigs attempted to limit William's prerogative by introducing a Bill instituting Triennial parliaments. William was dogmatically opposed to the move, believing that 'K Ch Ist lost his crown and life by consenting to pass such a bill'. He sought Temple's advice on the matter. Temple believed that William exaggerated the disadvantages of such a change, and sent Swift with a letter to tell him so.

> Whereupon Mr Swift was sent to Kensington with the whole account of that matter, in writing, to convince the King . . . how ill [he] was informed. . . . Mr Swift who was well versed in English history although he were then under twenty-one years old, gave the king a short account of the matter.

Despite the efforts of his earnest young pedagogue, William refused to pass the bill. Swift was, as Temple intended, suitably disillusioned.

This was the first time that Mr Swift had ever any converse with courts, and he told his friends it was the first incident that helped to cure him of vanity.[62]

After this Temple made little effort to find a post for Swift in either Church or State. In November 1692 Swift told his uncle William that although Temple promised him 'the certainty' of a prebendary, he was 'less forward than I could wish, because I suppose, he believes I shall leave him, and upon some accounts, he thinks me a little necessary to him at present'.[63] Years later he reminded Stella of the anguish and insecurity he had often experienced at Moor Park.

Don't you remember how I used to be in pain when Sir William Temple would look cold and out of humour for three or four days, and I used to suspect a hundred reasons?[64]

It is clear that Swift sought sympathy and amusement from this twelve-year-old girl, when his heart was full of a renewed sense of neglect. The tensions in his relationship with Temple are evident in the last poem which he wrote at Moor Park, 'Occasioned by Sir W.T.'s late Illness and Recovery'. The poem is less about Temple's illness than Swift's dejection. He presents himself as

. . . an abandoned wretch by hopes forsook;
Forsook by hopes, ill fortune's last relief,
Assigned for life to unremitting grief.[65]
(11. 108-10)

The repetitions here toll with self-pity. Congreve was the toast of London, but Swift had achieved nothing, and he complains of his Muse, since he cannot bring himself openly to complain of Temple.

To thee I owe that fatal bent of mind
Still to unhappy restless thoughts inclined;

62 *PW* v. 193-4.
63 *Corr.* i. 12.
64 *Journal,* i. 231.
65 *Poems,* 51-5.

To thee, what oft I vainly strive to hide
That scorn of fools, by fools mistook for pride.
(ll. 131-4)

Of course Temple was right; the world was full of fools and hypocrites. It could only be Swift's unhappy restless thoughts, his 'conjured spirit', that urged him to quit the happy valley of Moor Park.

Without Temple's assistance, Swift's best hopes of preferment lay within the Church of Ireland. However he had a scruple against entering the church 'merely for support'. His scruple was genuine, though it chimed in with his sense of being worth rather more than the obscurity of a country parish, with a congregation of a dozen narrow-minded, stiff-necked Anglo-Irish landowners. Above all, he told his cousin Thomas that he wished to keep his thoughts 'in a ferment, for I imagine a dead calm to be [the] troublesomest part of our voyage thro' the world'.[66] A dead calm was how Moor Park seemed to him now, the doldrums.

At last Temple, then Master of the Rolls in Ireland, found Swift a post in that office worth £120 a year; whereupon Swift, ever the casuist, 'told him, that since he had now an opportunity of living without being driven into the church for a maintenance, he was resolved to go to Ireland and take Holy Orders'.[67] The apparent perversity of Swift's behaviour is a good illustration of the lengths to which he would go to convince himself, and others, that he was exercising free choice, and not being driven to an expedient. In fact he was only doing what his family had always expected him to do, and what his cousin Thomas had already done. But it was of vital importance to Swift to feel that he was finally taking charge of his own life. He was declaring his independence of Temple in a gesture which caused a good deal of bitterness and recrimination on both sides. Swift left Moor Park in May 1694, and, as he told his cousin Deane, Temple 'was extreme[ly] angry I left him, and yet would not oblige himself any further than upon my good behaviour, nor would promise anything firmly to me at all, so that everybody judged I did best to leave him'.[68] Swift hoped to be ordained the following September, and set himself to 'make what endeavours I can to find

[66] *Corr.* i. 13. [67] *PW* v. 194. [68] *Corr.* i. 16.

something in the church'. He had escaped from the dead calm, and for a few weeks he felt quite heady with the sense of freedom. He even expressed an interest in becoming a chaplain in Lisbon where his cousin Willoughby was a merchant. After the doldrums he wanted storms.

In fact he settled for something less adventurous but equally demanding. The Church of Ireland, that is, the established Anglican church of that country, was in a parlous condition. Protestants generally were in a minority of less than 20 per cent throughout Ireland, and of those by far the greater number were Presbyterians of Scottish descent. In the eyes of all but a tiny minority of the population, the Anglican church was at best an irrelevance, at worst an alien imposition. With tiny congregations, pluralism became commonplace; churches, seldom used, fell into disrepair. Church tithes were often impropriated to the lay landowners who made up the church's sole membership, which caused bitter disputes between the spiritual and temporal wings of the Anglican ascendancy. Short of money, land, talent, and parishioners, the Church of Ireland was presided over by Archbishop Michael Boyle, a half-deaf, half-blind, senile octogenarian. This was the institution which Swift was to serve for the rest of his life.

The canons of the Church stated that Swift could not be ordained until he could certify that he had 'a living in readiness' so he set out from Leicester to Dublin to sound out his contacts there, including Narcissus Marsh, now Archbishop of Dublin. He was told that he would need to supply testimonials of his 'good life and behaviour' in the years since he had left university. It was a bitter condition, but one which he could not circumvent. Swallowing his pride he wrote to Temple in October 1694. Macaulay described the language of this 'penitential' letter as resembling that of 'a lacquey, or rather of a beggar'.[69] But this misses the point of the constrained humility that Swift adopts. Swift is like a teenager, fresh from home who needs simultaneously to assert his independence, while seeking a favour that requires humility.

> The sense I am in, how low I am fallen in your Honour's thoughts has denied me assurance enough to beg this favour till I find it impossible to avoid.[70]

[69] Ibid., p. 17. See Macaulay's *History of England,* 5 vols., 1849-61, ch. 19.
[70] *Corr.* i. 17.

When he declares that he stands 'in need of all your goodness to excuse my many weaknesses and follies and oversights' the language is strictly conventional. Pride struggles with self-abasement here. Swift is entirely in Temple's hands, but he will not apologize for the past or assume a false contrition. Luckily Temple was above spite, and sent the required testimonial by return. Swift was ordained deacon in October, and presented to the prebend of Kilroot in the diocese of Down and Connor the following January. The value of this living was estimated at £100, slightly less than the post which Temple had offered. Swift soon found that the price of independence was rather more than twenty pounds, however. Kilroot was a lonely run-down parish, part of a union of Kilroot, Templecorran, and Ballynure, the larger portion of whose tithes went to the lay impropriator, the Earl of Donegal. The church building at Kilroot was ruined, and the spiritual condition of the parish little better. Bishop Leslie reported on the tiny congregations of the Anglican churches of Antrim only three years before Swift arrived there.

The Nonconformists are much the most numerous portion of the Protestants in Ulster. . . . Some parishes have not ten, some not six, that come to church, while the presbyterian meetings are crowded with thousands covering all the fields.[71]

Kilroot itself, just a few miles from Belfast, was in the stronghold of Ulster Presbyterianism. Indeed Swift's predecessor at Kilroot had so neglected the parish that 'several considerable persons . . . were forced to frequent the presbyterian meetings for want of a fit minister to attend that cure'.[72] There was no such occasional nonconformity while Swift held the living. His original dislike of Dissent hardened to a fierce antagonism with his experience of the stern intolerance of the Presbyterians that he encountered at Kilroot. He became friendly with John Winder, vicar of nearby Carmoney, and was later instrumental in having Winder appointed as his successor at Kilroot. Despite this friendship, and the company of a few local landowners, Swift found even less to stimulate him in Ulster than at Moor Park, and was soon confessing his weariness. By the spring of 1696 both he and Temple were in conciliatory moods. Temple asked him to return and Swift was happy to agree, having stayed at Kilroot for barely

[71] Louis Landa, *Swift and the Church of Ireland,* 1954, pp. 20-1.
[72] Ibid., p. 19.

a year. He subsequently told Winder, 'had I been assured of your neighbourhood, I should not have been so unsatisfied with the region I was planted in'.[73] Yet the company of Winder alone would hardly have been sufficient to inure him to this inhospitable land. There was one person, however, who might have softened the hardships of that life and encouraged him to stay. Her name was Jane Waring.

V

THE WARING family was sufficiently distinguished to have lent their name to their native town, Waringstown, thirty miles south-west of Carrickfergus. Jane's father, formerly Archdeacon of Dromore, had died in 1692, and when Swift met her she was a frail young woman in need of guidance and protection. Two of Jane's cousins had been at Trinity with Swift and from these beginnings a friendship developed. The rift with Temple had left Swift with an emotional scar, and his voluntary exile from the comforts of a close family environment to exposure in the bleak Ulster landscape made him look for new diversions.Jane's situation may have reminded him of young Esther Johnson, now sixteen, who had humoured him out of his black moods in Moor Park. At twenty-one Jane was still immature, and it was easy for Swift to guide the relationship into the form that suited him best, in which he was part father, part lover, and part tutor. As in Leicester he was surrounded at Kilroot by a 'parcel of fools' and he sought this natural outlet for his conjured spirit. But Swift had now reached an age when he felt he should take charge of his destiny. His cousin Thomas had followed his ordination with marriage, and if Swift was to consolidate his new choice of life, then marriage was an obvious next step.

It was Temple who precipitated a decision. Typically, it was the invitation to return to Moor Park that prompted Swift to propose to Jane Waring. As with his decision to enter the church, he would not be driven into marriage, and would only risk a proposal when he had a secure alternative. His letter of proposal, so full of the contradictions of his character, makes fascinating reading.[74] It is a virtuoso display of impetuosity and

[73] *Corr.* i. 31. [74] Ibid., pp. 18-23.

high passion, and yet one senses that the rhetoric of love is used to conceal a lack of real feeling. As with Esther Johnson (Stella) and later with Hester Vanhomrigh (Vanessa), he Latinized Jane Waring's name to Varina. The device was a common enough feature of the amatory lyrics of the time, but in Swift's case his habit of renaming these fatherless girls is a distancing process that both elevates them to mock divinities and reduces them to pets. At all events, it stops him from having to confront them as women. He could not address Jane Waring in the voice of the courtly lover that he adopts towards Varina.

> Surely, Varina, you have but a very mean opinion of the joys that accompany a true, honourable, unlimited love; yet either nature and our ancestors have hugely deceived us, or else all other sublunary things are dross in comparison. Is it possible you cannot be yet insensible to the prospect of a rapture and delight so innocent and so exalted? Trust me, Varina, Heaven has given us nothing else worth the loss of a thought. Ambition, high appearance, friends and fortune, are all tasteless and insipid when they come in competition; yet millions of such glorious minutes are we perpetually losing, for ever losing, irrecoverably losing, to gratify empty forms and wrong notions, and affected coldnesses and peevish humour.

This is not a tone that we are accustomed to hearing from Swift, but it would be wrong to accuse him of insincerity on that account alone. He was clearly attracted to Jane Waring, and to the life that he might hope to live with her, and yet the impetuosity with which he writes, the casting aside of empty forms and peevish humours, is not simply the result of passion. There is also a peremptory note behind these declarations, an autocratic pride that wishes to command rather than to sue. He makes it clear that he has other options before him. 'I am once more offered the advantage to have the same acquaintance with greatness that I formerly enjoyed, and with better prospects of interest.' However, he is willing to sacrifice his career to his love, if she will take him now, immediately, without further question. If not, then 'in one fortnight I must take eternal farewell of Varina'. If she rejects him now, it will be forever, for Swift makes this melodramatic pledge.

> I here solemnly protest, by all that can be witness to an oath, that if I leave this kingdom before you are mine, I will endure the utmost

indignities of fortune rather than ever return again, though the king would send me back his deputy.

He deliberately dramatizes the choice that she must make. It is all or nothing. There is a quality of brinkmanship in this proposal that is part of the pattern of his life, a pattern that made disappointments and rejection almost inevitable. One suspects that for all their months of friendship and flirtation, Swift had never expressed himself so seriously before. Having possibly heard tales of his Leicester conquests, Jane Waring may have had doubts about his intentions. Swift's friends, he admitted, had reproached him for 'neglecting a close siege'. Now suddenly he confronted her with this ultimatum. One phrase in his letter is particularly revealing. 'Why was I so foolish to put my hopes and fears into the power or management of another?' Throughout his life Swift suffered agonies of resentment at finding himself in the power of others. He had already undergone the cold looks of Sir William Temple, and would not suffer 'affected coldnesses and peevish humour' from Jane Waring. She, confused, bullied, and astonished, could do nothing but refuse him, though she did so with reluctance. So be it. As far as Swift was concerned the episode was over. He returned to Moor Park a lonely but an independent man.

During his absence Temple had employed his cousin Thomas as secretary, but a comparison of the capabilities of the two cousins left Temple in no doubt of Swift's value to him. However Swift did not resign his living at Kilroot until January 1698 when a suitable moment occurred to provoke another dramatic gesture of renunciation. His main hopes of advancement at court rested upon Temple's friendship with the Lord Chamberlain, the Earl of Sunderland. But in December 1697 Sunderland resigned, without having found anything for Swift. It was in response to this midwinter blow to his hopes that Swift decided to burn his bridges at Kilroot, rather than be driven back there by circumstances. To Winder he wrote that '10 days before my resignation, my lord Sunderland fell, and I with him'.[75] Looking back on his life, Swift could portray this as one of those little ironies that had turned his career into a series of disappointments, but in fact one sees that he deliberately courted disaster, and even

[75] Ibid., p. 26.

made a boastful virtue of his defeat. It is as though he was seeking consciously to deride the 'paltry maxims' of prudence 'calculated for the rabble of mankind' that he had condemned in his letter to Jane Waring. Swift's sister claimed that it was Temple who 'made him give up his living in [Ireland,] to stay with him at Moor Park, and promised to get him one in England',[76] but there is no evidence to support this. On the contrary, Temple advised Swift to renew his licence at Kilroot. It was Swift's own ambition that prevented him from settling there. He was assiduous in promoting Winder's claims to the parish, and it pleased him when his imprudent behaviour became the subject of gossip. He assured Winder that he had no regrets about leaving.

> Since the resignation of my living and the noise it made amongst you, I have had at least 3 or 4 very wise letters unsubscribed, from the Lord knows who, declaring much sorrow for my quitting Kilroot, blaming my prudence for doing it before I was possessed of something else, and censuring my truth in relation to a certain lady. One or two of them talked of you as one who was less my friend than you pretended, with more of the same sort, too tedious to trouble you or myself with.[77]

He could feel at least a minor sense of triumph in confounding the narrow views of such anonymous well-wishers.

Swift now became Temple's official literary executor, and Temple added a codicil to his will granting Swift £100 for his labours. He began collecting, copying, and editing Temple's essays, memoirs, poems, and letters, although Temple would not allow his letters to be published while he was still alive. Swift enjoyed greater independence than formerly, but still no place was found for him, and there were times when Temple's aloofness continued to cause him pain. Many years later he repudiated the imputation of indebtedness to Temple. 'Being born to no fortune, I was at his death as far to seek as ever'.[78] Probably Swift made it as difficult for Temple to assist him, as he had made it for Jane Waring to accept him. He was a proud man, quick to take offence, and would have done so at the slightest hint of condescension on Temple's part.

Temple's health and spirits were gradually declining. His wife had died in February 1695, and his gout grew worse, though he

[76] Ibid., p. 32. [77] Ibid., p. 25. [78] Ibid., p. 125.

complained little. In September 1698 we have a colourful picture of him visiting the Somersets at Petworth, where he won twelve guineas at cards and was 'insufferably pert'.[79] He died on 29 January, quite suddenly 'and with him', wrote Swift, 'all that was good and amiable among men'.[80] In a later tribute he described Temple as 'a person of the greatest wisdom, justice, liberality, politeness, eloquence, of his age and nation; the truest lover of his country, and one that deserved more from it by his eminent public services, than any man before or since'.[81] Many have remarked on the extravagance and impersonality of this eulogy, and some have suspected irony here. In his latter years at Moor Park Swift came to distrust the casual superiority that Temple assumed in literary and political matters, and came to suspect that Temple's vaunted Epicureanism was merely a polite name for intellectual laziness. Nevertheless his affection and respect went deeper than such reservations, and he sincerely mourned the passing of his mentor and patron.

Swift stayed on at Moor Park to assist with the settlement of Temple's affairs. He also made a final attempt to capitalize on Temple's friendship with King William by petitioning the King for a promised prebend of either Canterbury or Westminster. The Earl of Romney promised to second Swift's petition but as he was, in Swift's words, 'an old vicious illiterate rake, without any sense of truth or honour'[82] he apparently said nothing to the King. In fact the only position that Swift was offered was in Ireland. Lord Berkeley was travelling to Dublin as a Lord Justice, and invited Swift to accompany him there as his secretary and chaplain. After the grandiloquence of his eternal renunciation of Ireland to Varina, it was a bitter prospect to contemplate returning. But there was worse to come, for after he had forced himself to accept this humiliating post, Swift discovered that Dublin promises could be as false as London ones. According to his own account he had served Berkeley as his secretary for the whole of the journey to Dublin when 'another person . . . so far

[79] See Julia C. Longe, *Martha, Lady Giffard, a Memoir,* 1911, p. 227.

[80] Reported by John Lyon in MS notes to a copy of Hawksworth's *Life of the Reverend Jonathan Swift,* 1755, in the Victoria and Albert Museum, Forster Collection, no. 579 (48.D.39).

[81] Inscribed on the flyleaf of Swift's Bible. The eulogy is discussed by George Mayhew in *Harvard Library Bulletin* 19: 404, n. 7; and by Elias, pp. 97-108.

[82] *PW* v. 195.

insinuated himself into the Earl's favour, by telling him that the post of secretary was not proper for a clergymen . . . that his lordship after a poor apology gave that office to the other'. There is a neatness to the way Swift shapes this story that has the ring of paranoia. For, having insinuated his way into the position of secretary with the argument that it would be of no value 'to one who aimed only at Church preferment', this scheming rival then used that position to block Swift's hopes of church preferments too.

> In some months the Deanery of Derry fell vacant; and it was the Earl of Berkeley's turn to dispose of it. Yet things were so ordered that the secretary having received a bribe, the Deanery was disposed of to another, and Mr Swift was put off with some other Church-living not worth above a third part of that rich Deanery, and at this present time, not a sixth. The excuse pretended was his being too young, although he were then 30 years old.[83]

With this thoroughly characteristic example of the perfidy of great men, Swift finally abandoned his fragment of autobiography, as though it were a case now proved beyond any further doubt or argument. The facts of the matter are, inevitably, less clear-cut than his painful memories and partial knowledge admitted. Berkeley had no reason to snub Swift, who was something of a favourite in his household, especially among the ladies. The Earl's appointment had been somewhat hurried, part of the King's response to a deteriorating political situation. In normal times Ireland was governed by a single Lord Lieutenant, but in his absence, a number of Lord Justices would preside. William was resisting Tory pressure to appoint a new Lord Lieutenant by persevering with a number of loyal, if undynamic Whig lords. In the hurry of preparations, misunderstandings about Swift's exact position in the entourage might easily have arisen, but with his tendency to assign causality to coincidence he presents the achievement of his rival, Arthur Bushe, in entering Dublin in Berkeley's own coach, as a masterpiece of guile. The attraction to Swift of the dual roles of chaplain and secretary was precisely that it further postponed the necessity of choosing between a literary or a clerical career. In the event he retained only the post of domestic chaplain, and since Berkeley

[83] Ibid., p. 195.

had at least one official chaplain, his duties were very light. When not leading family prayers Swift would spend his time cultivating contacts at Dublin Castle or playing at puns and riddles with the Earl's eldest daughter, Lady Betty, a lively and witty teenager, typical of the young women in whose company he always felt most relaxed.

Swift was also wrong to believe that the deanery of Derry was in Berkeley's gift, or that his own claims to the post were particularly strong. John Bolton, Berkeley's official chaplain who had been in orders for twenty years, was a much stronger candidate. The crucial figure in making the appointment was the Bishop of Derry, Dr William King, a man who was to have a significant influence on the whole of Swift's career in the Church of Ireland. On this occasion Swift's name simply never entered King's head. The post was first offered to Bolton who, being happily settled in his living at Laracor, close to Dublin, refused it. It was then offered to a second person who also refused it, but even now Swift was not considered. Instead attempts were made to tempt Bolton to accept the Deanery by allowing him to retain the living of Ratoath, adjacent to Laracor, in addition. This was at a time when strenuous efforts were being made, notably by King himself, to end such pluralism. The offer was too good to refuse, and Bolton thanked the authorities for their 'kindness and goodness' to him. Swift's part of the deal was to be offered the remainder of Bolton's original cure, after the living of Ratoath was detached. There could hardly be a more graphic illustration of his true status within the ecclesiastical hierarchy. Contrary to his belief that Bolton had offered a bribe to obtain the Deanery, he had actually been given a bribe to take it.

Swift found some relief from bitterness by voiding his resentment in the form of some smutty verses that can hardly be redeemed by the most sympathetic of Freudian analyses. In 'The Problem' the 'vicious illiterate rake' Lord Romney is shown farting contentedly amid a circle of admiring ladies, each of whom seeks to approach this 'back-way to his heart' while in 'The Discovery' the services that secretary Bushe performs for Berkeley are revealed as being as profound and intimate as an enema.[84]

[84] *Poems*, i. 64-7, and 61-3.

After ten years of unsuccessful attendance on great men, Swift had finally gained a living for himself. He soon developed a strong affection for Laracor, and held on to the cure there for the rest of his life. Even after the detachment of Ratoath, it was hardly a poor living, comprising a union of the three parishes of Laracor, Agher, and Rathbeggan which together brought in an income of £230 a year. In fact Swift had to petition the Primate for the right to continue to hold the three parishes in a union, arguing that together they offered 'but a comfortable support'. His petition was approved, and he was presented to the living in February 1700. In the autumn he was also appointed to Bolton's vacant prebend of Dunlavin, which brought him a few additional duties, a small additional income of some £15, but most significantly, his first entry into St Patrick's Cathedral.

VI

NOW A man of thirty-two, Swift's feet had finally touched ground. As a parish priest he had a house near Trim from which to oversee his little domain; as a prebendary had access to the ecclesiastical hierarchy of the Church of Ireland, and as chaplain to one of the three Lords Justices he kept in touch with political developments. Yet he could not conceal from himself that as a courtier he had proved a miserable failure. The man who, as the favoured protégé of Sir William Temple, had been offering advice to the King seven years before, was now easily outmanouevred by a self-important mediocrity like Bushe. Swift had an unmistakable sense of narrowing horizons. He was no longer young, and it seemed probable that a career as a parish priest–an eccentric, possibly even a literary parish priest, but still only a parish priest–was looming before him. He would have to be content with the slow and worthy path of ecclesiastical promotion, rather than the *carrière ouverte aux talents* of a courtly environment. It was when his thoughts were full of such thoughts as these that Jane Waring let it be known that she might now be prepared to reconsider his proposal. Maybe she thought that Swift's disappointments might have inclined him to a mood for compromise. If so she misjudged his character badly. His stomach was too full of humble pie and he seized this opportunity to assert a residual power, rather than accept her as consolation for his defeated hopes.

His sense of humiliation was increased by his sister's marriage to Joseph Fenton, a tanner, in December 1699. Whether from a snobbish distaste for Fenton's trade, or from personal dislike of the man, Swift thoroughly disapproved of the match, and felt betrayed by his sister. However, Jane Swift was thirty-three, no great beauty, and had virtually no dowry to offer. She may well have considered the widower Fenton a reasonable catch. But Swift felt she had let the family down by marrying this dunce, and he ensured that their mother should leave Jane's money in his safe keeping, with the stipulation that it was 'not to come to her husband'. In old age he remarked bad-temperedly that he did not care 'one straw' what happened to her, and 'did not employ one thought upon her except to her disadvantage' since she had 'during her whole life disobliged me in the most [material] circumstances of her conduct'.[85] Was this all that the Swifts were to become? Parish priests and tanners' wives? It was with a mind perplexed by such household thoughts that he received the 'imperious' and 'untractable' questions of Jane Waring concerning his own matrimonial intentions.

His reply is a disingenuous letter that completely lacks the impetuous bravado of his original proposal. Now he was trapped in Ireland and found himself subjected to the arguments and recriminations of a woman who showed every sign of becoming a 'clog'. He exerts all his rhetorical skills to manoeuvre the situation to his own advantage and deliver another triumphantly unacceptable proposal, but does not quite succeed. There is a cruelty of tone and hostility of intention in his letter.[86] He addresses her *de haut en bas,* as though she were guilty of some monstrous presumption in having taken the play-acting gallantry of his proposal seriously. His difficulty was an inability, which remained with him all his life, to acknowledge that he had simply changed his mind. No, he insists that he had been perfectly consistent, and to prove this he allows what was previously only a playful condescension to become open cruelty. Jane had cited her health as the main reason for her hesitation about marrying. Swift returns again and again to this weak point

[85] *Corr.* iv. 411. [86] Ibid. i. 32-6.

in her position, subjecting an emotional qualm to the rigours of logical cross-examination. 'I am extremely concerned at the account you give of your health', he begins, though concern is hardly the feeling that emerges. Her other stated reservation had been financial, and here again, Swift is anxious to know what she thinks has changed.

> The dismal account you say I have given you of my livings I can assure you to be a true one; and since it is a dismal one even in your own opinion, you can best draw consequences from it.

Swift's protestations of consistency did not in the least deceive Jane who knew that his tone had changed completely, and she asked him directly whether he had found someone else. The starchy solemnity of his reply reeks of insincerity.

> The other thing you would know is, whether this change of style be owing to the thoughts of a new mistress. I declare, upon the word of a Christian and a gentleman, it is not; neither had I ever thoughts of being married to any other person but yourself.

Yet when he left London in August 1699, Esther Johnson was nineteen and, as he later recalled, 'looked upon as one of the most beautiful, graceful and agreeable young women in London, only a little too fat'.[87] This image of Esther was much in his mind during those trying months in Ireland. It may be true that he had no thoughts of marrying her, for in the series of resolutions entitled 'When I come to be old' which he wrote at this time, the very first one is 'Not to marry a young woman'. But the tone of this, and of his declaration to Jane, smacks more of a self-denying ordinance, imposed with difficulty, than of a straightforward choice. Evidently the thought of marriage to a young woman, specifically to Esther Johnson, had occurred to him forcefully, and yet for some reason he, 'as a Christian and a gentleman' could not acknowledge it. The hostile tone of his letter to Jane results from being required to explain to her motives and emotions which he was unwilling or unable to explain to himself. His language sounds more appropriate to a couple contemplating divorce than marriage: 'All I had in answer from you, was nothing but a great deal of arguing, and sometimes in a style so very imperious as I thought might have

[87] *PW* v. 227.

been spared.' He is contemptuous of any arguments of hers based on her expectations in her present family situation, a 'sink' from which he had already advised her to move: 'No young woman in the world of the same income would dwindle away her health and life in such a sink, and among such family conversation'. It was to be a familiar feature of all Swift's relationships with women that they should be not only frail and fatherless, but should be prepared to give up any remaining family ties and rely entirely upon him. Willingness to make such a sacrifice was the guarantee that he needed of a woman's loyalty. They must also be prepared to be moulded and educated by him. Such demands might seem formidable and repugnant, but only those with a very superficial view of human nature will be surprised that at least two women found his dictated terms irresistible.

At the conclusion of his letter Swift details a set of marriage conditions that read like punishments for Jane's presumption in seeking to bandy terms. For a short while his happiness had seemed dependent upon her will; now, in revenge, he acts the tyrant, showing her just how dependent she must be on him. Expressed here as a cold and formal contract, without compassion or individuality, Swift's conditions seem like the rules of a penal institution. Yet the terms upon which Esther Johnson agreed to bind her life to his were scarcely less rigorous, though never expressed in so callous a manner.

> Are you in a condition to manage domestic affairs, with an income of less (perhaps) than three hundred pounds a year? Have you such an inclination to my person and humour as to comply with my desires and way of living, and endeavour to make us both as happy as you can? Will you be ready to engage in those methods I shall direct for the improvement of your mind, so as to make us entertaining company for each other, without being miserable when we are neither visiting nor visited? Can you bend your love and esteem and indifference to others the same way as I do mine?

The imperiousness of that 'bend' stimulates his imagination to assume the arrogance of a Petruchio. Just as in his proposal letter he had played the rake, so here he plays the shrew-tamer.

> Have you so much good-nature as to endeavour by soft words to smooth any rugged humour occasioned by the cross accidents of life? Shall the place wherever your husband is thrown be more welcome

> than courts or cities without him? In short, these are some of the necessary methods to please men, who, like me, are deep-read in the world; and to a person thus made, I should be proud in giving all due returns towards making her happy. These are the questions I have always resolved to propose to her with whom I mean to pass my life.

By becoming literary, Swift manages to wriggle out of addressing Jane Waring face-to-face. He is now being witty at the expense of Varina, and hence can conclude with an epigram which, if delivered *in propria persona,* would be an intolerable insult.

> Whenever you can heartily answer them in the affirmative, I shall be blessed to have you in my arms, without regarding whether your person be beautiful, or your fortune large. Cleanliness in the first, and competency in the other, is all I look for.

No reply from Jane Waring has survived.

It is worth taking a closer look at those resolutions 'When I come to be Old' which Swift wrote during this gloomy period when the world seemed to be closing in on him, and he set his features firmly against taking the Waring way out. They give the impression of a man no longer young anticipating, even welcoming the thought of a deprived and attenuated old age. Just as, many years later, Swift was to anticipate his own death, here he anticipates old age like a Beckett character, paring his life down to a bare minimum. In despair at his inability to expand his world according to his ambitions, he makes this gesture of regulated self-denial, by closing any further chinks of vulnerability in the armour of his self-sufficiency.

> Not to marry a young woman.
> Not to keep young company unless they really desire it.
> Not to be peevish or morose, or suspicious.
> Not to scorn present ways, or wits, or fashions, or men, or war etc.
> Not to be fond of children, or let them come near me hardly.
> Not to tell the same story over and over to the same people.
> Not to be covetous.
> Not to neglect decency, or cleanliness, for fear of falling into nastiness.
> Not to be over severe with young people, but give allowance for their youthful follies and weaknesses.
> Not to be influenced by, or give ear to knavish tattling servants or others.
> Not to be too free of advice nor trouble any but those that desire it.

To desire some good friends to inform me which of these resolutions I break or neglect, & wherein; and reform accordingly.
Not to talk much, nor of myself.
Not to boast of my former beauty, or strength, or favour with ladies, etc.
Not to harken to flatteries, nor conceive I can be beloved by a young woman. *et eos qui hereditatem captant odisse ac vitare.*
Not to be positive or opiniative.
Not to set up for observing all these rules, for fear I should observe none.[88]

The repeated reminder to avoid falling in love with a young woman is the cry of a damaged sensibility that wishes to draw attention to its own damage. Coupled with the references to children, it suggests a deep disturbance in Swift's affective life, and shows how much more easily, even at this stage in his life, he could don the mask of misanthropy than face up to the problems of involvement in human relationships. He is asserting ground rules for life, and among the first rules seems to be a resolution to trust no one. 'Why was I so foolish to put my hopes and fears into the power or management of another?' He would not willingly do so again. The bleak emphasis on cleanliness here, repeated in his letter to Jane, indicates the kind of austere minimum requirements that he has of life. Obviously there is an element of self-mocking exaggeration in these resolutions, but that should not blind us to their significance. Swift had evidently watched with distaste how vain old men would dote on children and make fools of themselves over young women. Murry suggests quite plausibly that something in Temple's 'infinite fondness' for children, and notably for Esther, had shocked Swift.[89] He sensed similar tendencies within himself, in the 'little language' which he reserved for her. He liked to believe that his admiration for Esther was based on her wit and understanding, her strength of mind and character, all of which he had helped to form. But he knew that his real tenderness for her had other sources, and feared that some day such weakness might get him by the throat. Hurt by Jane Waring, he made this gesture of setting his face against any further emotional disturbances. The child that he had educated and cared for would not get the

[88] *PW* i, p. xxxvii.
[89] J. Middleton Murry, *Jonathan Swift*, 1954, p. 66.

better of him by making him feel that she could love him. His resolutions indicate a deliberate attempt at self-mortification: not the hair shirt, but the reductive moral imperative.

VII

It seems appropriate that the composition of *A Tale of a Tub* can be assigned to no particular period in Swift's life. This work in which he walks the fence between ancients and moderns, was itself fairly antique by the time it appeared in 1704. In the 'Apology' which he added to the fifth edition of the *Tale* in 1710, Swift declared that 'the greatest part' of the book was 'finished above thirteen years since'. However, since the intention of this 'Apology' was to excuse 'several youthful sallies' in the *Tale*, 'which from the grave and wise may deserve a rebuke', this statement cannot be taken as definitive. The excuse of youth was the only concession that Swift was prepared to make to the many critics of the *Tale*. For the most part he was unrepentant, and demanded

> to be answerable no farther than he is guilty, and that his faults may not be multiplied by the ignorant, the unnatural, and uncharitable applications of those who have neither candour to suppose good meanings, nor palate to distinguish true ones.[90]

Parts of the *Tale* may well have been written at Trinity College, while the central narrative, with its fierce attack upon the Dissenters, no doubt derived from Swift's feelings of isolation among the Presbyterian hordes in Kilroot. The principal context for the *Tale*, and for *The Battle of the Books* however, was the dispute between the ancients and the moderns, or more specifically, between Temple and Boyle, the champions of ancient learning, and Bentley and Wotton, the defenders of the moderns. Temple's essay 'Of Ancient and Modern Learning' which appeared in 1690 is a polite and dilettante piece, with all the intellectual rigour of an after-dinner speech. Having already, in

[90] *PW* i. 2. In the interests of consistency, all references are to Herbert Davis's edition of *A Tale of a Tub* in the *Prose Works*, i, 1939. However, it is generally accepted that the edition of the *Tale* by A. C. Guthkelch and D. Nichol Smith (1958) represents a significant improvement on Davis's edition, and should be consulted by readers wishing to make a thorough study of the *Tale*.

his essay 'On the Gardens of Epicurus' (1685) poured scorn on the main achievements of modern learning, including Harvey's discovery of the circulation of the blood and Copernicus's discovery of the revolutions of the planets, Temple now chose to base his claims for the superiority of ancient culture upon the Epistles of Phalaris. His arguments were first politely challenged by William Wotton in his *Reflections upon Ancient and Modern Learning* (1694) and subsequently devastated by Richard Bentley in his *Dissertation* on the Phalaris epistles. Bentley, the foremost philologist and classical scholar of his age proved conclusively that these epistles were in fact spurious concoctions, and his contemptuous dismissal of Temple's feeble attempts at scholarship signalled a clear dichotomy between the professional philologist and the gentleman amateur.

When Swift returned to Moor Park in June 1696, it was natural that Temple should have encouraged him to participate in this controversy. However, neither the *Tale* nor the *Battle of the Books* is a simple vindication of Temple. Although many of the ideas and images in both works have their origins in Temple's writings, Swift's treatment of them is full of ironies. *A Tale of a Tub* is a bibliomaniac's fantasy, a virtuoso display of book-learning that draws attention to Swift's years of study at Moor Park. 'It exhibits', wrote Johnson, 'a vehemence and rapidity of mind, a copiousness of images, and vivacity of diction, such as he afterwards never possessed, or never exerted'.[91] Swift agreed, 'What a genius I had when I wrote that book,' he declared, forty years later.[92] It is this exuberant excess of Swift's verbal pyrotechnics that both dazzles and disturbs the modern reader. It is easy enough to sense the energy of the *Tale*. It is far less easy to detect the direction of its arguments.

One of Swift's favourite satiric techniques throughout his career was to mimic and parody the voices of his enemies. In the guise of an astrologer he undermined the astrologer Partridge; as a political economist he subverted the modest proposals of political economy. So here, to attack the hacks of modern culture he becomes a hack himself. The *Tale* embodies the vanity, pretensions, sycophancy, but above all, the ephemerality

[91] Johnson, iii. 51.
[92] See Swift's *Works*, ed. Walter Scott, 1814, 19 vols., i. 90.

of modern literature. Books had become commodities, to be sold when fresh, like hot pies or wet fish. 'I am living fast to see the time when a *book* that misses its tide, shall be neglected, as the *moon* by day, or like *mackerel* a week after the season'.[93] The market shortens all perspectives, of distance as well as time. 'Such a jest there is,' we are told in the Preface, 'that will not pass out of *Covent-Garden*; and such a one, that is nowhere intelligible but at *Hyde-Park* corner.'[94] The contrast between a classical culture, presenting the 'genuine progeny of common humanity', and a modern one calculated to satisfy the London fashion of the week, obsessed both Swift and Pope. In the Preface to his *Works* (1717) Pope wrote:

> They [the Ancients] writ in languages that became universal and everlasting, while ours are extremely limited both in extent and duration. A mighty foundation for our pride! when the utmost we can hope, is but to be read in one island, and to be thrown aside at the end of one age.[95]

In 1712 Swift published his *Proposals for Correcting, Improving and Ascertaining the English Tongue*, in which he asked:

> How then shall any man, who hath a genius for history equal to the best of the ancients, be able to undertake such a work with spirit and cheerfulness, when he considers that he will be read with pleasure but a very few years, and in an age or two shall hardly be understood without an interpreter?[96]

Yet paradoxically the excitement and animation of the *Tale* comes precisely from the hectic variety and novelty of the Hack's ephemeral talents. With the consummate knowingness of an impresario he tosses off prefaces, digressions, interpolations, epic similes, and moral *sententiae*. The work is full of the cant of the book-trade; in-jokes that deliberately devalue a culture to provide tit-bits for a metropolitan coterie.

It is usual for commentators to divide the *Tale* into 'narrative' and 'digressions', or between satire on religion and satire on learning. By such divisions the modern critic hopes to rule Swift's anarchic material. But there is a danger that such demar-

[93] *PW* i. 132.
[94] Ibid., p. 26.
[95] Pope, Preface to *The Works*, 1717. *The Poems of Alexander Pope*, ed. John Butt *et al.*, 11 vols., 1939-69, i, 7.
[96] *PW* iv. 18.

cations may block the imaginative currents that flow through the whole work. The tale proper begins with the words 'once upon a time . . .' and this phrase, together with the emphatically conventional narrative formula of the old man with his three sons recall the secure homiletic world of fables. However, the forty pages of satiric fantasy that precede this point should prevent us from interpreting the story of the three brothers with their magic coats as a simple moral allegory. The clothing metaphor itself is something which serves to bind together several different strands of satire. The brothers, Martin, Peter, and Jack clearly represent the three competing brands of Christianity in post-Reformation Europe, and it is customary for critics to note that Martin, the Anglican Church, is far less confidently characterized than either Peter (Roman Catholicism) or Jack (Calvinism). Yet the description of Martin's painstaking removal of the layers of ornamentation with which the brothers had decorated their coats on first coming to town, is a concise image of the Anglican compromise.

> Where he observed the embroidery to be worked so close, as not to be got away without damaging the cloth, or where it served to hide or strengthen any flaw in the body of the coat, contracted by the perpetual tampering of workmen upon it; he concluded the wisest course was to let it remain, resolving in no case whatsoever, that the substance of the stuff should suffer injury.[97]

Clothing is used as an effective medium for a parody of syllogistic reasoning when Peter seeks to justify his fashionable predilection for shoulder-knots by a casuistical deconstruction of their father's will (the Bible). Having failed to find any mention of shoulder-knots *totidem verbis* (in so many words), or *totidem syllabis* (syllable by syllable) he even fails to spell out the desired words *totidem literis* (letter by letter) since 'a K was not to be found'; Unabashed, he proclaims that 'K was a modern illegitimate letter, unknown to the learned ages',[98] and confidently assures his brothers that they can wear their shoulder-knots with a clear conscience. Ironically, this kind of redefinition is one of Swift's own most characteristic satiric devices, and the casuistry which he attacks here he would use elsewhere for his own purposes.

[97] Ibid. i. 85. [98] Ibid., p. 50.

The clothes 'philosophy' in the *Tale* which culminates in the worship of a tailor's dummy, is in reality no more than a restatement of a well-worked theme: that people should pay more attention to their souls than to their bodies. Yet in this triumphantly rhetorical strip-tease, Swift provides a new animation for old tropes.

> Is not religion a *cloak,* honesty a *pair of shoes,* worn out in the dirt, self-love a *surtout,* vanity a *shirt,* and conscience a *pair of breeches,* which, though a cover for lewdness as well as nastiness, is easily slipt down for the service of both?[99]

An analogy is an analogy is an analogy. Swift was never happier than when stringing together such homely metaphors as a more pedestrian preacher might drag in to prove a point, and turning them inside out. Those breeches lead us back to familiar Swiftian territory and pose a besetting problem. Is it better to cover nastiness or to cure it? To patch or to purge? When in the 'Digression on Madness' the Hack orders the carcass of a beau to be stripped, he affects surprise at finding 'so many unsuspected faults under one suit of clothes'. He 'justly' forms the conclusion that mankind is best served by those who contrive to conceal such unpalatable truths and endeavour to 'patch up the flaws and imperfections of nature . . . Happiness', he declares, 'is a perpetual possession of being well deceived.' Many of the ironies in this Digression can be traced back to Swift's own ambiguous attitude to Temple. Thus when, as part of his argument, the Hack commends those who can 'with Epicurus content their ideas with the films and images that fly off upon their senses from the superficies of things', this is clearly an allusion to the self-styled Epicurean, Temple. Similarly, in the Preface to the *Tale,* satire is described as 'a ball bandied to and fro, and every man carries a racket about him to strike it from himself among the rest of the company'. This has been plausibly interpreted as another oblique reference to Temple, whose love of tennis was well known.[100] In fact the Hack's arguments abound in borrowings from Temple, but few if any of them are used uncritically. While Swift had little sympathy for the egotistical modern pedantry of Bentley, he was intelligent enough to realise that Temple's complacent superficiality represented not only intel-

[99] Ibid., p. 47. [100] See Elias, p. 168.

lectual bankruptcy but also social snobbery and moral blindness. As a satirist Swift recognized that the process of flaying, stripping, and dissecting his victims might well produce no moral improvement in society at all, yet his preference throughout his life was for the instruments of correction, the lash and the knife, rather than the more civilized implements of his patron.

'Written for the universal improvement of mankind' is the *Tale*'s bold motto. Such confidently asserted reformism is, naturally, ironic. The Hack is the first of a series of Swiftian personae who offer their panaceas for the ills of mankind. In fact the *Tale* fairly bristles with good intentions. In the Preface, explaining the title of the work, he offers this analogy: 'That sea-men have a custom when they meet a whale, to fling him out an empty tub, by way of amusement, to divert him from laying violent hands upon the ship'.[101] This pleasant image of a whale with violent hands is quickly unscrambled as a reference to Hobbe's *Leviathan* 'which tosses and plays with all other schemes of religion and government', and Hobbesian materialism lies at the root of many of the Hack's physico-logical parodies. According to this explanation, the 'sole design' of the *Tale* was to 'employ those unquiet spirits' for an interim of some months. But, in yet another statement of intent, the Hack congratulates himself on having written 'so elaborate and useful a discourse without one grain of satire intermixt'. This is followed by one of those famous passages in which the man and the manner neatly contradict each other.

> I have observed some satirists to use the public much at the rate that pedants do a naughty boy ready horsed for discipline: First expostulate the case, then plead the necessity of the rod, from great provocations, and conclude every period with a lash. Now, if I know anything of mankind, these gentlemen might very well spare their reproof and correction: For there is not, through all nature, another so callous and insensible member as the world's posteriors, whether you apply to it the toe or the birch.[102]

'When you think of the world, give it one lash the more, at my request,' Swift wrote to Pope, many years later, no doubt recalling this passage with the rueful irony of an ageing pedagogue. No satirist, before or since, has delivered such chastisement to the

[101] *PW* i. 24. [102] Ibid., p. 29.

world's insensitive posteriors; and it was from their posteriors that the yahoos of the world voided back their indifference at his creature, Gulliver. For Swift, the sting is always in the tail. Like a good moralist, he spent his lifetime regarding the end.

One of the best ways to examine this 'physico-logical' system, according to which the Hack seeks to prove the interdependence of man's mental and physical faculties and which, coincidentally, links the satire on religion with the satire on learning in the *Tale,* is to consider briefly Swift's use of puns. Swift was a lifelong punster, a man whose love of riddles, verbal games, 'pun-ic wars', crambo, and Anglo-Latin doggerel demonstrated his adherence to the motto *'vive la bagatelle'.* Yet he shared with his fellow Dubliner Joyce an interest in pun at a deeper level, as a form of ironic revelation. Epiphanies, Joyce called them; acts of verbal magic that could transform old words into new ideas. Swift's puns are not so much epiphanies as incarnations, a constant process of words becoming flesh and spirit becoming substance. Swift transubstantiates words into things as a means of 'proving' the physical origins of all visionary phenomena. No word is used with greater ambiguity in the *Tale* than the word 'spirit'. Much of the Hack's ingenuity is devoted to presenting variations on the theme made explicit in the title of Swift's 'Fragment', *The Mechanical Operation of the Spirit.* There was a persisting uncertainty among philosophers as diverse as Bacon, Hobbes, and Berkeley, concerning the exact nature, status, and meaning of the term 'spirit'. For an age which had yet to discover the nervous system, 'animal spirits' were envisaged as providing the links between mind, body, and soul without entirely belonging to any one of these faculties. But the question of determining at what point mortal man contained an immortal, immaterial soul was infinitely perplexing, not least after Descartes confidently asserted the material location of the soul in the pineal gland. Swift presents the mechanical operations of the spirit like the changes in a volatile chemical, alternately evaporating and condensing in the human body. In doing so he reaffirms certain axioms: that the same operations link the fantasies of critics, fanatics, madmen, and tyrants; that polar extremes are frequently indistinguishable in their effects; and that public revolutions have their origins in private indispositions. Thus Louis XIV's imperial ambitions were all the result of a disorder of the bowels. 'The same spirits which in their superior progress would conquer a kingdom, descending upon the anus, conclude in a fistula.'[103] In Section VIII of the *Tale* the Hack presents the learned Aeolists who maintain 'the original cause of all things to be wind'. This is their interpretation of the concept of the '*anima mundi*; that is to say, the spirit, or breath, or wind of the world'. By literalizing puritan claims to *inspiration* in this way, Swift 'proves' them to be nothing but windbags. The wind/spirit pun is developed in a memorable cumulative fantasy. Since inspiration is prized above reason, 'the wise Aeolists affirm the gift of BELCHING, to be the noblest act of a rational creature'. In this they are at one with the 'modern saints' of the Introduction who have 'spiritualized and refined' their writings 'from the dross and grossness of sense and human reason'. To cultivate this noble art and render it 'more serviceable to mankind' the Aeolists used several methods.

At . . . times were to be seen several hundreds linked together in a circular chain, with every man a pair of bellows applied to his neighbour's breech, by which they blew up each other to the shape and size of a tun; and for that reason, with great propriety of speech, did usually call their bodies, their vessels. When, by these and the like performances, they were grown sufficiently replete, they would immediately depart, and disembogue for the public good a plentiful share of their acquirements into their disciples' chaps.[104]

In the *Mechanical Operation of the Spirit* the vapours of enlightenment flow through other bodily channels. Recalling his observations at Kilroot, Swift describes the atmosphere of a revivalist meeting whose members 'grow visionary' from the influence of a 'short pipe of tobacco', while their bodies move up and down until 'the reasoning faculties are all suspended' and 'a thousand deliriums' crowd their brains. The spiritual harangues of the dissenting preachers concentrate on sounds rather than sense.

Thus it is frequent for a single vowel to draw sighs from a multitude, and for a whole assembly of saints to sob to the music of one solitary liquid. But these are trifles; when even sounds inarticulate are observed to produce as forcible effects. A master-workman shall blow his nose so powerfully, as to pierce the hearts of his people, who are disposed to

[103] Ibid., p. 104. [104] Ibid., p. 96.

receive the excrements of his brain with the same reverence, as the issue of it.[105]

This droning delirium leads to a state of spiritual ecstasy that has an appropriate climax.

in the height and *orgasmus* of their spiritual exercise it has been frequent with them *****; immediately after which, they found the spirit to relax and flag of a sudden with the nerves, and they were forced to hasten to a conclusion.[106]

These techniques of literalization and *reductio ad absurdum* which are so characteristic of Swift's major satires carry with them interesting biographical and psychological implications. Taken together with his partiality for ironic masks, they indicate a mind which is happiest when parodying theories and ridiculing abstractions but seems unwilling or unable to tackle metaphysical concepts head-on. The robust literal-mindedness which instinctively equates inspiration with inflation, evangelism with cant, and enthusiasm with insanity is well equipped to satirize the follies, vices, and vanities of mankind. Throughout his works Swift plays his transubstantiation trick to good effect, reducing dreams of empire to an inflammation of the bowels and transforming the statistics of political economy into hecatombs of baby-flesh. It is in Swift's sermons that we see the negative side of this instinctive materialism, as he struggles to reduce the mystery of the Trinity to common-sense terms, or to reconcile his axiomatic belief in human self-interest with the morality of the Sermon on the Mount.

Like an expert impresario, the Hack of the *Tale* gives us a book of the world of the book, inviting us to indulge ourselves in the blissful ignorance of the Aeolist, the Bedlamite, and the thoroughly accomplished Modern. Yet for all the surface dazzle of the *Tale,* the truth is that we are never in any doubt about Swift's values. We may occasionally find ourselves misreading a parallel for an antithesis, or stumbling blindly through a short-cut out of one of his rhetorical mazes. We may mistake a positive for a negative charge in the battery of his allusions. But that is all part of the strategy of the work. To err is human, and the Renaissance humanists, whose heir Swift is in this work, exploited the human facility

[105] Ibid., p. 183. [106] Ibid., p. 189.

for error deliberately. As Rosalie Colie says of Erasmus's *Praise of Folly,* 'Mistake-making serves man well, since his salvation depends upon his ultimate realisation of his own folly'.[107] In the end it is perfectly correct to read the *Tale* as a defence of the established Anglican Church, of classical literary standards, of constitutional monarchy, of rationality and moderation. But Swift is both more entertaining and more convincing when he argues his case through paradoxes and parodies than when, as a reformist himself, he allows his authoritarian tendencies to get the better of his delight in *la bagatelle.*

[107] R. Colie, *Paradoxica Epidemica,* Princeton, 1966, p. 20.

Part Two
FIRST FRUITS

I

THE FINAL years of the reign of William III were marked by increasing political activity both inside and outside Parliament. The Tory party, enjoying a new surge of popularity, grew more confident in harrying the Whig ministry until in January 1699 they actually succeeded in forcing through a Bill to reduce the standing army to a mere 7,000 men. The main Tory grievance, however, was William's practice of making large gifts of the Irish lands confiscated from Jacobite rebels to his personal friends, most of whom were foreigners. An official commission, demanded by the Tories, revealed that William Bentinck and Arnold van Keppel had both received estates of over 100,000 acres, and that the King's former mistress had been granted James II's personal estate, also of 100,000 acres.[1] A vociferous campaign was mounted to oppose these gifts; impeachments were demanded and eventually the Whig grandees were forced to sell off the estates. This, though only a token victory, was greeted by bonfires and parades of the unruly Tory mobs through the streets of London.

Back in Dublin, Swift and Berkeley, both Whigs, kept an anxious watch on these developments. 'Our government sits very loose', Swift remarked to Jane Waring, 'and I believe will change in a few months.'[2] In fact the commissions of all three Lords Justices were revoked within the month, and the administration of Ireland entrusted to a Tory Lord Lieutenant, the Earl of Rochester. Having once tasted victory, the Tories were eager to retain the initiative, and renewed their clamour for impeach-

[1] *Report of the Commissioners into Forfeited Estates in Ireland,* 15 December 1699. For this and further details of political events leading up to Swift's publication of *A Discourse of the Contests and Dissensions,* see F. H. Ellis's excellent edition of the *Discourse,* Oxford, 1967. As before, however, my own references are to the edition of the *Discourse* in the *Prose Works,* i.

[2] *Corr.* i. 34.

ments, this time over the partition treaties which William had negotiated in 1698 and 1700 for the division of the Spanish empire. These treaties had been concluded without the knowledge or consent of Parliament, but solely on the advice of the 'four lords partitioners', Portland, Orford, Halifax, and Somers. When Swift arrived back in London in April, he found the capital buzzing with rumours that the four lords were to be impeached, and on the 16th the Commons urged the King to remove these men 'from your council and presence for ever'. Tory mobs were countered by Whig petitions in defence of the four lords, ostensibly sent in by loyal subjects up and down the country, but actually written by Defoe and his associates in back rooms in the City. Throughout May an energetic paper war was waged in which the Whigs began gradually to recover their nerve, as the Tory onslaught seemed to ebb away. When on 17 June the House of Lords solemnly assembled to hear the impeachment proceedings against Somers, the drama of the situation rapidly turned to farce as the Commons, his accusers, simply failed to attend. The Tories were forced to concede defeat.

The experience of so much political volatility had a profound effect on Swift, who was engaged in editing Temple's *Miscellanea* for the press. The Third Part of Temple's work begins with an essay 'On Popular Discontents' in which Temple traced all 'factions, seditions, convulsions and fatal revolutions' of state back to 'a certain restlessness of mind and thought, which seems universally and inseparably annexed to our very natures.[3] Swift could hardly fail to draw comparisons between Temple's theories and the scenes which he witnessed in London, and he set out his own thoughts on these matters in his *Discourse of the Contests and Dissensions . . . in Athens and Rome.* Though largely composed during the summer, this work was not published until October, well after the impeachment proceedings had been dismissed. It is therefore less a defence of the Whig lords than a theoretical discussion of political discontents through the ages, though its starting point is Swift's conviction that 'the liberties of Athens and Rome' had been undermined by just such actions as had been recently proposed. The *Discourse* is an example of the popular genre of parallel history, which

[3] Temple, *Miscellanea, The Third Part,* ed. Swift, 1701. First Essay, 'On Popular Discontents', p. 7.

allowed political criticism to masquerade as historical analysis. However, Swift's historical framework is more than a camouflage for polemic. He offers his classical perspective as a corrective to the short views and narrow opinions of party strife, and begins by defining the balance of powers between the one, the few and the many which, he claims, exists within all societies. In English political terms these figures translate naturally into the king, the lords and the commons. Swift realizes his image of a balance by emphasizing the vital third element, the fulcrum which keeps the other two forces in equilibrium. Once such an equilibrium is destroyed, the result is tyranny. Much of the first section of the *Discourse* is devoted to proving that such tyranny need not be vested in a single dictator, but can equally well be exercised by the few–such as the thirty tyrants of Athens–or the many. In Carthage at the time of the second Punic war, he writes, 'some authors reckon the government to have been then among them a *dominatio plebis*'.[4] Lacking the phrase 'dictatorship of the proletariat', Swift is in no doubt that such a dangerously democratic state could exist, and that it would be a fatal denial of the 'mixed government' which he believed was both natural and rational. It is noticeable that, although he claims to deal impartially with the balance of power, all the examples which Swift chooses from Roman history illustrate the dangers of such a *dominatio plebis*. 'The people are much more dextrous at pulling down, and setting up,' he complained, 'than at preserving what is fixed.' His fears of over-mighty patricians were evidently less acute. He agreed with Temple in identifying the restlessness of post-lapsarian Man as the root cause of political disorders. 'Endless and exorbitant are the desires of men,' he lamented. Like Temple he believed that innovations should always be resisted, and that when they could not be avoided 'large intervals of time must pass between every . . . innovation, enough to melt down, and make it of a piece with the constitution'. Now and again he descends from these philosophical heights to draw specific lessons from more recent events.

The orators of the people at Argos (whether you will style them in modern phrase, *great Speakers in the House*, or only in general, representatives of the people collective) stirred up the *Commons* against the

[4] *PW* i. 199.

Nobles; of whom 1600 were murdered at once; and at last, the Orators themselves, because they left off their accusations; or to speak intelligibly, because they *withdrew their impeachments*; having it seems, raised a spirit they were not able to lay. And this last circumstance, as cases have lately stood, may perhaps be worth noting.[5]

Towards the end of the *Discourse* Swift ventures the conclusion that popular disturbances characteristically occur 'when the state would, of itself, gladly be quiet'. It is an indication of the shallowness of Swift's political philosophy here and elsewhere that he can write of the state having a will 'of itself' quite separate from the will or interests of the one, the few, or the many. It was this belief which allowed him, as the *Examiner*, to present the 'national interest' as a unified and uncontroversial goal, and to denounce all opposition as factional and self-seeking. He added that he conceived it 'far below the dignity, both of human nature, and reason, to be engaged in any party'. According to this high-minded view, the party member sacrificed the basic human right of thinking for himself, and became as much of an automaton as an acolyte among the Aeolists, moved by the spirit within him. Swift reinforces the point with language that echoes *A Tale of a Tub*.

He hath neither thoughts, nor actions, nor talk, that he can call his own; but all conveyed to him by his leader, as wind is through an organ. The nourishment he receives hath been not only *chewed,* but *digested,* before it comes into his mouth.[6]

Looking back, in 1714, Swift claimed that it was shortly after this, in conversations with Lord Somers in 1702, that 'I first began to trouble myself with the difference between the principles of Whig and Tory; having formerly employed myself in other, and, I think, much better speculations.'[7] This is disingenuous bluster to gloss over the embarrassing fact of his support for the Whigs in the *Discourse.* What he remembers telling Somers of himself was this:

Having been long conversant with the Greek and Roman authors, and therefore a lover of liberty, I found myself much inclined to be what they called a Whig in politics; and that, besides, I thought it impossible,

[5] Ibid., p. 199.
[6] Ibid., p. 234.
[7] *PW* viii. 120, 'Memoirs relating to . . . The Queen's Ministry'.

upon any other principle, to defend or submit to the revolution: But as to religion, I confessed myself to be an High-Churchman, and that I did not conceive how anyone, who wore the habit of a clergyman, could be otherwise.[8]

Swift saw no contradiction in being a Whig in politics and a Tory in religion. On the contrary, he saw it as a positive indication of his independence. But he goes further: he cannot conceive how anyone who is loyal to both Church and State can think other than he does. This belief, that his own position is the only one logically and loyally tenable, lies behind his presentation of the national interest as something rational, natural, and indivisible. He could never envisage the two-party system, as Burke did, as the most effective check upon an over-powerful executive. And yet, when the one-party state arrived with the carefully managed Parliaments of Walpole, and with it a period of unparalleled political stability, it was Swift himself who led an opposing 'faction'.

The ambiguities of Swift's position were the result of his combining a distrust of the mob with a desire to appeal to majority opinions and 'common sense'. This unacknowledged contradiction can be seen in many of his writings. In the *Discourse* he envisages an ideal House of Commons, comprised entirely of independent members, without any party labels. He allows that 'many pernicious and foolish overtures would arise', but believes that they 'would die, and disappear,

because this must be said in behalf of human kind; that common sense and plain reason, while men are disengaged from acquired opinions, will ever have some general influence upon their minds; whereas, the species of folly and vice are infinite, and so different in every individual, that they could never procure a majority, if other corruptions did not enter to pervert men's understandings and misguide their wills.[9]

There is an untypical woolliness about this. The assertion that reason is unitary, while folly is made up of infinite variations, has the superficial appeal of Tolstoy's opening aphorism in *Anna Karenina*: 'All happy families are alike, but an unhappy family is unhappy after its own fashion';[10] but it is equally false. Swift never

[8] Ibid., p. 120.
[9] *PW* i. 232.
[10] Tolstoy, *Anna Karenina*, tr. Rosemary Edmonds, Penguin Books, 1954.

ceased to expose the fallacy that common sense is actually common, but where else could he repose his trust, in his struggles with the virtuosi, the enthusiasts, and the party bloodhounds, if not in the large averageness of common humanity? It is a highly un-Swiftian contention that the common denominator of a group of five hundred men would be reason. Nor are those superadditions that he mentions, the 'acquired opinions' and 'other corruptions', so easily dismissed. The curious limpness here betrays an opinion which Swift cannot seriously support. In *A Tale of a Tub* he says something which seems similar, but makes a very different point.

> . . . all the virtues that have been ever in mankind, are to be counted upon a few fingers, but his follies and vices are innumerable, and time adds hourly to the heap.[11]

To conclude from this numerical discrepancy between the virtues and the vices of mankind that the virtues must be stronger, because each is shared by a greater number of individuals, is a curious piece of pseudo-logic that defies rational analysis.

Swift sent the *Discourse* 'very privately to the press, with the strictest injunctions to conceal the author'. He returned to Ireland in late September, and soon heard that the book was a great success, being 'greatly bought, and read'. It was an exciting opening to his career as a political writer, and soon 'the vanity of a young man prevailed with me, to let myself be known for the author'.[12] His time in London was not entirely taken up with politics, however, for Esther Johnson was much in his thoughts. She and Rebecca Dingley were serving in Lady Giffard's house in Dover Street, and it was now that he suggested that they should both come over with him to Dublin. There is no way of knowing how Esther understood this invitation, or whether she knew of his resolution 'never to marry a young woman'. In Swift's subsequent narration of the episode there is naturally no mention of any emotional tie, and financial considerations are given as the main reason for the move.

> going to visit my friends in England, I found she was a little uneasy upon the death of a person on whom she had some dependence. Her fortune at that time was in all not above fifteen hundred pounds, the

[11] *PW* i. 30. [12] *PW* viii. 119.

interest of which was but a scanty maintenance, in so dear a country for one of her spirit. . . . Money was then at ten per cent in Ireland, besides the advantage of turning it, and all necessaries of life at half the price.[13]

As Temple had been dead for two years, Esther's 'uneasiness' was hardly a new thing. Decoding this rather guarded recollection, one can sense that Swift wished to assume Temple's place in Esther's life, as the person on whom she depended. By asking her to leave her friends and family in England, he was imposing the same kind of test of her feelings as he had applied to Jane Waring. She had been unwilling to leave the 'sink' of her own family circumstances without a firm offer of an advantageous marriage, but Esther was prepared to make herself dependent upon Swift in a way 'very much for my satisfaction, who had few friends or acquaintance in Ireland'. It naturally seemed rather a curious arrangement for an attractive twenty-year-old girl to place herself under the protection of an unmarried thirty-four-year-old clergyman. Swift admitted that 'the adventure looked so like a frolic, the censure held, for some time, as if there were a secret history in such a removal'.[14] Whereas he usually liked to fly in the face of such gossip, preferring, as Sheridan wrote, to employ 'no other shield in his defence, but that of conscious integrity',[15] on this occasion he sought to minimize, not dramatize the incident. He declared that rumours soon died down 'by her excellent conduct', but this is not entirely true. Six years later his cousin Thomas showed that gossip still persisted when he wondered whether Jonathan had been 'able to resist the charms of both those gentlewomen that marched quite from Moor-Park to Dublin (as they would have marched to the north or anywhere else), with full resolution to engage him'.[16] 'Both those gentlewomen' is important. Esther's move to become Swift's Stella in Dublin would have been inconceivable without the fussy, forgetful, compliant Mrs Dingley to act as her chaperone. Dingley was the vital third element, the fulcrum in the balance of powers between Swift and Stella. There is a tradition that Swift and Stella were never alone together, but that a third person, usually Dingley, witnessed all their meetings. Contemptuous of the skin-deep morality of the world, Swift nevertheless

[13] *PW* v. 228.

[14] Ibid., p. 228.

[15] Sheridan, 'Introduction', fol. A2.

[16] *Corr.* i. 56.

clamped this strait-jacket of self-discipline upon his own deepest feelings. The pretence that his concern was for 'both those gentlewomen' equally was not merely a fiction to prevent gossip. It was also a deception which Swift practised upon himself, to prevent him from having to acknowledge the nature of his real feelings for Stella. Already in 1701 he had begun the habit of compressing their names to Ppt (poppet?) for Stella, Dd (Dear Dingley?) for Dingley, and most commonly MD (my dears?) for the two of them together. The Latin names Varina, Stella, Vanessa, still allowed too strong an erotic charge to filter through his defensive grid. These stern ciphers seem almost to defy personality. The series of almost daily letters which he sent to Stella from 1710 to 1713, and which are known as his *Journal to Stella* ought more accurately to be called his *Journal to MD,* since each letter is addressed to this composite persona. It was not sufficient that he should never see her alone. He did not even correspond with her alone, but wrote only such things as could be addressed equally well to Dingley.

Having arranged for the ladies' removal, Swift was keen to return to Ireland, but Lord Rochester was in no hurry to take up his post there, preferring to remain in London, where he could exert some political influence. Consequently the ladies were forced to travel to Ireland in August without the help and advice of their protector, and 'were very discouraged to live in Dublin, where they were wholly strangers'.[17] Eventually the Lord Lieutenant's entourage sailed from Holyhead on 17 September, and the following day Swift entered Dublin to a welcome of cannons, cheering crowds, and assembled civic dignitaries. Ducking out of the official reception as soon as he could, he hastened to visit the ladies in their new lodgings. Apart from one short trip to England in 1708, Dublin was to remain home for Stella and Dingley for the rest of their lives.

II

LORD ROCHESTER did not have long to settle in Dublin before fresh political changes sent him on his travels once more.

[17] *PW* v. 228.

Throughout the summer of 1701 the indications of a new European war became stronger and when, in September, James II died, Louis XIV alienated opinion in England completely by recognizing his son as the rightful heir to the crown. In the face of mounting war pressure the Tories were forced on the defensive and their triumphant gesture of cutting the army back to 7,000 men now seemed little short of treason. William took the opportunity to call an election which resulted, as expected, in a more compliant House of Commons. Rochester hurried back from Dublin, but was only in time to hear that the new Parliament had voted the King 80,000 men as England's contribution to the Grand Alliance.

Swift, who had no great liking for Rochester, did not accompany him back to London, preferring to stay near the ladies and help make their first winter in Dublin a happy one. In February he took his degree of Doctor of Divinity at Trinity. It cost him £44, about a fifth of his annual income. That same month King William was thrown from his horse and shortly afterwards died. Immediately a check was given to the recent surge of Whig successes. Princess Anne was not only a Stuart, but a woman of renowned piety, and a friend to the High Church party. From the beginning she let it be known that she would be especially active in protecting the interests of the Church. For an ambitious young clergyman it seemed an opportunity to be seized. Swift quickly completed the final volume of Temple's *Letters* for publication and made arrangements to visit England in April.

What Swift did not sufficiently acknowledge was that his position as a Whig in politics and a High Churchman in religion, put him on both sides of an increasingly sharp divide. The Whig Lords on whose behalf he had written his *Discourse* were happy to welcome him, but were powerless. If he really wished to gain the Queen's good opinion, then he should have severed his contacts with the Whigs, and cultivated new friends among Lord Rochester's acquaintance. In fact, with the same self-destructive instinct that had led him to throw in his lot with Sunderland in the week that Sunderland fell, he now became intimate with the Whig lords, in the month that they lost any power to further his career. Somers and Halifax both 'lamented that they were not able to serve me since the death of the king, and were very liberal in promising me the greatest preferments I could hope for, if it ever came in their power'.[18] Swift wrote this in 1714, as a further illustration of his theme of the unreliability of courtiers' promises. At the time he was more realistic, recognizing that 'nothing is so civil as a cast courtier'.

Swift's occasional annotations to the volume of Temple's *Letters* published in 1702, confirm that he was still instinctively a Whig at this time. In one place he notes that Temple had once informed him that Charles II had finally refused the bribe of a French pension when Louis XIV had insisted that England should never keep more than 8,000 men in its standing army.[19] This striking example of parallel history is a clear indictment of the Tories.

While in England Swift visited his mother in Leicester, and in August went to Berkeley Castle, where he found the Earl bemoaning the fresh crop of Tory successes in the recent elections. Swift quickly fell in with the friendly family atmosphere at the castle which he describes in his charming colloquial poem 'The Humble Petition of Frances Harris'. Composed in irregular, rambling couplets and full of bathetic rhymes, this poem in which Swift imitates the breathless, garrulous voice of Lady Betty's waiting-maid is the first of his many evocations of the muddled animation of below-stairs life. Swift's *Meditation on a Broomstick* also apparently began life as a Berkeley family joke. The Countess, a very pious lady, was fond of Boyle's *Meditations,* a partiality which Swift did not share. According to legend, once, when the Countess asked him to read aloud from Boyle's work, Swift slipped his own *Meditation* into the volume, and solemnly announced the title. 'Meditation on a broomstick!' exclaimed Lady Berkeley, 'bless me, what a strange subject! But there is no knowing what useful lessons of instruction this wonderful man may draw from things apparently the most trivial.'[20] Sheridan goes on to describe how Swift read the piece 'in the same solemn tone' and how Lady Berkeley, not in the least suspecting any trick, interrupted him to express 'her admiration of this extraordinary man, who could draw such fine moral reflections from so contemptible a subject'. This parody of the homiletic style, which is one of Swift's most accomplished

[18] *PW* viii. 119.

[19] Temple, *Letters to the King . . . and Other Persons,* ed. Swift, 1703, pp. 355-6.

[20] Sheridan, pp. 43-4.

squibs, concludes with an erotic innuendo which must have passed unnoticed by the Countess. Comparing man to a broomstick, Swift concludes:

His last days are spent in slavery to women, and generally the least deserving; till, worn to the stumps, like his brother besom, he is either kicked out of doors, or made use of to kindle flames for others to warm themselves by.[21]

The Countess apparently took the joke in good part, and Swift was spared any further readings from the works of Robert Boyle.

In October Swift left Berkeley Castle and returned to Ireland, where he remained for over a year. His account book for that year gives a detailed picture of the material conditions of his life. He passed much of his time at Laracor, where he spent large sums to improve the glebe, the church, and the surrounding area. Throughout the rest of his life he spoke with great affection of the river walk where he would fish for trout and pike, of the willows along the canal, and of the holly bank and cherry trees. His care of the parish was such that in 1723 the church was officially described as 'in very decent repair . . . handsome, well-built . . . and furnished with all conveniences except a surplice and carpets'.[22] The 'good garden' and 'exceedingly well-enclosed' glebe were also mentioned with approval. All this was a shining exception to the general rule of ruin and disrepair in most of the churches of Ireland.

Although Swift preached only intermittently at Laracor himself, he took care that the curate to whom he entrusted the spiritual duties of the parish was suitable in character and outlook, and was prepared to pay him £57, well above the usual salary. Parochial duties were hardly arduous, since Rathbeggan contained no Protestants, and Laracor only sixteen Anglican families, including two Members of Parliament. After paying his curate, Swift's next largest expense was the £50 annual allowance that he paid to Stella throughout her life.[23] His servant, 'the rogue John Kemp', cost him £4.10*s.* in wages, £10 in board and lodgings, and further incidental expenses for clothes and shoes.

[21] *PW* i. 240.

[22] Landa, p. 39.

[23] Swift's account books, detailing his income and expenditure for several years, are to be found in the Forster Collection in the Victoria and Albert Museum: 48.D.34/1-9.

Swift's own clothing was another substantial item in his expenditure. In January he spent £1.12*s.* on seven pairs of shoes, and a pair of galoshes for the wet weather. His passion for long walks made this a recurrent expense. In July he spent £5 on a shirt, nightgown, cap, and slippers. A fortnight's winter ale cost him 17*s.* 8½*d.*, and in February he spent 7*s.* 6*d.* on coffee. He carefully recorded each item of income and expenditure, down to the last farthing, in his account book. Christmas boxes to servants, and gifts to relatives and friends are all listed. In the spring he spent 11*s.* 3*d.* on some little 'treats' for Stella. He also kept a careful tally of exactly how much he won or lost at cards, and, in the early years at least, it is noticeable that he usually managed to end the year on the winning side. Horses form another large item in Swift's accounts. In January he bought a bay nag for £4.5*s.*, and in March a chestnut horse for £3.15*s.* Hay, oats, barley, saddles, etc. were all costly items, and altogether he spent over £20 of his year's salary on his horses.

During 1703 two of Swift's superiors in the Church of Ireland were promoted to positions which they would occupy for much of his career. William King began his long reign as Archbishop of Dublin in March, upon the elevation of Narcissus Marsh to become Primate. Both men were erudite and conscientious in their different ways, but neither had the kind of mind to appreciate the eccentric talents of the vicar of Laracor.

Towards the end of the year Swift began to prepare for another visit to England. He was careful to introduce the ladies to a number of his Dublin friends to ensure that they should not feel lonely in his absence. Among those they met was William Tisdall, a facetious fellow-collegian from Trinity, with a high opinion of his own literary skills. Tisdall was a man whom Swift indulged rather than respected, but who might prove an adequate diversion while he was away in London.

III

SWIFT LEFT Dublin on 11 November and landed in England two days later. As usual he spent a few days with his mother in Leicester, and arrived in London at the end of the month. There he found that a new political storm was raging over the Sacramental Test which excluded all but communicants of the

Anglican Church from holding public office. During the previous winter the Tories had forced a Bill through the Commons to end 'occasional conformity', the tolerated deceit by which Dissenters qualified themselves for office by taking communion once a year in the Church of England. That Bill had failed in the Lords, but now the Tories were eager to renew their campaign. The government, led by Marlborough and Godolphin found its measures to prosecute the war with France hampered by a High Church parliament more interested in doing battle with Dissenters. Queen Anne quickly realized that she could not promote this High Church interest without seriously damaging the war effort, and in her speech from the throne she advised Parliament to 'avoid any heats and divisions'.[24] This was a clear indication that she was not prepared to support the Tories' Bill, which the Lords promptly rejected. It was a situation which placed further strains on Swift's position of neutrality, though he pretended otherwise in the facetious account of events which he sent to Tisdall. The previous ten days had demonstrated 'the highest and warmest reign of party and faction that I ever knew', he declared. Even 'the dogs in the street' were more contumelious and quarrelsome than usual' and 'a committee of Whig and Tory cats had a very warm and loud debate upon the roof of our house'.[25] Swift's apparent high spirits may be the result of a sense of relief, for he had been 'mightily urged' by the Whig lords to follow up the success of his *Discourse* with an attack on this latest manifestation of Tory demagoguery. Their invitation faced him with a clear clash between his principles and his hopes of preferment. His visits to London, not easily financed out of his clerical salary, were all aimed at securing just such a foothold on the wider political landscape of England as the Whig lords seemed to offer. Yet he could not blind himself to the fact that they did not regard the Test in the same light as he, for whom it was the cornerstone of the relationship between Church and State. In his letter to Tisdall he tries, unsuccessfully, to demonstrate that opposition to the Bill was not the same as opposition to the Church.

I cannot but think (if men's highest assurances are to be believed) that several, who were against this bill, do love the church, and do hate or

[24] Ehrenpreis, ii. 118. [25] *Corr.* i. 38-9.

despise Presbytery. I put it close to my Lord Peterborough just as the bill was going up, who assured me in the most solemn manner, that if he had the least suspicion that rejecting this bill would hurt the Church, or do kindness to the Dissenters, he would lose his right hand rather than speak against it. The like professions I had from the Bishop of Salisbury, my Lord Somers, and some others; so that I know not what to think, and therefore shall think no more.[26]

It is a damning indictment of the position into which Swift is attempting to talk himself that he prefers to 'think no more' about it. His face-saving parenthesis, 'if men's highest assurances are to be believed', betrays far more doubts than certainties. Peterborough's colourful phrase is a typical piece of soldierly hyperbole, and Somers had ulterior reasons for wishing to retain Swift's journalistic skills. One cannot help suspecting that these assurances were given with a hollow insincerity that enjoyed watching Swift squirm in his dilemma. 'Climbing', as Swift later observed, 'is performed in the same posture with creeping.'[27]

Eventually he did write against the Tory Bill, 'but it came too late by a day, so I would not print it'.[28] How lucky for Swift's conscience that day was, and how revealing is his refusal to print. His *Discourse* came too late by several months to defend the Whig lords against impeachment, but he was still happy to publish it, to gain their friendship, and to offer some general remarks on politics. This time he was only too glad to seize on the excuse of being late to suppress a work that conflicted with some of his deepest instincts.

'Pox on the Dissenters and Independents', he wrote to Tisdall in February, in the same strained tone of affected flippancy.[29] His attitude to Tisdall in these letters is one of barely concealed condescension, and he mocked Tisdall's earnest desire to join him in the world of political pamphleteering.

I look upon you as under a terrible mistake, if you imagine you cannot be enough distinguished without writing for the public. Preach, preach, preach, preach, preach, preach; that is certainly your talent.[30]

Swift, who saw it as a slightly comic presumption for Tisdall to

[26] Ibid., p. 39.
[27] 'Thoughts on Various Subjects', *PW* i. 245.
[28] *Corr.* i. 44.
[29] Ibid., pp. 43-4.
[30] *Corr.* i. 43.

attempt to emulate his own hard-won achievement, used the younger man as a puppet to be manipulated at long range for the ladies' amusement, an actor for his own jests and anecdotes. In December he sent Tisdall a script to perform, teaching him how to outwit Stella in the latest court game of 'bites' or riddles: 'You must ask a bantering question, or tell some damned lie in a serious manner, and then she will answer, or speak as if you were in earnest: and then cry you *"Madam,* there's a *bite."* '[31] Evidently 'biting' was all the rage at the ladies' house in William Street that Christmas. Stella conspired with Swift in leading Tisdall on, pretending to admire his wit, gallantry, and skills in argument. Yet while enjoying the joke, Swift sought to keep a firm hold on his puppet. 'You seem to be mighty proud', he wrote, '(as you have reason if it be true) of the part you have in the ladies' good graces, especially of her you call the *party*; I am very much concerned to know it.'[32] Swift's arch humour here conceals more than a germ of unease. Part of the joke between himself and Stella was to watch Tisdall's attempts to supplant his puppet-master in her affections. But even Swift could not be entirely sure, and no doubt would often mock Tisdall in more basic ways. 'Do his feet stink still?' he asked her some years later.[33] It sounds like an aside that had been long shared.

Tisdall had been indiscreet enough to show one of Swift's letters to Primate Marsh. Swift promised to put an end to such liberties by including in all future letters something that would make them unfit to be seen by any but Tisdall himself. Thus, commenting on the 'evil' of Tisdall's new-found favour with the ladies he includes this tale.

> A cast mistress went to her rival, and expostulated with her for robbing her of her lover. After a long quarrel, finding no good to be done; 'Well', says the abdicated lady, 'keep him and stop him in your a--.' 'No,' says the other, 'that will not be altogether so convenient; however to oblige you, I will do something that is very near it.'–*Dixi.*[34]

Swift was confident that Stella enjoyed seeing Tisdall as piggy-in-the-middle as much as he did, and could not take him seriously as a rival, yet this sudden descent into smuttiness is a characteristic indication that a nerve has been touched. Like Dissenters

[31] Ibid., p. 40.
[32] Ibid., pp. 41-2.
[33] *Journal,* ii. 671.
[34] *Corr.* i. 42.

and Independents, Tisdall was no more worth writing against than a louse or a flea, but this, beneath the patronizing banter, is a warning shot.

A few weeks later Swift heard from Tisdall that he wished to marry Esther Johnson. Suspecting that Swift had similar, though unavowed intentions, Tisdall wrote asking for a clarification. This was the last thing that Swift could supply. His relationship with Stella was another of those sensitive areas about which he preferred to think no more. The present situation was for him so ideal that he chose to think it natural and permanent. Yet clearly, in the eyes of the world it was a highly irregular relationship. To cover his confusion he replied with a mixture of bluster and attack. 'You talk in a mystical sort of way, as if you would have me believe I had some great design, and that you had found it out.[35] In the absence of Tisdall's letter it is difficult to be certain, but there is every reason to think that his intentions and enquiries were perfectly clear. It is Swift who weaves a mystical veil around the answer which is, he asserts 'upon my conscience and honour . . . the naked truth'. With similar protestations of completed sincerity he had prefaced his disingenuous letter to Jane Waring.

Stella was probably quite happy to receive Tisdall's proposal, believing that it would finally provoke Swift into proposing himself. It was to Swift, and not her mother, that she turned for advice on the matter. But she was mistaken. Her appeal to him merely confirmed Swift's conviction that her own good sense and taste could be relied on to reject Tisdall. It is important to recognize that Swift believed he was behaving with complete candour and impartiality in this matter. His own deep self-interest did not allow him to perceive that a relationship which seemed ideal to him might be less than ideal for her. 'What they *do* in heaven we are ignorant of; what they do *not* we are told expressly; that they neither marry nor are given in marriage';[36] this was a 'thought' which he composed shortly after this.

As usual, his main arguments against the match were financial. 'I did not conceive', he wrote to Tisdall in April, 'you were then rich enough to make yourself and her happy and easy.'[37] This did not satisfy Tisdall who had just received the promise of a good

[35] Ibid., p. 45. [36] *PW* i. 244. [37] *Corr.* i. 45.

living in Donegal and Swift was forced to withdraw this objection when Tisdall protested.

The objection of your fortune being removed, I declare I have no other; nor shall any consideration of my own misfortune of losing so good a friend and companion as her, prevail on me, against her interest and settlement in the world, since it is held so necessary and convenient a thing for ladies to marry; and that time takes off from the lustre of virgins in all other eyes but mine.[38]

The juxtaposition of the cold, formal language of interest and settlements with the maudlin wistfulness of the virgins' lustre, is an obvious piece of emotional blackmail. Even as Swift makes his gesture of renunciation, it must have been clear to Tisdall that he was not renouncing Stella at all.

I think I have said to you before, that if my fortunes and humour served me to think of that state, I should certainly, among all persons on earth, make your choice; because I never saw that person whose conversation I entirely valued but hers; this was the utmost I ever gave way to . . .

That phrase 'the utmost I ever gave way to' is highly revealing, confirming again Swift's horror at any form of emotional dependence. Once more he gambled with disaster and rejection, but this time he won. Stella accepted that these guarded and defensive remarks were the strongest testimonies he could give of his affection for her. She refused Tisdall who quickly recovered from his disappointment, and married Miss Eleanor Morgan of Co. Sligo. Meanwhile in London Swift grew tired of living off the 'good words and wishes of a decayed ministry'. It was clear that the Whigs could do little to promote his ambitions at present, and 'therefore I am resolved suddenly to retire, like a discontented courtier, and vent myself in study and speculation'.[39] Having found himself caught out in the contradictions of his position both in London and Dublin he decided to hasten home to cling on to the security that he needed most.

He left behind him the manuscript of *A Tale of a Tub* which finally appeared on 7 May 1704. By dedicating the work to Somers, Swift sought, somewhat quixotically, to align himself and Somers with a work written ostensibly, though eccentrically, in defence of the Anglican Church. He chose badly, however. For

[38] Ibid., pp. 45-6. [39] Ibid., p. 46.

those readers who considered the book an affront to religion, this dedication merely confirmed their suspicions. It was from desperation at such reactions to his work that Swift added in 1710 a prefatory 'Apology' in which he protests that *A Tale of a Tub* 'celebrates the Church of England as the most perfect of all others in discipline and doctrine, it advances no opinion they reject, nor condemns any they receive'.[40] He points out, with exasperation 'that there generally runs an irony through the thread of the whole book . . . which will render some objections that have been made very weak and insignificant'. But irony is the flag of convenience invariably hoisted by satirists who intend sailing close to the wind. If Swift had delayed publication of the *Tale* much longer, he might never have published it at all. The time would never be right for such an idiosyncratic and provocative work. Without it his career might well have been smoother, but his achievement very much less.

Leaving Leicester on 29 May he found a favourable wind at Neston, and reached Ireland on 1 June. It was three and a half years before he returned to England. There were several reasons for this prolonged residence in Ireland. The first and most important was the shock of Tisdall's proposal to Stella. Swift needed the secure home base of the ladies in Dublin and was prepared to acknowledge the fact by his presence, if not by a proposal. Secondly he was living beyond his means, finding it impossible to maintain a life near the court in London on the revenues of an Irish country parish. Moreover Archbishop King took a dim view of absentees, and having failed to persuade the Whig lords to further his career as a man of affairs, Swift could not afford to antagonize his superior as a man of the cloth. He went straight to Laracor after landing and was in his parish by 3 June. Later that year he very pointedly headed a letter to King, 'Trim', to assure him that his tiny flock there were receiving all the care to which membership of the Anglican establishment entitled them.

Swift recognized that further trips to England might most conveniently be undertaken if he could travel as an official representative of the Irish Church. On her birthday Queen Anne had consented to remit the Crown's right to the 'First Fruits and

[40] *PW* i. 2.

Twentieth Parts' of the Church of England, and had agreed to use the revenues for augmenting some poorer livings.[41] The Convocation of the Church of Ireland quickly petitioned for a similar remission to be made in Ireland. The Queen made them a 'gracious answer' but no action followed. What was needed, Swift believed, was an envoy from the Church of Ireland to solicit the ministry in London to give effect to this remission. Moreover, if her Majesty would consent to give up not only the First Fruits, but also the Crown rents due on Church properties in Ireland, the value of the remission would be doubled for rather less than the cost of the grant made to the Church of England. He lost no time in seeking to ingratiate himself with King. Hearing that the Archbishop was about to leave for England, Swift wrote to ensure him that when he had been in London that winter 'at a great man's table who hath as much influence in England as any subject can well have'[42] he had constantly spoken up on King's behalf. Many of his letters to King have an obsequious tone that reflects little credit on Swift or the Archbishop. But King had no very high opinion of wit, nor any great skill in detecting it, and Swift exaggerated his humility for fear of having it pass unnoticed. He went on to say that he was confident 'with some reason' that remission of the First Fruits and of the Crown rents 'would be easily granted'.[43] King was less sanguine, thanking Swift for his support but declining to 'meddle' with the First Fruits yet. He was particularly pessimistic about the Crown rents and was, he confesed, 'a little afraid to ask too much'.[44] In February the Dean of St Patrick's died, and in the little constitutional struggle which followed, Swift was careful to appear as a model of orthodoxy.[45] Following King's own precedent the canons decided to elect a new Dean, and nominated John Stearne, chancellor of the cathedral. However, Edward Synge, one of the men who had refused the Deanery of Derry when Swift had sought it, now sought the Deanery of St

[41] The First Fruits and Twentieth Parts were ancient fees paid to the Crown by clergymen. On her birthday in 1704 Queen Anne agreed to remit these fees, a decision subsequently known as Queen Anne's Bounty.

[42] *Corr.* i. 48.

[43] Ibid., p. 49.

[44] Ibid., p. 50.

[45] William Monck Mason, *The History . . . of the . . . Cathedral Church of St Patrick's,* Dublin, 1820, pp. 217 ff.

Patrick's, and applied directly to the Crown to pre-empt any election. Both sides moved quickly. With a nice irony the canons chose Synge's brother Samuel, together with Swift, to represent their case to the Lord Lieutenant and the city council. On 9 March Edward Synge's patent was approved by the Crown. On the 20th Swift organized an election, while acting as Stearne's campaign manager. 'I was the most busy of all your solicitors', he reminded him many years later,[46] when Stearne revealed himself as another promise-breaking friend. Stearne was duly elected and Swift lost no time in communicating the result to King in London, urging him to complete the appointment with all possible speed. He need not have worried. King, who referred to Stearne as his 'bosom friend' had actually confirmed the appointment on the day before he received the election result.

Stearne's promotion left vacant his rectory at Trim which was soon occupied by Anthony Raymond, a man nine years younger than Swift. Over the years Raymond and his wife were to prove trusted and useful friends to both Swift and Stella. Little else at Laracor changed, with its congregation of between ten and fifteen 'most of them gentle, and all simple'.[47]

In 1703 the Convocation of the Church of Ireland assembled for the first time in almost forty years, and its debates, which took place in the chapter room of St Patrick's, only increased Swift's anxieties about the relationship between Church and State. As in England, Convocation was divided between an Upper house where the bishops sought to maintain good relations with the administration which had appointed them, and a Lower house acutely aware of the deplorable conditions of unpaid tithes, unrepaired churches, and unviable livings in many country parishes. For many Anglican landlords in Ireland, membership of the Church was little more than a ticket to power. They resented the tithes, which were the price of that ticket, and disapproved of occasional conformity which made the temporal benefits of their faith only too obvious. Swift remarked of his wealthiest parishioner, John Percival, that 'he would not lessen his rent-roll to save all the churches of Christendom'.[48] Nothing brought out this conflict between the temporal and spiritual wings of the Anglican community more

[46] *Corr.* iv. 182. [47] *Corr.* i. 163. [48] Ibid., ii. 236.

clearly than the different interpretations which both parties put on their shared desire to prevent the further growth of popery. The dispossession and disenfranchisement of the Catholic majority had concentrated wealth and power in the hands of Anglican landlords, and provided them with a native force of labourers, tenants, and servants bereft of civil rights. The last thing they wanted was for the Church to take on an evangelistic role, converting these subservient papists to Anglicanism and civil liberty. When Convocation discussed the idea of encouraging the use of the Irish language in churches, it was paid only the scantest of lip-service. For their part, the landlords were keen to introduce a Bill for improving the manufacture of hemp and flax, which would decrease the amount of tithes which they paid. Debates on these matters were conducted in a climate of such suspicion and jealousy, and reached such a pitch of intensity, that the Lord Lieutenant Ormonde was forced to adjourn both Parliament and Convocation. Amid all these arguments the plan for the remission of the First Fruits was entirely neglected. 'Some men are very dextrous at doing nothing,' remarked Archbishop King, sardonically.[49]

In June 1705 Ormonde departed for England, leaving Ireland in the care of two Lords Justice, Lord Cox and General Cutts. The General's reputation for bravery, or bravado, had earned him the nickname 'salamander', a soubriquet which Swift borrowed for a satirical attack on Cutts in his poem 'The Description of a Salamander'.[50] Savagely adapting the terms of Pliny's description of this fiery creature in his *Natural History,* Swift depicts the general as 'a battered Beau, by age and claps grown cold' who spews out a contagion of leprosy and the pox.

Later that year Dubliners observed with some vexation the preliminary moves towards a union of the kingdoms of England and Scotland, and reflected ruefully on the discrepant treatment accorded by England to her two neighbours. Ireland was a loyal Anglican country with a legislature and economy subservient to England's demands, while Scotland flaunted its independent Parliament, Presbyterian religion, and freedom of trade. Yet it was Scotland that was being offered a free partnership with England, while Ireland was being steadily reduced to the status of a

[49] Richard Mant, *History of the Church of Ireland,* 2 vols., 1840-1, ii. 178.
[50] *Poems,* i. 82-5. See Pliny, *Natural History,* x. 67.

colonial possession. Throughout 1706, as the *Dublin Gazette* reported the details of the proposed Anglo-Scottish treaty to its Irish readers, feelings of envy spread, which were represented by Swift in his *Story of the Injured Lady*. This short pamphlet, written before the completion of the Act of Union in May 1707, but not published in Swift's lifetime, presents Ireland's grievances as the complaint of a cast-off mistress against her former suitor who is now engaged to marry a cunning gold-digger. Swift's arguments, though sharpened by the specific political context, show the clear influence of Molyneux's treatise. *The Case of Ireland* (1698), which had argued that Ireland was not a dependent colony, but a sister kingdom with as many independent rights under the Crown as England. A friend and admirer of Locke, Molyneux had asserted the 'universal law of nature' which the American colonists were later to adopt, 'of being governed only by such laws to which they give their own consent by their representatives in parliament'.[51] The injured lady's story allegorizes Molyneux's thesis that Ireland had submitted only to English protection, not to conquest. She confesses 'with shame, that I was undone by the common arts practised upon all easy credulous virgins, half by force and half by consent, after solemn vows, and protestations of marriage'.[52] Now, as she grows 'pale and thin with grief and ill-usage' she finds her servants dismissed, and her household run by her lover's domineering steward. Meanwhile her rival, a sluttish, malodorous, and ill-shaped creature, who had once set her lover's house on fire, receives a proposal of marriage. Swift's own recent matrimonial hesitations give an added piquancy to the allegory, which is coherent and clever but lacks the crusading anger of his later Irish tracts.

In April 1707 the Duke of Ormonde was replaced as Lord Lieutenant of Ireland by the Earl of Pembroke. Ormonde's Anglo-Irish roots had inclined him to sympathize with Irish grievances, but Pembroke, who added this Irish post to Presidency of the Privy Council, could be relied upon to be more responsive to the ministry in London than to Dublin city council. Swift had serious reservations about the new appointment, but he soon found that Pembroke, who combined an amateur

[51] Molyneux, *The Case of Ireland,* 1698, p. 48.
[52] *PW* ix. 5.

interest in arts and antiquities with a relish for humorous and witty conversation, could be a most congenial companion. Pembroke brought with him Sir Andrew Fountaine, a young man who might have sat as the model for the virtuoso of the age. He had done the Grand Tour, dabbled in Anglo-Saxon antiquities, collected drawings by the Italian masters, published a monograph on Danish and Saxon coins, and had a fashionable smattering of the 'new science'. In addition he shared with Pembroke a liking for puns and riddles, which quickly endeared them both to Swift.

Indeed Dublin Castle soon became the setting for extended logomachic exchanges between the Ashe brothers, Swift, Fountaine, and Pembroke. A short extract from one of their joint productions, the *Petition to my Lord High Admiral* (i.e. Pembroke) illustrates the remorseless quest for puns that so delighted Swift, and in which he was surpassed only by his fellow Dubliner, James Joyce.

> I have a *Deal* of stories to tell your lordship, and tho' you may have heard them before, I should be glad to *Chatt'em* over again; but I am now sick, tho' I hope not near *Graves'end*.[53]

In a similar vein of intellectual exercise, Swift compiled at this time his first collection of *Thoughts on Various Subjects*, modelled on the *Maximes* of La Rochefoucauld. From his first 'thought'

> We have just religion enough to make us *hate*, but not enough to make us *love* one another

to his last,

> Censure is the tax a man pays to the public for being eminent.[54]

he maintains a sardonic assault upon any assumptions of human benignity. Twenty years later he boasted that La Rochefoucauld was his favourite author, 'because I found my whole character in him'.[55] He never tired of reading such finely turned cynical epigrams as these:

> la modération des hommes dans leur plus haute élévation est un désir de paraître plus grands que leur fortune.

[53] Ibid., iv. 261. [54] Ibid., i. 241-5. [55] *Corr.* iii. 118.

Or,

> si on juge de l'amour par la plupart de ses effets, il resemble plus à la haine qu'à l'amitié.[56]

That last thought must have often entered his head during the vicissitudes of his long relationship with Vanessa.

Although Swift found Pembroke personally congenial, he was firmly opposed to him politically. For the new Lord Lieutenant had been sent to Ireland with strict instructions to have the Sacramental Test abolished. Throughout 1707 a war of slogans was waged on the issue. 'Unity' was the slogan of the abolitionists, meaning a unity of all Protestants against the Jacobite menace. 'Loyalty' was the watchword of the testers, loyalty to the Church, in the face of those whose 'disaffection to the public good' made them dangerous. This was a conveniently vague designation, meant to include both Dissenters and Catholics. The result was never in much doubt, however, since the Testers enjoyed a large majority in Parliament. Archbishop King observed that 'upon trial, it proved that nothing was more averse to the universal inclination of the parliament' than abolition.[57] Dodington, Pembroke's secretary, was forced to agree, and wrote home to Sunderland in rueful tones. 'Believe me, this country is priest-rid near as much as the Portuguese and Spaniards'.[58] Probably the very size of the majority helped Swift to remain on good terms with Pembroke, though political differences always threatened to disrupt their amicable relationship. 'Whig and Tory has spoiled all that was tolerable here, by mixing with private friendship and conversation, and ruining both',[59] Swift complained, although he claimed to think it as useless for a private man to argue about the relative merits of Rochester and Godolphin, as about Copernicus and Ptolemy. No doubt it was with such bluff disclaimers as this that he turned the edge of any dispute with Pembroke, and converted it into a less contentious trial of wit.

In fact, despite his involvement in public affairs, Swift spent much of this three and a half years in Ireland enjoying the simple

[56] La Rochefoucauld, *Maximes*, 1678, 18, 72.

[57] Mant, ii. 186.

[58] J. A. Froude, *The English in Ireland in the Eighteenth Century*, 3 vols., 1872-4, i. 321-2.

[59] *Corr.* i. 55.

pleasures of an unambitious private life. In particular he enjoyed the company of Stella and Dingley, conversation, cards, and games of wit in William Street, or riverside walks among the willows and cherry trees at Laracor. It was during this period that his cousin Thomas wrote to ask whether Swift were now married, having perhaps heard from John Temple that Swift was so content in Ireland that he could not be persuaded to return to Moor Park. Actually Swift had told Temple that it was mere poverty that kept him in Ireland, though he somewhat belied this by admitting to having put on weight. When he was back in the political turmoil of London, the memory of these idle days would often recur to him, like a breath of fresh air.

> Oh, that we were at Laracor this fine day! The willows begin to peep, and the quicks to bud . . . I was a-dreamed last night that I eat ripe cherries. And now they begin to catch the pikes, and will shortly the trouts . . . [60]

In July Swift was elected proctor of the Lower House of Convocation, some small recognition of his three years of clerical orthodoxy. This elevation came at just the right moment, for Convocation was again considering making a concerted effort to gain remission of the First Fruits. By the end of the year he was arranging to return to England, but this time as the Church's official representative to negotiate this business. King informed Pembroke that Swift was commissioned to 'put his Excellency in mind of the business of the First Fruits', and for his part Pembroke 'desired that it might be so'.[61] As usual, however, the Church bureaucracy failed to supply Swift with the necessary certificates of his status before his departure, and he complained that he was left "in the dark' about the exact extent of his powers. However, on the eve of his fortieth birthday he set sail once again for England in a wartime convoy of three vessels. He was on his way back to the centre of the political world, with a commission to the Great Men in London.

IV

AFTER AN absence of almost four years, Swift was astonished by the newness of things in England. Compared with Dublin, even

[60] *Journal*, i. 220. [61] *Corr.* i. 60.

Leicester had the appearance of a brave new world, as he told King: 'The buildings, the improvements, the dress and countenance of the people put a new spirit into one. . . . Here is a universal love of the present government'.[62] When he reached London, he stayed first with Andrew Fountaine in Leicester Fields, and then went to visit the Berkeley family again. He bubbled over with an infection of high spirits, and wrote back to Archdeacon Walls whose wife Dorothy had caught the punning bug from him before he left, with the latest flashes of his wit.

A gentleman was mightily afraid of a cat; I told him it was a sign he was pus-ilanimous, and Lady Berkeley talking to her cat, my lord said she was very impertinent, but I defended her & said I thought her ladyship spoke very much to the poor-pus.[63]

He was particularly pleased when Stella and Dingley decided to come over to England to visit him for a few months. Although he would tease Stella, claiming that she 'cannot make a pun if she might have the weight of it in gold', she no doubt entered into the spirit of repartee with Fountaine and the others. Together with the ladies Swift did the round of the London sights, listened to nightingales and made a trip to Greenwich to see the new naval hospital, where Stella's dog Pug chased deer in the park. They visited Lady Giffard's house to see Stella's mother and may also have called upon another Dublin family, recently arrived in London, the Vanhomrighs. Mrs Vanhomrigh, the widow of an émigré Dutch merchant who had acted as commissary-general to the army in Ireland, had just established herself in lodgings near St James's Square. She had four children of whom the eldest was the vivacious twenty-year-old Hester. Swift had first encountered the family at an inn at Dunstable where it seems Hester had initiated a conversation by spilling a cup of coffee.

Swift had not been long in London before he became an accepted member of a literary group which included Addison, Steele, Congreve, and Ambrose Philips. Addison, five years younger than Swift, had a reputation for aloofness, but seems to have struck up an immediate and close friendship with him. Yet when Swift contrasted his own career with that of this elegant young man, it would have been remarkable if he had not felt a

[62] Ibid., p. 58. [63] Ibid., p. 66.

certain envy. From an ecclesiastical background similar to his own, Addison's career had followed a smooth and continuous ascent. At every point where Swift had encountered checks and reverses, Addison had been fêted with success. From Charterhouse he had gone on to distinguish himself at Magdalen College, Oxford. Early verses, dedicated to Somers and Halifax, had resulted in a government pension of £200 which enabled him to make the Grand Tour. When he returned he was offered first a lucrative position in the Excise office, and subsequently promoted to an Under-Secretaryship, at £400 per year. Addison had the perfect blend of diplomatic and literary skills to succeed in an age when poets needed to be not only courtiers, but preferably bureaucrats as well. Richard Steele frequently made a third at dinner with Addison and Swift throughout 1708. Born in Dublin, he had left Christ Church without a degree, and had joined the army, dedicating a poem to Lord Cutts which promptly gained him a commission in the Coldstream Guards. Needless to say it was very different in tone from Swift's poem to Cutts. Where Addison was cautious, reserved, and pompous, the perfect civil servant, Steele was thriftless, impetuous, and indiscreet. Addison had a bank manager's respect for money, Steele a Micawberish fecklessness.

Swift's friendship with these men began as Swift was developing the most elaborate of his many April Fool's Day hoaxes.[64] In mid-February, writing under the name of Isaac Bickerstaff, Swift published a pamphlet called *Predictions for the year 1708*. In this parody of an astrological almanac the first prediction, casually described as 'but a trifle', is the key to a complex satiric strategy.

> It relates to Partridge the almanac-maker; I have consulted the star of his nativity by my own rules, and find he will infallibly die upon the 29th of March next, about eleven at night, of a raging fever.[65]

John Partridge, who published annually his almanac *Merlinus Liberatus* was the best-known astrologer of the time, and his practice flourished as speculation and gambling of all kinds increased. Even Marlborough's officers would regularly lay bets

[64] The 'Bickerstaff Papers' are in volume ii of the *Prose Works* (1939), pp. 141-64.

[65] Ibid., p. 145.

on the outcome of future battles, and those without the benefit of such inside information often invoked the aid of the stars to get ahead of the odds. A former cobbler, Partridge had turned to this more lucrative trade early in life, but it was not merely for exploiting the credulity of the gullible that Swift attacked him. Partridge combined astrological beliefs with a fierce brand of Nonconformist zeal, which found expression in such unsubtle doggerel as this:

> High Church! the common curse, the nation's shame.
> 'Tis only Pop'ry by another name.[66]

To Swift, Partridge presented the perfect image of the modern fanatic. However, recognizing the irrationality of attempting to discredit irrationality by an appeal to reason, Swift pretends not to undermine Partridge, but rather to blow him up until he bursts. Bickerstaff is a super-inflated Partridge, a 'true astrologer' who seeks to rescue his art from the 'gross abuses', the 'nonsense, lies and folly' foisted on the public by Partridge. By dating Partridge's 'death' on 29 March Swift not only primed his elaborate April Fool's Day hoax, but also allowed seven weeks for the 'bite' to take. Having served notice on this overmighty cobbler, Bickerstaff proceeds like an epidemic to cut a swathe through the enemies of Protestantism. On 4 April the Archbishop of Paris will die; on the 11th the young Prince of Austria; on 7 May the Dauphin; on 29 July Louis XIV, and on 11 September the Pope himself will fall.

Swift's plan succeeded beyond his wildest expectations. His own penny pamphlet sold in multitudes and pirate publishers were soon offering halfpenny reprints, replies, and imitations. At least one of these has been plausibly attributed to Swift himself though it was not published at the time. In it the supposed author, a 'person of quality' purports to see through Bickerstaff's pose, asserting that he is, in reality,

> without question a gentleman of wit and learning, although the piece seems hastily written in a sudden frolic, with the scornful thought of the pleasure he will have, in putting this great town into a wonderment about nothing: Nor do I doubt but he and his friends in the secret,

[66] 'The Englishman's Humble Thanks', ibid., p. x.

laugh often and plentifully in a corner to reflect how many hundred thousand fools they have already made.[67]

Here, argues Ehrenpreis, 'is Swift pretending to be a man who sees through a man whom Swift is pretending to be. Some passages come so close to autobiography that to read them gives one a dizzy, happy sense of infinite regress.'[68] Capitalizing on this careful build-up, Swift launched *The Accomplishment of the first of Mr Bickerstaff's Predictions* on 30 March. This is written in the person of a former revenue officer who gives a conscientious account of the fulfilment of Bickerstaff's prediction. On his death-bed Partridge, incredulous at feeling the effects of the prediction, is brought to a desperate confession that he had always believed that 'all pretences of foretelling by astrology are deceits . . . and none but the poor ignorant vulgar give it any credit'. It is a neat ironic coup to present him admitting the falsity of astrology while apparently succumbing to its power. He expires with a 'fanatic preacher' as his spiritual guide, while the punctilious narrator, whose careful prose nevertheless betrays a certain shudder at what he has witnessed, records some minor inaccuracies in the prediction. Partridge in fact died at five past seven, he reports, 'by which it is clear that Mr Bickerstaff was mistaken almost four hours in his calculation.'[69] Yet this apparent discrepancy only serves to guard the satiric fantasy. Let cardinals, kings, and popes beware.

Swift's hoax had now become a fashionable craze, and the whole of Grub Street strove to celebrate Partridge's 'death'. The most inventive variation on the common theme, *Squire Bickerstaff Detected,* is variously attributed to Congreve or Rowe.[70] It is a superb comic lament, in the name of poor Partridge, as his home is besieged by lawyers, undertakers, and creditors. Coming downstairs on the morning after his 'death' Partridge finds a man mounted on a table measuring the walls and windows for mourning drapery. Partridge announces his name, but the man is unsurprised. 'Oh the Doctor's brother, belike, cries he', and goes back to his task. 'The stair-case, I believe, and these two apartments hung in close mourning will be sufficient.' Partridge chases him out with a cudgel, but shortly afterwards is visited by

[67] Ibid., p. 195.
[68] Ehrenpreis, ii. 202.
[69] *PW* ii. 155.
[70] Ibid., pp. 217-23.

the sexton, anxious to know whether the grave is to be plain or bricked. He is understandably indignant when Partridge protests that no grave is needed. 'It looks as if you had a design to defraud the church,' he complains. 'What is become of the freedom of an Englishman,' cries out the Partridge persona pathetically, for 'although I produce certificates . . . I am alive . . . truth is borne down, attestations neglected, the testimony of sober persons despised, and a man is looked upon by his neighbours as if he had been seven years dead, and is buried alive in the midst of his friends and acquaintance'. Ironically, life imitated art in this case. The Stationers' Company struck the 'deceased' Partridge's name from their rolls, and gained the exclusive right to issue the *Merlinus Liberatus* themselves. The almanac lived, only the almanac-maker had been killed.

The Bickerstaff hoax was a stunning example to Swift of the manipulative powers of the press. What had begun as a simple practical joke grew into a full-blown fantasy until more people knew of Partridge's death than had ever heard of him alive. The press was now littered with rival 'Partridges'; how could one possibly tell the real Partridge from all the impostors? When the real Partridge did attempt to hit back at that 'impudent lying fellow' Bickerstaff, in his almanac for 1709, he sounded, inevitably, like yet another figment of Swift's imagination. Swift seized on the opportunity to lay the ghost of Partridge, having already killed the man, with his *Vindication of Isaac Bickerstaff*. By now the rest of Bickerstaff's original predictions had been disproved, of course, and Swift makes little attempt to maintain that part of the hoax. He does note, however, that we have only a Frenchman's word that the Archbishop of Paris is alive, and 'how far a Frenchman, a papist, and an enemy is to be believed . . . I shall leave to the candid and impartial reader'.[71] The main purpose of the *Vindication* is to disprove Partridge's claims to be alive. Yet 'Bickerstaff' speaks of him throughout as a living antagonist, while advancing several distinct 'proofs' that he is dead. Swift's strategy is deliberately bewildering. having created, through his extended hoax, a monument to public gullibility, he now sought to draw attention to that gullibility. His 'proofs', facetious, semantic, or arithmetical, are a virtuoso

[71] Ibid., p. 161.

display of false trails down which he might lead us, if he had a mind to. We are in his power to create fact or fiction, history or polemic, life or death. What remains of Partridge is 'an uninformed carcass' and if such an object pleases to call itself Partridge 'Mr Bickerstaff does not think himself any way answerable for that'. The irony of Swift's triumph in these pamphlets is that he could only undermine Partridge by proving himself to be an even greater conjuror, exerting an even stronger hold over public credulity. In the *Vindication* he draws attention to the hoax, almost as if trying to shake off part of his success. The power of the ironist over human credulity was just a measure of the failure of the moralist to reassert rationality.

The most notable by-product of Swift's Bickerstaff pamphlets was the *Tatler*, the first issue of which was produced by Steele in April 1709 under the pseudonym of Isaac Bickerstaff. Steele claimed that the popularity of Swift's character had 'created an inclination in the town towards anything that could appear in the same disguise'. In the Preface to the first collected edition of the *Tatler* he made a gracious acknowledgement to Swift, whose company had been 'very advantageous to one whose imagination was to be continually employed upon obvious and common subjects, though at the same time obliged to treat them in a new and unbeaten method'. In reality Swift's influence on the *Tatler* was minimal. He contributed hints and paragraphs to some early numbers, but they are slight things, self-consciously trivial, offering gossip in the guise of allegory, and with none of the ironic edge that gives a paradoxical longevity to the self-consciously ephemeral productions of the hack of the *Tale*.

Pressures of government business and deadlines meant that Addison, Steele, and Swift met increasingly in pairs, rather than as a threesome. 'The triumvirate of Mr Addison, Steele and me come together as seldom as the sun, moon and earth,' Swift told Ambrose Philips in July 1708.[72] He was not unhappy at this, declaring that whenever he and Addison dined together 'they neither of them ever wished for a third person, to support or enliven their conversation'. During one such meeting he showed Addison a draft of a poem he had written on the fable of Baucis and Philemon. Addison's advice was that he should 'blot out

[72] *Corr.* i. 91.

fourscore [lines], add fourscore, and alter fourscore'.[73] The symmetry of this judgement is entirely Addisonian. However the poem exists in two forms, before and after Addison's alterations, and modern readers have usually preferred the original version.

The story, from Ovid, tells how Jupiter and Apollo, coming to earth in disguise to test human hospitality, are shunned by all except a poor peasant couple who share their homely fare with the strangers. In reward, the gods create a temple for the couple, and, as they grow old, transform them into trees. Swift's poem retains all the essentials of the story, only substituting medieval saints for Olympian gods. Written in a jog-trot octosyllabic metre, 'Baucis and Philemon' has the humorous charm of a Chaucerian fabliau, and demonstrates Swift's ear for the rhythms and idioms of colloquial speech.[74] In his original version there is a detailed comic setting as the two saints go from door to door like mendicants, and are chased away by the villagers.

> One surly clown looked out, and said
> 'I'll fling the piss pot on your head;
> You sha'n't come here, nor get a sous
> You look like rogues would rob a house
> Can't you go work . . .'[75]
>
> (ll. 37-41)

All this, even without the piss-pot, was too low for Addision, who expressed his distaste by objecting that these vulgar details impeded the narrative flow of the poem. In the revised version, Swift's thirty lines of entreaty are reduced to just three.

> They begged from door to door in vain;
> Tried every tone might pity win,
> But not a soul would let them in.[76]
>
> (ll. 12-14)

Addison's instinct throughout is to dignify and generalize Swift's descriptions. If these alterations are typical of the lines which Addison blotted, those which he added are designed to place the fable in an ironic framework, with a modern knowingness that recalls Dryden's adaptations of Chaucer. The portrait of the parson Philemon is extended to include these epigrammatic lines.

[73] Delany, p. 19.
[74] *Poems,* i. 88-96; 110-17.
[75] Ibid., p. 91.
[76] Ibid., p. 111.

Against Dissenters would repine,
And stood up firm for Right Divine.
(ii. 133-4)

Philemon ceases to be a simple idealized figure in a Gothic fable and becomes a contemporary 'type' of the country parson, treated with comic condescension. One senses a little ribbing here between Addison and the country parson Swift, who would also repine against Dissenters. Part of the charm of the poem consists in the fact that it is a fable written for a newly scientific age. The metamorphosis of the cottage into a church is seen in slow-motion detail, with a Disneyesque plasticity that seems to parody the close observation by members of the Royal Society of the transformation of a caterpillar into a butterfly.

The groaning chair began to crawl
Like a huge insect up the wall,
There stuck, and to a pulpit grew,
But kept its matter and its hue,
And mindful of its ancient state
Still groans while tattling gossips prate.[77]
(ll. 105-10)

Despite this buoyancy in his literary and social life, Swift soon found that his clerical ambitions were doomed to further disappointments. Only a month after his departure from Ireland the Bishop of Waterford died, and Swift understood, from Fountaine, that both Pembroke and the Archbishop of Canterbury were prepared to support his application. When he did not get the post, he managed to convince himself that it was because Fountaine had been away from London at the crucial time. Worse disappointments followed. In return for his support in the recent Deanery election, Swift understood that Stearne had promised him one of two new curacies that were to be created by dividing the parish of St Nicholas Without. As time went on Swift sent an indirect reminder to Archdeacon Walls to ask Dr Smith to ask Dr Synge 'to enquire whether Dr Stearne designs really to give me the parish that has the church'.[78] If Stearne was offering him the parish without a church, Swift declared he would 'not accept it, nor come to Ireland to be deceived'. In fact he was offered neither parish. According to King there were

[77] Ibid., p. 93. [78] *Corr.* i. 66-7.

insufficient funds to erect a second church, and he decided that Stearne should continue to hold the livings of both parishes, putting the revenues aside until enough had been accumulated to build one. Swift was furious and twenty-five years later he was still angrily reminding Stearne of this betrayal. 'You absolutely and frequently promised to give me the curacy of St Nicholas Without. But you thought fit, by concert with the Archbishop, to hold it yourself'.[79] For this reason, he declared, though he respected Stearne, he would 'never hope for the least friendship from you'.

Swift nevertheless continued to adopt an ingratiating manner towards King. In April he declared that King's letters were 'full of everything that can inspire the meanest pen with generous and public thoughts',[80] and in December pretended to chide King for his incorrigible itch for doing good. 'Your time of sleep is mis-spent in perpetual projects for the good of the Church and kingdom.'[81] If this was irony, it went totally unsuspected by King, who confessed that this section of Swift's letter pleased him 'the best of any part'. While offering such flatteries, Swift must have smarted at the pharisaical disingenuousness of the Archbishop, who continued to sympathize with him for the loss of livings which he had contrived to deny him. Declaring that 'Necessity has no law', King sought to convince him that all was for the best in the loss of St Nicholas Without, 'for to deal ingenuously with you, I do conceive that you would have no prospect of temporal interest if you were tomorrow put into one of those parishes'.[82] Three years later, King sent Swift some platitudinous advice, designed to improve his temporal interests. Observing that 'it is with men as with beauties, if they pass the flower, they grow stale', he urged him to tackle 'some serious and useful subject'. Should Swift be stumped for an idea, he gives him this tip. 'If you look into Dr Wilkins' Heads of Matters . . . you will be surprised to find so many necessary and useful heads . . . a good genius will not fail to produce something new and surprising on the most trite.'[83] There is something almost comic about such a total misunderstanding of the talents of a man who, in *A Tale of a Tub* and a *Meditation on a Broomstick* had made a virtuoso

[79] Ibid., iv. 182. [80] Ibid., i. 79. [81] Ibid., p. 59.
[82] *Corr.* i. 75-6. [83] Ibid., pp. 254-5.

display of parodying such trite themes and homiletic heads as are contained in Dr John Wilkins's *Gift of Preaching*. 'I know you are not ambitious,' King added, 'but it is prudence and not ambition to get into a station that may make a man easy.' How galling such sanctimonious advice must have sounded to Swift, whose literary output had not included the one piece of pious inanity that would earn him respectability and the respect of his superiors. 'Assure yourself, that your interest as well as your duty requires this from you,' King concluded, and a clear threat is implied. Dullness guarded the gates of preferment, and Swift must succumb, or waste his life away among the willow trees of Laracor.

No progress had been achieved in the matter of the First Fruits since the Irish clergy were reluctant to provide accurate information about the value of their livings. 'We know not (say they) what use will be made of it,' King reported.[84] Exasperated, Swift recognized this as an endemic Irish fault 'who seldom understand their true interest or are able to distinguish their enemies from their friends'.[85] Speed was essential, he believed, for ugly rumours were already circulating in the lobbies. On New Year's Day he reported to King a 'whisper' that he had heard, that remission might have to be 'purchased by a compliance with what was undertaken and endeavoured in Ireland last session, which I confess I cannot bring myself yet to believe nor do I care to think or reason upon it'.[86] In other words, the Whigs might demand repeal of the Test as the price of remission. King found the rumour 'hardly credible', but the power of the Dissenters grew steadily as the pressures of war made the cry of Protestant unity all the more relevant. In January a clerk in the office of Secretary Robert Harley, called William Greg, was arrested for treason. This seriously damaged Harley's reputation at a time when he, as leader of a broad 'Country-Church-Tory' group, had been gaining a hold on the Queen's confidence. Greg pleaded guilty of espionage, but the House of Lords chose to make an elaborate enquiry into the case, and appointed an examining committee of seven leading Whig peers. The aim was to discredit Harley as much as possible, but he, in retaliation, finally persuaded the Queen to dismiss Godolphin in the first

[84] *Corr.* i. 60. [85] Ibid., p. 62. [86] Ibid.

week of February, and to allow him to form a ministry. The plan misfired, for Marlborough (the indispensable figure for maintaining a credible war policy) promptly resigned as well. Without his support, Harley was unable to form a government, and the Marlborough-Godolphin partnership resumed power. Swift related these events to King with a detachment that betrays no inkling of his future involvement with the man whose collapse he watched.

Mr Harley had been some time, with the greatest art imaginable, carrying on an intrigue to alter the ministry, and began with no less an enterprise than that of removing the Lord Treasurer, and had nearly effected it, by the help of Mrs Masham, one of the Queen's dressers, who was a great and growing favourite, of much industry and insinuation.[87]

Seven years later, when writing his *Memoirs* of this period, he told the story very differently, picturing Harley as the victim of a treacherous plot by Marlborough and Godolphin who at first 'had concerted with them and their friends upon a moderating scheme, wherein some of both parties should be employed' but who 'were secretly using their utmost efforts with the Queen to turn Mr Harley . . . and his friends, out of their employment'.[88] In this revised version, Swift insisted that the Queen had 'a great opinion of Mr Harley's integrity, and was 'determined to remove the Earl of Godolphin'. It was only when Marlborough delivered his ultimatum that she was 'unwillingly forced to yield'. That was how Swift wished posterity to understand the incident. In his letter to King, however, he correctly anticipated the harsh view that the Archbishop would take of Harley's manoeuvres at a time of national emergency. News had already arrived of a French invasion fleet that was assembling at Dunkirk. 'The great cry is that this was Harley's plot,' wrote King, 'and if he had continued three days longer in his place, the French would have landed at Greenwich.'

In fact this fleet of thirty ships, carrying the Old Pretender, failed to effect a landing, but the invasion threat had serious political consequences. The Dissenters became increasingly confident that the Test Acts in both England and Ireland would be removed, to enable them to play a full part in national

[87] Ibid., p. 69. [88] *PW* viii. 113.

defence. In Dublin, Alan Brodrick, Speaker of the Irish Commons, encouraged the city council to present an address to the Queen with a strong emphasis on unity. 'The hearts of all your Protestant subjects of this city are entirely devoted to your Majesty's service, though the hands of some be restrained from serving in the commission.'[89] Swift's relations with his Whig friends Addison and Steele came under severe pressure at this time, since Steele, as editor of the *Gazette*, reproduced this loyal address at length. Swift would occasionally suggest modifications to such pieces, but Steele would seldom trim his editorial line. Addison had recently published a bland piece entitled *The Present State of the War*, arguing the Whig case for continuing the war against French tyranny, a case which Swift came increasingly to distrust. Swift soft-pedalled his reservations, content to present himself as a rather old-fashioned country parson. Yet doubts nagged him that he might be drawn into compromising his principles; or, worse still, that he might compromise himself, and still be left disappointed, and discredited. He had acquired an unconscious tendency to regard failure as a guarantee of integrity. Both his grandfather and William Temple had suffered defeats in their careers, rather than deny their principles. Now that a certain kind of literary success seemed almost within his grasp, he feared that it might be a poisoned bait. No wonder that his account book for the year commencing 1 November 1708 is headed 'in suspense'.

In April he set about the task of soliciting the remission of the First Fruits, and was treated to a truly expert display of double-talk and buck-passing. He told Somers that he understood the Queen had assented in principle to remission, and that 'it wanted nothing but solicitation'. He then asked Somers to reintroduce him to Sunderland, 'with whom I had long been acquainted',[90] but who seemed too busy lately to receive 'common visits' from petitioners on the slim pretext of a previous acquaintanceship. Somers replied that Sunderland would gladly see Swift and introduce him to Lord Treasurer Godolphin to plead his case in person. However, the invasion threat delayed Swift's meeting with Godolphin, or at least, that was the excuse. 'For it is the method here of Great Ministers, when any public

[89] *London Gazette,* 8 April 1708. [90] *Corr.* i. 80.

matter is in hand, to make it an excuse for putting off all private application.'[91] Then it proved impossible to find a time when both Godolphin and Sunderland–who was to effect the introduction–were free. However, Sunderland assured Swift that he had prepared Godolphin 'to think well of the matter'. Eventually, one day in early June, Swift was ushered into a private room where he made his long-rehearsed application. Godolphin heard him out in silence, but then with a complacent shrug said that he was passive in this business. It was not really his affair, but he would be glad, as a favour to Swift, to mention the matter to the Lord Lieutenant, Pembroke. Swift bit back his irritation at this transparent evasion, explained that he already knew Pembroke and indicated that he had judged it best to apply directly to the head of the administration. But Godolphin insisted that he could do nothing and advised him to go back to Pembroke. He assured Swift that he would say nothing to the Lord Lieutenant about this ill-conceived attempt to go over his head. However, the interview did not end there. Godolphin went on to say that Swift exaggerated the importance of the remission and told him that since the grant of the First Fruits in England 'not one clergyman in England was a shilling the better'.[92] Well, if the sum were so inconsiderable, it would not harm the Crown to grant it, Swift countered. True enough, Godolphin agreed. He doubted if the grant to Ireland would cost more than £1,200, 'which was almost nothing for the Queen to grant. Upon two conditions.' Godolphin, whose nickname in the Tory press was Volpone, dropped the word with mock-innocence. Conditions? What conditions? First that the sum would be well disposed of. That was no problem. Swift was confident that 'the bishops would leave the methods of disposing it entirely to her Majesty's breast'. But the sting was in the second condition: 'That it should be well received with due acknowledgements; in which cases he would give his consent, otherwise . . . he never would.' At first Swift pretended not to understand this, assuring Godolphin that the Queen 'might count upon all the acknowledgements that the most grateful and dutiful subjects could pay to a prince'.[93] But then he asked whether Godolphin had any *particular* acknowledgements in mind. The grin on Volpone's face never altered. 'I

[91] Ibid., p. 84. [92] Ibid., p. 85. [93] Ibid.

will so far explain myself to tell you I mean better acknowledgements than those of the clergy in England.' Again Swift refused to be lured into naming the Test but, feigning ignorance 'begged his lordship to give me his advice, what sort of acknowledgement he thought fittest for the clergy to make'. 'Such acknowledgement as they ought', replied the perfect diplomat and the interview was at an end. Swift left the room, angry, his nerves worn down, and no further advanced than he had been six months before.

Later that evening he was forced to go through the charade of asking Pembroke's permission to approach Godolphin for the interview which had just taken place. Pembroke too was evasive. It was nothing to do with him either, apparently, for the matter 'was entirely with the Queen', and therefore his permission 'was needless'. The whole episode was a perfect example of bureaucratic evasion. 'Thus your Grace sees that I shall have nothing more to do in this matter, further than pursuing the cold scent of asking his excellency once a month how it goeth on?' Swift signed off his letter to King with a bitter memory that 'if only' he had been allowed to negotiate the matter with Ormonde four years earlier 'it might have been done in a month'. But now–'God knows when'.[94]

V

IN AN ATTEMPT to settle the doubts which nagged at him, Swift began to set down some thoughts on the Sacramental Test. The trickle soon turned to a torrent as day after day he sat in his chamber near Leicester Fields, pouring out a series of 'speculations' on this crucial issue. Though not published until 1711, his *Sentiments of a Church of England Man* was probably among the first of these new pamphlets. As the Church of England man of his title, Swift gives a self-portrait whose defiant stance of impartiality indicates the conflicting pressures that seemed to be tearing him apart.

> I believe I am no bigot in religion; and I am sure I am none in government. I converse in full freedom with many considerable men of both parties.[95]

[94] Ibid., p. 87. [95] *PW* ii. 2.

The pamphlet is one of those perennial appeals to the middle-ground of politics. No one, Swift declares, who 'hath examined the conduct and proceedings of both parties' can conceive it possible 'to go far towards the extreme of either without offering some violence to his integrity or understanding'. Moderation is Swift's watchword, and he seeks to convince us of his impartiality by indicating a number of faults on both sides. But where exactly is Swift's middle ground? In another pamphlet, a *Letter from a Member of the House of Commons in Ireland . . . concerning the Sacramental Test,* Swift says of the Irish Commons that whoever venerates the glorious memory of King William; whoever is firmly loyal to the Queen; whoever abhors the Pretender, approves of the Hanoverian succession, and supports the Church of England 'may be justly allowed a Whig; and I believe there are not six members in our House of Commons, who may not fairly come under this description'.[96] The only people to oppose such institutions are 'a set of men not of our own growth', Presbyterians and their associates, whose ideas are foreign and noxious. In other words, we are all Whigs now. Swift manages to assert his own fidelity to Whig principles by pretending that no such things as Tories exist. Having redrawn the political map in this way, it is not surprising to find that Swift saw most to criticize on the 'left' or Dissenting wing of politics since the High-Tory, Jacobite wing was apparently invisible to him.

Swift had no doubt that Dissent was a dangerous evil. As such, he was prepared to tolerate it, within limits, but not to encourage it. It would not be right for the Government to use 'violent methods against great numbers of mistaken people'. The Church of England must have no truck with the kind of methods that Louis XIV had employed against the Huguenots, and Swift offers a modest proposal, not a shortest way for dealing with the Dissenters. But Occasional Conformity marks the outermost bound of his toleration. If he will not approve methods to eradicate Dissent, he nevertheless thinks it unreasonable to allow the infection to spread. To this end he makes a distinction between 'liberty of conscience' and 'freedom of opinion', which might be glossed as the distinction between freedom *from*

[96] Ibid., p. 118.

prosecution and freedom *to* proselytize. The apparent balance here seems to confirm Swift's moderation. Yet, having started from the axiom that the Church of England has a monopoly of truth and light, while Dissenters are evil schismatics, Swift's middle position has the inevitable unease of all bargains struck between God and the Devil.

Many of the same arguments reappear in the *Letter on the Sacramental test,* written in 1708 following the alleged 'persecution' of two Presbyterians in Drogheda. The Church insisted that no Dissenting community existed in Drogheda, and therefore that the Presbyterians had been proselytizing. One of the two was gaoled, and quickly hailed as a martyr by Whigs in both England and Ireland. To Swift in this *Letter* the papist threat appears as a paper tiger, or rather a bound lion, compared with the danger posed by the Presbyterians:

> if a man were to have his choice, either a lion at his foot, bound fast with three or four chains, his teeth drawn out, and his claws pared to the quick, or an angry cat, in full liberty at his throat; he would take no long time to determine.[97]

Severe civil disabilities rendered the Catholics harmless, whereas the Scottish Presbyterians who had occupied Ulster were a real threat, not least because they possessed so many of those virtues of self-discipline and industry that Swift admired.

> These people by their extreme parsimony, wonderful dexterity in dealing, and firm adherence to one another, soon grow into wealth from the smallest beginnings, never are rooted out where they once fix, and increase daily by new supplies.[98]

Give such determined people as these access to political power and, Swift believed, it would not be long before the Kirk, not the Church would rule in Ireland, and he would be 'forced to keep a chaplain disguised like my butler, and steal to prayers in a backroom, as my grandfather used in those times when the Church of England was malignant'. As he thought of his congregation of half a score in Laracor, they seemed easy meat for the advancing zealots from Antrim. However the real anger in this *Letter* is directed against Speaker Brodrick's subterfuge in seeking to have the Test in Ireland repealed by an Act of the English Parlia-

[97] Ibid., p. 122. [98] Ibid., p. 116.

ment, which might then be made binding on the Irish Parliament. In Swift's view, Brodrick was not only attacking the true Church; also, as Speaker of the Commons, he was seeking to subvert the institution which he represented, by relying upon its impotence to resist legislation forced upon it by England. As he reacts to these combined attacks on his country's Church and his country's Parliament. Swift's arguments rise above doctrinaire self-interest to enunciate certain general principles. Here, for the first time, we find Swift's ringing denunciation of England's callous exploitation of her neighbour.

> If your little finger be sore, and you think a poultice made of our vitals will give it any ease, speak the word, and it shall be done; the interest of our whole kingdom is, at any time, ready to strike to that of your poorest fishing town; it is hard you will not accept our services, unless we believe at the same time, that you are only consulting our profit, and giving us marks of your love. If there be a fire at some distance, and I immediately blow up my house before there be occasion, because you are a man of quality, and apprehend some danger to a corner of your stable; yet why should you require me to attend next morning at your levee, with my humble thanks for the favour you have done me?[99]

The conundrum which faced defenders of the established Church was to decide which of the two, Church or State, was finally supreme. Was the Church a department of the civil service, or the State the administrative machinery of a theocracy? For Sancroft the issue had been clear; the Church was supreme, or as the young idealistic Swift wrote, 'the weathercock of state/Hung loosely on the Church's pinnacle'. The Church of England man, however, reveals some of the inconvenient loopholes in the subsequent Erastian compromise:

> although he will not determine whether episcopacy be of divine right, he is sure it is most agreeable to primitive institutions . . . [and] he would defend it by arms against all the powers on earth, except our own legislature.

Swift's assertions are frail rope-walks across yawning chasms here. The interdependence of Church and State was one of those mystical unions, like the Trinity, which were above understanding, but not against it. Also, like the Trinity, it was a subject upon which, he felt, fewest words were best.

[99] Ibid., p. 114.

Among the Dissenters who believed that the institutions of primitive Christianity were very different from those of the established Church was Matthew Tindal, who published in 1706 *The Rights of the Christian Church Asserted.* In this he argued that 'the Church of England, being established by Acts of Parliament, is a perfect creature of the civil power'. In a rare display of unity both houses of Convocation deplored this attack on the Church, and hastened to refute Tindal's subversive doctrines. Swift also began making notes to 'show the silliness' of Tindal's book, but never completed the task. There are several probable reasons for this: too many replies would merely magnify Tindal's reputation; besides, many of those attacking Tindal had previously attacked *A Tale of a Tub* and Swift felt no eagerness to be their ally. Most importantly though, Swift found he was unable to deliver a good knock-down argument to demolish his opponent. In many places the style of his *Remarks* is vigorous, and employs a colourful vein of invective. Yet as he piles on the parodies, one senses a certain staleness in the attack. Swift scores a number of ironic hits, but they read like clever student exercises.[100] Tindal remains substantially unharmed. Swift is a guerrilla, sniping from cover, but Tindal is engaged in a war of attrition. Swift's *Remarks* finally degenerate into a series of one-liners: Thus when Tindal declares 'the love of power natural to churchmen', Swift adds, 'truly, so is the love of pudding'.

The last two pamphlets which Swift composed at this time concerning the relationship between the Church and State indicate, in their contradictions, something of his own uncertainties in this area. Among the range of his satiric personae, none is more familiar or effective than that of the projector, the benevolently blinkered ideologue with his pet scheme for 'the universal improvement of mankind'. Yet Swift's own *Project for the Advancement of Religion and the Reformation of Manners* is a scheme as narrow, naïve, and Utopian as any that he satirizes, and totally without a trace of irony. His language throughout has the authentic tone of confidence of all enthusiasts.

For, as much as faith and morality are declined among us, I am altogether confident they might, in a short time, and with no very great trouble, be raised to as high a perfection as numbers are capable of receiving.[101]

[100] *PW* ii. 65-107. [101] Ibid., p. 44.

Like a true projector, Swift has hit on the one infallible method whereby this miraculous transformation 'may be easily put into execution'. It is that the Queen should only promote to positions of power men of proven piety and religion. If it were in a man's interest to be virtuous, the temptations of vice would be considerably diminished.

If religion were once understood to be the necessary step to favour and preferment: can it be imagined, that any man would openly offend against it, who had the least regard for his reputation or his fortune? . . . How ready would most men be to step into the paths of virtue and piety, if they infallibly led to favour and fortune.[102]

Naïve as this seems, it plays upon the central chord in Swift's view of human psychology, that self-interest was the ruling passion of human life. This is his attempt to link self-interest to social responsibility, and to graft a morality on to a survival instinct. In the years since the Revolution the 'reformation of manners' had become a popular theme, and Swift's scheme is as thorough as anything envisaged by the evangelists of SPCK. The *Project* has running through it, in the best millenialist style, a mechanistic 'wave-theory'. One impulse from the centre, Queen Anne, will send out ripples through the court, the town, the provinces, the whole country. The list of Swift's targets makes depressing reading. The stage is 'undecent' and 'prophane' and Swift anticipates the imposition of censorship by the Lord Chamberlain when he sees himself in the role of Censor morum. Surely, he writes, 'some man of wit, learning and virtue' should be employed 'who might have power to strike out every offensive or unbecoming passage from plays'. The press too should be censored, 'at least so far as to prevent the publishing of such pernicious books, as under pretence of free-thinking, endeavour to overthrow . . . religion'. If you cannot beat Tindal, at last you should be able to ban him. Swearing, lewdness, drunkenness, and gaming are all to be subject to new punitive measures. And to police his virtuous Utopia, Swift plans a new commission of officers,

appointed to inspect everywhere throughout the kingdom, into the conduct (at least) of men in office, with respect to their morals and

[102] Ibid., p. 50.

religion as well as their abilities; to receive the complaints and informations that should be offered against them; and make their report here upon oath, to the court or the ministry.[103]

Robespierre himself could not have formulated a more thoroughgoing apparatus for ensuring the tyranny of virtue. There is something distasteful about finding Swift as the author of a tract of which any puritan fanatic, or dictator, might have been proud, with its dedication to narrow principles of orthodoxy, and its paraphernalia of thought-police and informers. Yet it is a fact that in Swift's responses to the vices and vanities of his age, authoritarianism and libertarianism alternate with a disconcerting frequency. He is torn between tolerating vanities with the rueful scepticism of a Montaigne, and seeking to eradicate them with the righteous indignation of a Cromwell. Swift's sermons, his *Proposals for Ascertaining the English Tongue,* and his frequent calls for curbs on press freedom all indicate radical, authoritarian, even Utopian intentions. They seek to impose a fundamental moral improvement upon society by a central change in policy–a change towards greater state regulation and control.

There is a further disquieting element in this pamphlet. Swift anticipates the objection that if virtue is rewarded it may encourage people to feign, but not necessarily follow virtue.

Making religion a necessary step to interest and favour might increase hypocrisy among us: And I readily believe it would. But if one in twenty should be brought over to true piety by this, or the like methods, and the other nineteen be only hypocrites, the advantage would still be great. Besides, hypocrisy is much more eligible than open infidelity and vice: It wears the livery of religion, it acknowledgeth her authority, and is cautious of giving scandal. Nay, a long continued disguise is too great a constraint upon human nature, especially an English disposition. Men would leave off their vices out of mere weariness, rather than undergo the toil and hazard, and perhaps expense of practising them perpetually in private. And, I believe, it is often with religion as it is with love; which, by much dissembling, at last grows real.[104]

One scans this passage in vain for irony. Swift means exactly what he says here. This is a form of occasional conformity which is not merely a constitutional convenience, but the next best thing

[103] Ibid., p. 49. [104] Ibid., p. 57.

to true godliness. Hypocrisy ceases to be a vice, and becomes instead a form of utilitarian virtue. The impact of this *Project* with its vision of a society of nominal Christians all conforming to a strict state apparatus is so entirely different from that of Swift's other pamphlet on nominal Christianity written at this time, that one is astounded by a mind that can veer so violently between ideals of freedom and control.

Swift's *Argument to prove that the abolishing of Christianity in England, may, as things now stand, be attended with some inconveniences* . . . is, by common consent, one of the show-pieces of his irony. Composed as a mock-response to a non-existent proposal, it adopts a tone of modest temerity in presuming to contradict 'the general humour and disposition' for abolishing Christianity. He cites with respect a 'fundamental law' often mentioned by Swift, 'that makes this majority of opinion the voice of god'. However, it is a voice that declares itself with delphic ambiguity; and a modern reader, interpreting 'abolition' as a rhetorical exaggeration for 'neglect', may, at the end, still be left wondering whether Swift's irony actually confronts or endorses the majority view. A vital concession is made early on. The author indicates that he is no wild-eyed puritan.

> I hope no reader imagines me so weak to stand up in defence of *real* Christianity . . . that would indeed be a wild project.[105]

The *Argument* is offered 'in defence of nominal Christianity' only, or of the hypocrisy recommended in the *Project.* To attempt the restoration of real Christianity

> would be to dig up foundations; to destroy at one blow all the wit, and half the learning of the kingdom; to break the entire frame and constitution of things; to ruin trade, extinguish arts and sciences . . .

One cannot help feeling that the ambiguity here comes close to cynicism. The 'irony' tells us that this is an attack on hypocrisy; yet the dismissal of the primitive Utopianism of Tindal or Whiston is convincingly total. Real Christianity would make for unreal politics; and the mention of Horace, 'where he advises the Romans . . . to leave their city, and seek a new seat in some

[105] *PW* ii. 27.

remote part of the world', does nothing to endorse a life of ascetic virtue on either the Roman or the Christian model.

It is suggested that one beneficial consequence of the abolition of Christianity might be the disappearance of political parties, following the removal of 'factious distinctions of high and low church'. The anti-abolitionist has little trouble refuting this.

Will any man say, that if the words whoring, drinking, cheating, lying, stealing, were, by act of parliament, ejected out of the English tongue and dictionaries; we should all awake next morning chaste and temperate, honest and just, and lovers of truth.[106]

On the contrary, as Gulliver finds, if one cannot divide on the issue of high or low church, one can argue just as fiercely over the issue of high or low heels. Yet, in his *Proposals for Correcting . . . The English Tongue* (1712) Swift advanced exactly the opposite view. His motivation in that pamphlet was moral, not simply aesthetic. Like money, language was a dangerously fluid currency, and Swift's explicit aim of 'fixing our language for ever' was intended to preserve the culture, advance learning, and resist barbarity by controlling the linguistic tap. In other words, he did apparently believe that control of the language supply could have beneficial social effects. This was an enigma to which he returned many times. The Houyhnhnms have no words for a whole range of human vices; and they lack the vices too. But which came first? In the beginning, was there a word or a thing? The arguments of the anti-abolitionist are entirely pragmatic. The effects of Christianity are shown to be good; the anticipated consequences of its abolition, bad. Like Swift, the author shows a striking disinclination to tangle with ontology. Religion provides a harmless Aunt Sally.

Great wits love to be free with the highest objects; and if they cannot be allowed a God to revile or renounce; they will speak evil of dignities, abuse the government, and reflect upon the ministry.[107]

God is a perennial tub, thrown out to divert restless leviathans that might otherwise disturb the ship of state. Christianity is an invaluable social cement.

[106] Ibid., p. 32. [107] Ibid., p. 29.

I conceive some scattered notions about a superior power to be of singular use for the common people, as furnishing excellent materials to keep children quiet, when they grow peevish; and providing topics of amusement in a tedious winter night.[108]

The restrictions imposed by Christianity are themselves purely nominal, and mainly serve to add a 'wonderful incitement' to certain pleasures, by designating them as forbidden fruits. Neither business nor pleasure suffers any abatement on the Lord's Day. 'Are not the taverns and coffee-houses open? . . . Are fewer claps got upon Sundays than other days?' Not only God but also the humble country parson would need to be reinvented in another guise if Christianity were abolished since it was surely necessary that 'in certain tracts of country, like what we call parishes, there should be *one* man at least, of abilities to read and write'. And so on; the social uses of religion are listed with a playful exaggeration that does not conceal the implied necessity for political conformism. At this point, Swift suddenly alerts our critical faculties with a deliberate bathos.

Nor do I think it wholly groundless, or my fears altogether imaginary; that the abolishing of Christianity may perhaps bring the Church in danger.

This is merely a rhetorical trick. Throughout the foregoing argument 'the Church' has been used metonymously for Christianity. Not for the first time, Swift shows himself a perfect Thwackum in doctrine, identifying Christianity, Protestantism, and the Anglican Church. Hence one of the most dangerous consequences of the end of Christianity might be a resurgence of popery. But for his final, clinching argument, the author relies on an appeal to our pockets.

To conclude: Whatever some may think of the great advantages to trade by this favourite scheme; I do very much apprehend that in six months time, after the Act is passed for the extirpation of the gospel, the Bank and East-India stock may fall at least one per cent. And since that is fifty times more than ever the wisdom of our age thought fit to venture for the *preservation* of Christianity, there is no reason we should be at so great a loss, merely for the sake of *destroying* it.[109]

[108] Ibid., p. 34. [109] Ibid., p. 38.

Read superficially, the *Argument* attacks hypocrisy by describing and 'defending' the residual utilitarian functions of religious institutions in a society whose ethos and impetus are to be found elsewhere. Yet, when we describe the work as ironic, what exactly do we mean? The society described here is one in which occasional conformity or the nominal Christianity of the *Project* have been universally accepted as governing principles. Swift gives us a parody of nominalism, yet at the same time recommends it to us. It is not simply this ambiguity that may disturb us. What is more worrying is the complete failure to suggest or imply any other reasons for accepting Christinity. Nowhere is there a hint of faith, or a whisper of a hope of salvation.

Swift's deep aversion to public displays of faith is well known. His early biographers agree in presenting him as a man of great piety who nevertheless shunned all religious ostentation. 'During his residence in London . . . he was seldom seen at church at the usual hours that pretenders to religion show themselves there'.[110] However commendable this dislike of display, it may have had deeper roots in doubt and confusion. The evidence suggests that Swift's ambivalence towards father-figures in his life extended also to God, and inhibited him from truly believing in the personal love, for him, of God the Father. His instinct for playing the hypocrite in reverse may well have been an attempt to project outwards a deep sense of hypocrisy that he felt within himself. We find the same tussle between conscience and conformity fought over again and again in his sermons. Where Hamm in Beckett's *Endgame* expresses the incoherent rage of a godfearing atheist, 'God . . . the bastard! He doesn't exist!' Swift reveals the tormented pessimism of an orthodox clergyman who cannot really believe in God–certainly not in the loving God of the New Testament. The most positive interpretation that one can offer of the *Argument* is that Swift is attempting to force us to recognize and acknowledge our own hypocrisy, our failure to live up to the name of Christians. In doing so he expects no changes. The *Argument* is not designed to inspire sudden conversions: But it may at least instil a little self-knowledge and humility.

[110] Sheridan, 'Introduction', A2 verso.

VI

AFTER BEING treated like a shuttlecock in a game of ministerial battledore, Swift made little further mention of the quest for the First Fruits in his correspondence with King. The Archbishop affected not to understand Godolphin's 'oracular sayings', declaring that he had tried to puzzle out what was meant by 'acknowledgements' but that the two or three meanings he had arrived at seemed either 'so trifling or so wicked that I can't allow myself to think that I have hit right'.[111] His caution may have been dictated by a fear of his letters being intercepted, but Swift was heartily sick of all these little games. He replied candidly that he had 'not stirred a foot further' in the matter. Since he had no doubt that Godolphin's allusion was to the repeal of the Test, he saw no point in pursuing the subject.

Stalemated in both his personal and ecclesiastical ambitions, Swift made sardonic observations on the fickleness of fortune, as Godolphin 'gave a young fellow a friend of mine, an employment sinecure of £400 a year added to one of £300 he had before'.[112] He was sorely tempted to return to Laracor, where he would be free to 'talk morals and rail at courts'.

He spent the 'fag-end of the summer' in Kent, and at the spa town of Epsom with the Berkeleys. Refreshed by this little jaunt, he contemplated new plans. There was a rumour that Berkeley might be sent on a diplomatic mission to Vienna, and Swift was mentioned as a suitable secretary. Alternatively Addison suggested that he might like to contact their friend Robert Hunter, recently appointed as Lieutenant-Governor of Virginia with a view to becoming Bishop of the colony. If all else failed, Swift took the precaution of reminding King that the living of St Nicholas Without was still in his thoughts. He returned from the country on the night of 28 October, to find that Prince George, the prince consort, had died, and that there was 'a new world'. Pembroke assumed the Prince's position as Lord High Admiral, allowing Somers to take over one of his previous posts as Lord President of the Council, and Wharton to fill the other, as Lord Lieutenant of Ireland.

[111] *Corr.* i. 93. [112] Ibid., p. 98.

Despite his severe reservations about Wharton, Swift was excited by the changes, 'not knowing how far my friends may endeavour to engage me in the service of a new government'.[113] Still, he confided inevitable doubts to Charles Ford, a young Irishman with an estate in Co. Meath whom he had recently met in London, fearing that his old friends might 'turn courtiers every way, which I shall not wonder at, though I do not suspect'.[114] As a further insurance against disappointment, he asked Walls to mention to the Archbishop that if his ailing kinsman, Thomas King, vicar of Swords, should unfortunately die, he would be very glad to have the post; 'for I like it, and he told me I should have the first good one that fell and you know, great men's promises never fail'.[115] That last refrain sounds so constantly throughout Swift's correspondence that one begins to suspect that 'great men' may have been tempted to play up to his paranoia, so as not to disappoint his expectation of being disappointed.

King made no mention of Swords in his next letter, but approved Swift's plans for Vienna, fearing only that he might 'lose too much time in it'. King's main recommendation was that Swift should seek a post as chaplain to the new Lord Lieutenant. However, that position was quickly snapped up by Ralph Lambert, who had preached a Whiggish sermon on the virtues of Protestant unity in October 1708. Swift informed King that he had made no application for the post 'for reasons left you to guess'.[116] Instead, almost as a conscious riposte to Lambert, he went ahead with publication of his *Letter concerning the Test.*

In November he was 'often giddy. God help me'[117] and paid 11*d.* for a vomit to ease his condition. Christmas came and went with still nothing settled. Berkeley, sixty years old and in poor health, was having second thoughts about Vienna. Swift paid several visits to his impoverished cousin Patty Rolt, even buying her a three-shilling coffee roaster so she could entertain him with his favourite beverage when he called. He also spent a good deal of time with Charles Ford, an unstrenuous companion whom he

[113] Ibid., p. 105.
[114] Ibid., p. 110.
[115] Ibid., p. 106.
[116] Ibid., p. 114.
[117] Account Book for 1708/1709 (48 D.34/2).

tended rather to patronize. 'I use him to walk with me as an easy companion, always ready for what I please, when I am weary of business'.[118] Anthony Henley, a wealthy aristocrat with a dilettante interest in the arts and a lucrative government pension was another new acquaintance to whom he was introduced by Addison. With friends such as these Swift toasted the society beauties of the season, Anne Long and Catherine Barton; played piquet, traffic, and ombre; debated the rival merits of the various Italian castrati at the opera, and whether the opera craze would or would not be the ruin of the nation. 'A good old lady, five miles out of town asked me the other day what these *uproars* were that her daughter was always going to,' wrote Swift to Philips.[119] At St James's coffee-house they discussed the state of poetry, ancient and modern, while criticizing each other's latest efforts, 'Your versifying in a sledge', Swift wrote of Philips's latest effusion, 'seems somewhat parallel to singing a psalm upon a ladder'.[120] The thing is not done well, but one is surprised to see it done at all, he might have added, but preferred to spare the younger man's feelings. The sledge was a natural thought, however, since the winter was so severe that another favourite diversion that Swift enjoyed was to eat ginger bread in a booth by a fire on the frozen Thames. He addressed gallant or facetious verses to a number of ladies, including Anne Finch, Lady Betty Germain, and her companion Biddy Floyd.

> Jove mixed up all, and his best clay employed;
> Then called the happy composition Floyd.[121]
> (ll. 11-12)

Addison was the most eminent, and most constant of his companions. Time and again his name appears in Swift's careful accounts of visits to coffee-houses and taverns. There is an element of name-dropping in his casual remark to his Irish friends that he and Addison would often 'steal to a pint of bad

[118] *Journal,* i. 210.
[119] *Corr.* i. 129.
[120] Ibid., p. 128. Philips was spending the winter in Copenhagen, hence the appropriateness of the 'sledge'. But, on another level, a traitor is drawn on a sledge to be hanged at Tyburn; before being 'turned off', he stands on the 'ladder' while the hanging psalm is sung.
[121] *Poems,* i. 117-18.

wine, and wish for no third person'[122] (the war made good wine difficult to find); but their friendship was still firm. For the family atmosphere that he so much enjoyed, however, he turned again to the Berkeley household. He records losing £1 1*s.* 6*d.* at cards there, without complaint, in November. Still other happy hours were spent with Pembroke and Fountaine in word-play and riddles, while enjoying the 'costly dainties' of Pembroke's table. Only occasionally would the conversation turn to politics, and then most often when the ladies were present to ensure a light-hearted tone. 'The best intelligence I get of public affairs is from ladies,' he joked, 'for the ministers never tell me anything.'[123] Neither the prolonged cold of the winter, nor the continued frustration of his ambitions, extinguished his pleasure in Bickerstaffian pranks.

On 31 March he placed an advertisement in the *London Post Boy* for an auction of esoteric books, prints, and medals at 'Mr Doily's in the Strand' to be held the following afternoon–April Fool's Day. Presumably he and Fountaine amused themselves by watching the crowd of eager bibliophiles gather outside the shop, for this totally fictitious auction. Next day the *Post Boy* admitted 'that there was no such auction designed'. Tricks of this kind reveal a curiously childlike pleasure in deception that lies at the heart of many of Swift's more sophisticated ironic strategies. Among his happiest hours were those spent with the Vanhomrigh family. Mrs Van, a widow since 1703, was a handsome, vivacious, but improvident hostess, who had already run through a good deal of the £16,000 which her husband had left her, in seeking to maintain a position in London society for herself and her children. Since her arrival in London in December 1707 she had kept something of a modest *salon* at their home near St James's Square, and was highly delighted by the visits of Swift and Fountaine who came with witty conversation and anecdotes of state affairs.

Swift soon fell into the agreeable habit of spending his evenings, when not otherwise engaged, at Mrs Van's. Often he would talk books with the elder daughter, Hester, over coffee or a special confection of oranges and sugar; but just as often they would chat idly of nothing in particular or play a few hands of

[122] *Corr.* i. 121. [123] Ibid., p. 133.

ombre. He was now far more confident and mature than the insecure youth who had first encountered and befriended the lonely Esther Johnson. He easily assumed the authority of Hester's dead father, and no doubt Mrs Van was happy to encourage their developing friendship.

While Londoners held fairs on the frozen Thames, the French were shivering in good earnest. Years of war, a poor harvest, and the severe winter caused widespread famine. As a consequence, Louis XIV was finally persuaded to negotiate with the Allies about peace. Meetings took place at The Hague, but the Allies pressed their advantage too fiercely, and although Louis conceded several points he refused to deliver Spain to the Habsburg Charles III. The negotiations, which had begun optimistically in the spring, dragged on until the summer, when they collapsed in time for a new campaign to begin.

In the New Year Swift abandoned his hopes of Vienna, and resumed his quest for the First Fruits. It appeared that Pembroke, as his final act as Lord Lieutenant, had gained the Queen's permission for the grant to be made. In January Swift, writing with an acute headache, told King that there was 'word of the affair of the First Fruits being performed'. He couldn't resist a little shrug of self-congratulation.

> I have a fair pretence to merit in this matter . . . two great men in office, giving me joy of it, very frankly told me, that if I had not smoothed the way . . . it would have met with opposition.[124]

Well pleased with himself, Swift went off for a week at Cranford with the Berkeleys, only to find when he returned that the matter was, once again, not so straightforward as he had thought. 'It is wonderful a great minister should make no difference between a grant and the promise of a grant.[125] It was just another of those court promises. After two months of fruitless enquiries, Swift wrote to Addison, the secretary to the new Lord Lieutenant, to find out what had happened to this particular promise. He was told that no order for the grant had ever reached the Treasury. Eventually Swift tracked down Pembroke, who was enjoying the pleasures of newly married life and proved 'very hard to see'. Pembroke shuffled and prevaricated, but at last admitted 'that he had been promised he should carry

[124] Ibid., p. 117. [125] Ibid., p. 137.

over the grant when he returned to Ireland',[126] but, since he wasn't returning to Ireland . . .

Both King and Addison urged Swift to seek preferment from Wharton, 'The Deanery of Down is fallen, and application has been made for it to my lord Lieutenant, but it yet hangs.'[127] It did not hang in Swift's direction. As might have been expected, it fell neatly into the outstretched palm of Ralph Lambert, further proof of the tangible rewards for loyal Whiggism.

Repressing any jealousies, Swift continued his pursuit of the First Fruits, by now rapidly acquiring the quality of a Holy Grail, and went to attend on Wharton. His lordship was 'in a great crowd' and treated Swift with apparent coldness. Swift told him his business, which Wharton put off 'with very poor and lame excuses, which amounted to a refusal'. Swift left, suspecting that Alan Brodrick was receiving a far more favourable welcome. However, either at Wharton's or at Somers's suggestion, Swift returned to press his case once more. He was received 'as drily as before' as Wharton tried to fob him off. Annoyed at this, Swift began refuting Wharton's objections with some vigour, when Wharton 'rose suddenly, turned off the discourse, and seemed in haste'.[128] Swift was left standing there, furious and still empty-handed. 'I never expected anything from Lord Wharton,' he wrote later, 'and Lord Wharton knew that I understood it so.'[129] He was not now disinclined to heed Addison's promptings to return to Dublin. 'He hath half persuaded me to have some thoughts of returning,' he wrote in January, when most of his other prospects seemed blasted. He proposed to leave in the summer, 'being not able to make my friends in the ministry consider my merits'. However, it was not a move that he contemplated with relish. In March, as Addison busied himself with removal arrangements on the official yacht, Swift compared himself to 'a Gentleman I know, who, having eat grapes in France, never looked up towards a vine after he came back for England'.[130] He warned Ford that his Irish friends would find him morose and discontented, 'but the fault will not be Ireland's, at least I will persuade myself so'. The sour grapes stuck in his belly and not surprisingly he complained of a 'pain in my stomach'. As

[126] Ibid., p. 136. [127] Ibid., p. 125. [128] Ibid., p. 137.
[129] *Journal,* i. 15. [130] *Corr.* i. 126.

with many of Swift's ailments, we can diagnose at least a partially psychosomatic cause. In the same month he complained of a cold, a cough, and shortness of breath. Returning to Ireland meant, at least, a Horatian retreat and the company of the ladies. 'I desire to know', he asked, 'whether a man may be allowed to sit alone among his books.'[131] Yet, highly as Swift regarded this stoic ideal, he was not yet ready to bury himself in meditation. It was too bad to be leaving London when the *Republica Grub-Streetaria,* to which Swift confessed himself 'a small contributor', was 'never in greater altitude';[132] when there were a new set of spring beauties, including Mrs Chetwynd and Mrs Worsley; when there were new operas, new pranks, and the newly launched *Tatler* to enjoy.

Addison arrived in Dublin on 21 April and wrote to Swift immediately 'I long to see you'.[133] Before leaving, Swift paid a final round of visits. On 1 April, a date deliberately chosen, he not only set up his hoax in the Strand, but went on a fool's errand to Halifax. A month later he paid another visit to Halifax. This time, seeing a volume of French poetry on a table, he begged it as a gift. Halifax graciously agreed, and Swift retorted that he should remember that this was 'the only favour he ever received from him or his party'.[134] They both laughed, and the interview was terminated.

He spent longer than usual in Leicester on his journey to the coast. His mother's health was failing, and he may well have suspected that this was to be the last time he would see her alive. He too was in poor health, suffering from piles; 'an ailment incommodious for riding', but not for punning. In letters to Fountaine and Pembroke he piled *pila* on *pilis.*[135] He amused himself studying Roman mosaics and visiting friends in the neighbourhood. His delay was partly caused by lingering hopes of a recall to London, and on the day he left Leicester he sent the first of two ingratiating letters to Halifax, acknowledging 'all your favours to me while I was in town'.[136] As he flatters Halifax

[131] Ibid., p. 126.
[132] Ibid., p. 133.
[133] Ibid., p. 138.
[134] John Forster, *The Life of Jonathan Swift,* vol. i, 1875, p. 252.
[135] *Corr.* i. 139-41.
[136] Ibid., p. 142.

as the new Maecenas, each phrase teeters on the brink of parody, indicating that narrow partition separating panegyric from satire. When he told Halifax that he had 'fifty times more wit than all of us together' he is speaking in the false hyperbole of a courtier. Many years later he gave a more candid opinion of Halifax's wit: 'I never heard him say one good thing, or seem to taste what was said by another.'[137] It was several months before Halifax replied, but when he did so he was full of magnificent promises.

> Mr Addison and I are entered into a new confederacy, never to give over the pursuit, nor to cease reminding those, who can serve you till your worth is placed in that light where it ought to shine. . . . I will be your constant solicitor, your sincere admirer, and your unalterable friend.[138]

This letter bears Swift's cynical endorsement: 'I kept this letter as a true original of courtiers and court promises.' Yet one wonders whether he ever asked himself if the insincerity of Halifax's reply was any more gross than that of his original request.

Swift finally embarked a little before dawn on the 29th. After lying for a night in the bay of Dublin, he landed at Ringsend early the following morning and set out directly for Laracor, needing a couple of days alone in the country before he could wear his public face in Dublin. His final task, before leaving England, had been to see through the press the third and last part of Temple's *Memoirs.* In April he had received £40 for the 'original copy' of this work and late in June, as he sailed for Ireland, it went on sale.

That £40 was poor recompense for all the trouble that this volume caused Swift. In the Preface he speaks of having postponed publication since certain sections of the *Memoirs* 'might give offence to several who were still alive'. He was quite right—they did. Temple had criticized the late Earl of Essex for corruption and duplicity. The Earl's widow, however, was a friend of Lady Giffard and Lady Somerset. Swift had asked Lady Giffard for her copy of the *Memoirs,* but she had refused. He therefore went ahead and published from his own copy, made under Temple's personal direction. Yet as soon as the volume appeared, Lady Giffard inserted an advertisement in the London

[137] *PW* v. 258. [138] *Corr.* i. 150.

Postman, declaring that Swift had acted wholly without authority and from an 'unfaithful copy'. Her aim clearly was to acquit herself of any complicity in the publication in the eyes of her aristocratic friends. In this she was successful. The Duchess of Somerset sympathized with her, calling Swift 'a man of no principles either of honour or religion'.[139] Swift was back in Ireland by the time that Lady Giffard's disclaimer appeared, but he soon got to hear of it and took it as a slur on his integrity both as a man and an editor. He complained of this attempt to ruin his reputation.

I pretend not to have had the least share in Sir Wm Temple's confidence above his relations, or his commonest friends; (I have but too good reason to think otherwise). But this was a thing in my way; and it was no more than to prefer the advice of a lawyer or even of a tradesman before that of his friends, in things that related to their callings. Nobody else had conversed so much with his manuscripts as I, and since I was not wholly illiterate, I cannot imagine whom else he would leave the care of his writings to.

I do not expect your ladyship or family will ask my leave for what you are to say; but all people should ask leave of reason and religion rather than of resentment. And will your ladyship think indeed that it is agreeable to either to reflect in print upon the veracity of an innocent man? Is it agreeable to prudence or at least to caution, to do that which might break all measures with any man who is capable of retaliating?[140]

Swift is very much on his dignity here, and all his suppressed resentment at the condescending treatment he had so often received from the Temple family breaks out. The 'tradèsman' reference is a piece of inverted snobbery and there is more than a hint of menace in that mention of retaliation. He vowed that he would not see Lady Giffard again until she apologized to him for her behaviour. In this way he would either notch up a notable coup for his pride or sever a connection that had become tedious.

VII

IT WAS Swift's firm intention to shun Dublin Castle and bury himself in his country parish, but Somers, anxious to prevent him from such a sulky retreat, had asked him to carry over a letter to

[139] Longe, p. 248. [140] *Corr.* i. 156.

Wharton. When Swift refused, Somers ordered the letter to be left at Swift's lodgings. Annoyed at this attempt to manipulate him, Swift at first determined to send the letter by post. But after only three days in the country, the 'incessant entreaties' of his friends, notably of Addison, drew him to Dublin. However, having delivered the letter, Swift returned to Laracor. 'During the greater part of his [Wharton's] government', he later recalled, 'I lived in the country, saw the Lieutenant very seldom when I came to town, nor ever entered into the least degree of confidence with him.'[141] His thoughts of his native land at this time were quite straightforward. 'I reckon no man is thoroughly miserable unless he be condemned to live in Ireland,'[142] he wrote, as he set about 'cultivating half an acre of Irish bog'. His London friends were unconvinced by this misanthropic pose, and Henley drew a whimsical picture of Swift in exile as a proto-Crusoe.

You are now cast on an inhospitable island, no mathematical figures on the sand, no *vestigia hominum* to be seen, perhaps at this very time reduced to one single barrel of damaged biscuit, and short allowance even of salt water. What's to be done.[143]

Swift played the part of the castaway properly, living in a cabin and sleeping on a field-bed on an earthen floor. After a year of this silence, Andrew Fountaine exploded with exasperation.

I neither can nor will have patience any longer, and Swift you are a confounded son of a – May your half acre turn to a bog, and may your willows perish; may the worms eat your Plato, and may Parvisol break your snuff box!

Swift recorded the activities of his parishioners at Laracor thus:

Mr Percival is ditching, Mrs Percival in her kitchen, Mr Wesley switching, Mrs Wesley stitching, Sir Arthur Langford riching . . .[145]

Langford's support for a Presbyterian chapel at nearby Summerhill was a continuing grievance, but otherwise his care of these

[141] *PW* viii. 121-2.
[142] *Corr.* i. 154.
[143] Ibid., p. 149.
[144] Ibid., p. 164. Isaiah Parvisol was Swift's steward and tithe-collector, and frequently castigated by Swift for inefficiency and negligence. He died, still in Swift's service, in 1718.
[145] Ibid., p. 163.

gentle souls was not an arduous task. In the evenings he might dine with Percival, or with Raymond at Trim, enjoying a game of ombre or whisk. He also visited the Ashe brothers at Clogher, where St George was bishop, or at Finglas where Thomas lived. While at Clogher he renewed his acquaintance with the young poet and archdeacon there, Thomas Parnell. Charles Ford's estate at Wood Park was another favourite retreat. Much as he railed against this country life, it seemed beneficial to his health, and during the autumn his headaches and giddiness disappeared.

Stella would often accompany Swift on visits to Clogher and Wood Park. During the summer Swift also introduced her to Addison. According to Swift, they were both pleased by the meeting. Stella enjoyed Addison's politeness and easy manner, and for his part, Addison assured Swift that had he remained in Ireland, 'he would have used all endeavours to cultivate her friendship'.[146] One suspects the partiality of an anxious tutor behind Swift's description of the encounter. He was keen that his favourite pupil should acquit herself well before an influential friend. Despite these pressures, or perhaps because of them, Stella was not inhibited, and passed this test, like all the others that Swift had set her, in a dutiful and deferential manner. When in town Swift lodged at Mrs Curry's in Capel Street, very close to the ladies. Social evenings in Dublin consisted of card parties with Archdeacon Walls and his punning spouse, with John Stoyte, a city merchant and his wife, and with Stella and Dingley. He would often recall such evenings with affection when he was back in England.

The Irish Parliament had another tempestuous session. A new bill for further preventing the growth of popery sought to close loopholes in previous legislation, and to regulate further the denominational apartheid in the country. Wharton vetoed a proposal that Catholic conversions should be bought at the rate of £5 per family, arguing that 'little trust could be reposed in such converts'.[147] In August Lord Abercorn and the Bishop of Killaloe delivered fierce speeches in a familiar but doomed attempt to reassert the rights of the Irish Parliament and judica-

[146] *PW* v. 229.
[147] See *The Letters of Joseph Addison*, ed. W. Graham, Oxford, 1941. p. 163.

ture. Wharton weathered the storms, and as soon as the session ended made plans to return to London where the political situation was even more volatile. There were strong rumours that his enemies were gathering to impeach him. Addison too had reasons to hurry back, having been unseated as member for Lostwithiel. He lost no time in having himself nominated for Malmesbury, a pocket borough controlled by Wharton. It was after Addison's return to England in September that Swift received Halifax's fulsome letter. It came enclosed in a letter from Steele, equally full of promises. Both men assured him that great men at their tables in London never ceased to upbraid themselves for permitting him to languish in Ireland. His aching desire for recognition is evident in his request to Philips at the time: 'When you write any more poetry, do me honour, mention me in it . . . that Prince Posterity shall know I was favoured by the men of wit in my time.'[148] There's something touchingly unguarded about this request, and the fear of neglect that it implies. He replied to Halifax in November, and the letter is another masterpiece of disingenuousness. He enumerates Halifax's virtues and achievements: his good taste, his wisdom, his power to decide 'the welfare of ten million'; yet each commendation has an ironic edge, as when he speaks of Halifax's reputation in Ireland. 'I must inform you, to your great mortification, that your lordship is universally admired by this tasteless people.'[149] He mentions that the Bishop of Cork was ill with the spotted fever, but to no avail. When the Bishop died, the appointment went, inevitably, to a friend of Lambert's.

In England, as the long war grew increasingly costly and unpopular, the Whigs were forced back on the defensive. When Marlborough committed the blunder of seeking to be appointed Captain-General for life, opposition reached a new pitch of intensity. On 5 November, the High Churchman Henry Sacheverell preached a sermon that so flagrantly challenged the Revolution Settlement of 1688 that the ministry had little choice but to impeach him. The sermon was widely distributed; perhaps as many as two hundred thousand copies were circulated, and the Tory mob appeared again in the streets. Sacheverell's show trial, in Westminster Hall in March, was a

[148] *Corr.* i. 154. [149] Ibid., p. 159.

contest between the ministry and a newly exultant High Church party. Sacheverell was found guilty, but received so light a sentence that it was a clear moral and political victory.

Anticipating his return to Ireland, Addison wrote to Swift of his hopes 'of waiting on you very suddenly at Dublin . . . I heartily long to eat a dish of bacon and beans in the best company in the world.'[150] The picture he gives of Swift's domestic life at Laracor, 'quarrelling with the frosty weather for spoiling my poor half dozen blossoms', is charming. Only Swift's company, he insisted, made Dublin tolerable. 'I love your company and value your conversation more than any man's.'

Only three days after Addison's arrival in Ireland, Swift suffered a severe blow. Late in the evening of 10 May he received at Laracor the news of his mother's death. Though she had been ill for most of the year, the news still came as a shock. Whatever the gulf between them in Swift's infancy, since reaching manhood he had been an attentive son, never failing to visit her on his journeys between London and Dublin. The valediction which he wrote for her was sincere, pious and utterly without affectation.

> I have now lost my barrier between me and death; God grant I may live to be as well prepared for it, as I confidently believe her to have been! If the way to heaven be through piety, truth, justice and charity, she is there.[151]

The particularity of the expression of his grief as the loss of a barrier is a typical confirmation of Swift's view of friends and loved ones as securities against the bleakness of life and the terror of death. That final conditional clause, '*if* the way to heaven be . . .' hints at those dark shadows of doubt that lurk beneath all Swift's attempts at assertions of faith.

When Swift and Addison did meet, their conversation was still dominated by Swift's hopes of preferment. In August he gave Addison his current list of vacant situations.

> You ordered me to give you a memorial of what I had in my thoughts. There were two things. Dr South's prebend and sinecure; or the place of Historiographer.[152]

This is the first mention of the Historiographer's post, after which Swift was to hanker for a good many years, convinced that

[150] Ibid., p. 161. [151] *PW* v. 196. [152] *Corr.* i. 170.

one of his main vocations in life was to provide Prince Posterity with an impartial record of the political events of the age.

Wharton's position in Ireland became increasingly precarious as the Whig ministry came under mounting pressure in London. In the post-Sacheverell Tory euphoria several ministerial changes were forced through, and finally, at the beginning of August, the Queen dismissed Godolphin himself. A general election could not be far off and Addison, anxious about his seat at Malmesbury, obtained leave from Wharton to scurry home. Wharton was equally anxious to return, but it was not until the end of August, with his wife in the later stages of pregnancy, that the Queen gave her permission. The Irish Parliament was prorogued and three days later Wharton left for England. With him went Swift, promise-crammed but pessimistic as ever. He wrote to Addison:

My Lord Lieutenant asked me yesterday whether I intended for England. I said I had no business there now, since I suppose in a little time I should not have one friend left that had any credit and his excellency was of my opinion.[153]

This is hardly an ingenuous comment, even discounting the irony of Swift identifying his own friends with Wharton's. He had written to the publisher Benjamin Tooke a few weeks earlier in a rather different vein. 'It is like to be a new world', he declared, 'and I have the merit of suffering by not complying with the old.'[154] So, he was not entirely without hopes, though with no settled plan for his advancement.

Moreover he had a reason for coming to England, the familiar old reason of the First Fruits. For a few weeks he feigned reluctance to undertake the commission again, and indeed declared that he 'ne'er went to England with so little desire.'[155] Yet we can probably discount this as a white lie to spare Stella's feelings. It took some dispatch to obtain his official signed commission to negotiate for the First Fruits within three days of the adjournment of Parliament, and it was finally signed on the very day, 31 August, that he and Wharton sailed for England. In fact, waiting for the document to be completed involved him in the additional expense of two crowns for a boat to carry him to the viceregal yacht that had already set sail.

[153] Ibid., p. 169. [154] Ibid., p. 166. [155] *Journal,* i. 4.

Part Three

THE LIFE OF A SPIDER[1]

I

SWIFT LANDED at Parkgate on the river Dee on 1 September. Five days later, weary from riding, bruised with a fall, but full of renewed excitement, he was back in London. The Whigs were in full retreat. On the 20th the Lord President, Somers, together with the Lord Steward, Secretary of State Boyle, and the Duke of Devonshire were all sacked. 'I am almost shocked at it,' wrote Swift, adding in an insouciant tone, 'though I did not care if they were all hanged.'[2] He claimed to have washed his hands of the Whigs, who were nevertheless 'ravished to see me, and would lay hold on me as a twig while they are drowning'. It pleased him to report that 'every whig in great office will, to a man, be infallibly put out, and we shall have such a winter as hath not been seen in England'.[3] Only one Whig grandee retained sufficient composure to treat Swift with customary condescension. On his second day in London, still travel-weary, he was greeted by Lord Treasurer Godolphin 'with a great deal of coldness, which has enraged me so, I am almost vowing revenge'.[4] He returned to his lodgings in Pall Mall to take his revenge in the form of the bawdy lampoon 'The Virtues of Sid Hamet's Rod'.

Despite this ostentatious detachment from the Whigs, Swift slipped back into a circle of friends, few of whom would have repudiated that description. On Sunday, 10 September he sat with Addison and Steele, though when he notes that 'Steele will certainly lose his Gazetteer's place, all the world detesting his engaging in parties', it is clear that he endorses the world's opinion.[5] Suddenly Swift was much sought after. Lady Lucy and

[1] 'It is a miserable thing to live in suspense; it is the life of a spider.' *Thoughts on Various Subjects, PW* i. 244.

[2] *Journal,* i. 24.

[3] Ibid., p. 7.

[4] Ibid., p. 6.

[5] *Journal,* i. 13.

Lady Betty Germain plied him with invitations, and he was so 'caressed by both parties' that he was able to boast after a month in town that 'it has cost me but three shillings in meat and drink since I came here'.[6] This was a proud boast indeed to a man as thrifty as Swift. As if to confirm his declared intention of remaining an 'indifferent spectator' of these state revolutions and of returning 'very peaceably to Ireland, when I had done my part in the affair I am entrusted with', he spent much of his time with Irish friends, visiting Patty Rolt, Francis Stratford, and Charles Ford in that first month. He refused to see Lady Giffard, but contrived to visit Stella's mother and her sister Anne, 'a good modest sort of girl'.[7] He paid close attention to Stella's reports of domestic life in Dublin, arranging for the ladies to occupy his own vacant lodgings and sending instructions that a favourite horse that Stella loved to ride should not be sold. He remarked that 'everybody asks me, how I came to be so long in Ireland, as naturally as if here were my being; but no soul offers to make it so'.[8] He was careful to draw Stella's attention to the fact that he did not speak of 'we' in England and 'you' in Ireland,[9] but failed to point out his gradual neglect of this scrupulous use of pronouns as the winter drew on.

Swift's *Journal to Stella* was the most obvious symbol of his determination to retain his connection with Dublin. By virtue of this 'I shall always be in conversation with MD, and MD with Presto.'[10] Every day, night and morning, usually while still in bed, Swift would cover the pages with his tiny writing, pouring out all the news of the day, mingled with riddles and puns, admonitions and anecdotes, hopes and fears, all in a teasing tone, half lover-like, half avuncular.

[6] Ibid., p. 37.

[7] Ibid., p. 39.

[8] Ibid., p. 7.

[9] Ibid., p. 47.

[10] Ibid., p. 8. For a detailed discussion of the editorial problems presented by the *Journal to Stella,* see pp. xlvii-lix of the *Journal.* For the texts of letters II-XL we are dependent upon Deane Swift's *Essay* (1755), whereas for letter I, and letters XLI-LXV the original MSS exist (BM Add, MSS 4804-6). In the MSS Swift's name for himself is 'Pdfr', pronounced podefar, and meaning, perhaps, 'poor, dear, foolish rogue'. Deane Swift amended or translated most of the terms of Swift's 'little language' in the *Journal,* substituting 'Stella' for 'ppt', and 'Presto' for 'pdfr'. It was the Duchess of Shrewsbury, Deane Swift explains, 'who, not recollecting the Dr's name, called him Dr Presto, (which is Italian for Swift)'.

Here is such a stir and bustle with this little MD of ours; I must be writing every night; I can't go to bed without a word to them; I can't put out my candle till I have bid them goodnight.[11]

The *Journal to Stella* is not a diary, but one side of a dialogue. Part of its appeal lies in guessing at the words, the tone, the frowning face of the unseen interlocutor, Stella, as Swift postures, boasts, rails, teases, and preaches before her. It is not a dialogue of equals. The paraphernalia of Swift's 'little language' does not allow that, and he destroyed all Stella's replies, to present the relationship entirely from his perspective. Nor is it always truthful; but the inaccuracies of the *Journal* are less significant than the simple fact that every night he would unburden himself of the hopes, cares, and experiences of the day in childlike banter with her.

Part of the pleasure of Swift's new experiences resided in their capacity to be relived and adorned in the *Journal*. In this way he took a double delight in the diversions of the town. He sat to have the portrait, begun on a previous visit, retouched, and given 'quite another turn' by Charles Jervas. He went with Addison to see the draw for the million lottery at Guildhall where 'the jackanapes of blue-coat boys gave themselves such airs in pulling out the tickets, and shewed white hands to the company to let us see there was no cheat'.[12] He renewed acquaintances at Robin's and the St. James's coffee-houses. He visited china-shops in search of gifts for MD, and promised Dingley 'the finest piece of Brazil tobacco that ever was born'.[13]

The impending elections gave greater animation to the London streets. On a morning visit to the celebrated portraitist Kneller, Swift's coach was jostled by an election mob. Being afraid of 'a dead cat, or our glasses broken',[14] he demonstrated a prudent impartiality, declaring himself 'always of their side'. It fell in with his mood to wish a plague of dead cats on both their houses. He spent his first fortnight in town in lodgings in Pall Mall, but moved on the 21st to Bury Street, where he rented a dining-room and bedchamber for the 'plaguy deep' sum of eight shillings a week. Still he consoled himself, 'I spend nothing for eating, never go to a tavern, and very seldom in a coach'.[15] Part of

[11] *Journal*, i. 27. [12] Ibid., p. 19. [13] Ibid., p. 31.
[14] Ibid., p. 42. [15] Ibid., p. 34.

the attraction of Bury Street was the presence 'but five doors off' of the Dublin widow Mrs Vanhomrigh and her family.[16] In October Swift's circle of friends was completed when Andrew Fountaine returned to town. Meanwhile his clerical mission languished, since nothing could be achieved until the present political 'hurry' was over. He had even heard from Archbishop King that the government had no money to pay for it. 'I am told', wrote King on the 16th, 'there is only £223 in the treasury, and the army unpaid'.[17]

The holiday atmosphere of these early weeks in London did not last long. At the end of September he dined at Addison's 'country place' near Chelsea. Returning home in high spirits he set about answering MD's latest letter in a rallying style before informing her that 'tomorrow I go to Mr Harley'. He continued to insist that 'I don't think any further than the business I am upon', and hoped 'in God, Presto and MD will be together this time twelvemonth',[18] but already the anticipated period of his stay in England was lengthening.

The meeting with Harley which was to change so much in Swift's life was carefully stage-managed by the artful politician. He did not see Swift on the day appointed, but put him off with an excuse till the following Wednesday. As Swift waited, a hint of anticipation crept into his tone. The Whigs were loud in their regrets for their previous 'ill usage of him', but he 'minded them not'.

> I am already represented to Harley as a discontented person, that was used ill for not being Whig enough; and I hope for good usage from him. The Tories dryly tell me, I may make my fortune, if I please; but I do not understand them, or rather, I do understand them.[19]

Worried at the prospect of Swift's defection, the Whigs made belated attempts to flatter his vanity. Two days later he dined with Halifax at Hampton Court where he walked in the gardens, and was shown the Raphael cartoons. But he could not be bought so easily, and would only agree to toast the 'reformation' not the 'resurrection' of the Whigs, informing Halifax that he was 'the only Whig in England I loved'.[20] On the eve of the postponed appointment with Harley, late at night, after Swift had

[16] Ibid., p. 24 and n. [17] *Corr.* i. 176. [18] *Journal,* i. 33-4.
[19] Ibid., p. 36. [20] Ibid., pp. 38-9.

gone to bed and extinguished his candle, there was a knock at the door, and a servant of Lord Halifax stood there with a rival invitation. It must have given Swift all the satisfaction in the world to reply as he did. 'I sent him word I had business of great importance that hindered me.'[21]

Harley continued to play his man, however. He received Swift privately 'with the greatest respect and kindness imaginable' but put off discussing serious matters till the following Saturday. However the interview left Swift in an undisguised mood of elation. 'I will open my business to him,' he told Stella, then checked himself (the phrase was a sexual euphemism), 'which expression I would not use if I were a woman. I know you smoked it, but I did not till I writ it.'[22] He sat up into the small hours writing a Memorial of the Irish Church's claims to present to Harley at their next meeting.

On Saturday, Swift was all but baulked at the door by Harley's notorious lying porter, Read, who solemnly informed him that his master was at dinner and not to be disturbed. Luckily Harley then appeared and ushered Swift in to meet his other guests, with whom he sat for several hours drinking wine. Afterwards he and Harley withdrew in private to discuss Swift's business.

> He . . . entered into it with all kindness, asked for my powers and read them; and read likewise a memorial I had drawn up, and put it in his pocket to show the Queen; told me the measures he would take; and, in short, said everything I could wish; told me he must bring Mr St John (secretary of state) and me acquainted; and spoke so many things of personal kindness and esteem for me, that I am inclined half to believe what some friends have told me, that he would do everything to bring me over.[23]

After the frigidity of his reception by Wharton and Godolphin, Swift's pleasure at such treatment is understandable. He was not so flattered, however, that he failed to observe some of Harley's little ploys. 'He knew my Christian name very well . . .' Writing to King he affected his customary cynicism. 'I never yet knew one great minister who made any scruple to mould the alphabet into whatever words he pleased'[24]; but privately he was convinced that Harley was an exception to this rule. He was particularly

[21] Ibid., p. 41.
[22] Ibid., p. 41.
[23] Ibid., pp. 45-6.
[24] *Corr.* i. 185.

pleased when Harley asked him to call often, not at the general levee, which 'was not a place for friends', but at home, among the family. Only three days after pocketing Swift's memorial Harley told him that he had shown it to the Queen and fully expected the matter to be settled within the week. At a dinner on 15 October Swift's verses on 'Sid Hamet's Rod' were read, and although everyone maintained the fiction of their anonymity Swift reported that 'Mr Harley bobbed me at every line to take notice of the beauties'.[25] Harley did more than merely flatter however, and a week later Swift confided to Stella that the Queen had granted the First Fruits. The Crown rents too, he believed might 'follow in time'. 'Never anything was compassed so soon,' he boasted, 'and purely done by my personal credit with Mr Harley.' Indeed it seemed that he was finally to gain some reward for his efforts. 'The Queen designs to signify it to the bishops in Ireland . . . that it was done upon a memorial from me, which Mr Harley tells me he does to make it more respectful to me.'[26] He really seemed to be 'forty times more caressed' by the new men in power than the old. How galling therefore to hear from King that the Irish bishops believed quite otherwise.

I am not to conceal from you that some expressed a little jealousy that you would not be acceptable to the present courtiers, intimating that you were under the reputation of being a favourite of the late party in power.[27]

In fact the bishops were arranging a new representation to be made to the new Lord Lieutenant as soon as one was appointed. Swift's irritation was made all the worse by being sworn to secrecy by Harley. All he could hope was that the matter would be safely settled before the new Lord Lieutenant took office. Two days later he prevailed with Harley to allow him to tell King that 'the Queen hath granted the First Fruits. . . . It was done above a fortnight ago, but I was then obliged to keep it a secret.'[28] Even this was not official, however. There were further delays while letters patent were drawn up, and the Irish clergy were sceptical of assurances they had heard before. Finally Swift's anger exploded, stung by the 'very different usage I meet

[25] *Journal*, i. 60.
[26] Ibid., p. 66.
[27] *Corr.* i. 189.
[28] Ibid., p. 190.

with' in the two countries.[29] 'If my lords the bishops doubt whether I have any credit with the present ministry, I will if they please, undo this matter in as little time as I have done it.'[30] This petulant outburst is absurd in its suggestion that Swift could make or unmake the Queen's decision at will, but indicates the strength of his resentment at this affront to his dignity. He complained to Harley in more moderate terms, 'and rallied him for putting me under difficulties with his secrets'. But King was still not convinced, and told Swift he wanted no more 'mouthfuls of moonshine' but documentary proof.[31] Swift's position was further complicated since, despite his repeated protests that he desired no personal credit for the remission, he was bitterly hurt when he received none. In the event the Irish House of Lords and Convocation thanked the Queen, the new Lord Lieutenant Ormonde, and Harley. There was no mention of Swift.

Harley's speedy resolution of this matter was not motivated by friendship for Swift or a pious regard for the temporal conditions of Ireland's spiritual guardians. He had his own reasons for enlisting Swift's support, and as soon as he had presented Swift's memorial to the Queen, he approached him about writing the Tory weekly, the *Examiner*. This paper had been launched in August by the new Secretary of State Henry St John as an antidote to the attacks of the Whig press. But Harley was anxious that the *Examiner* should express his own, rather than his Secretary's views.[32] As the new Parliament assembled Swift reported that 'the Tories carry it among the new members six to one', but Harley deliberately kept several Whigs in office, to temper Tory extremism. 'He would not let the Tories be too numerous, for fear they should be insolent, and kick against him,' wrote Swift, approvingly.[33] St John, however, was a root-and-branch man, who wished to chase every Whig from office and 'fill the employments of the kingdom, down to the meanest, with Tories'. He wanted to impeach the Whig Junto and to pillory the 'hireling scribblers' of the Whig press. An eloquent, ambitious thirty-two-year-old, St John contrasted in style,

[29] Ibid., p. 202.
[30] Ibid., p. 193.
[31] Ibid., p. 198.
[32] See J. A. Downie, *Robert Harley and the Press*, Cambridge, 1979, pp. 126-30.
[33] *Journal*, i. 44.

temperament, and outlook with the older, more cautious and cunning politician Robert Harley.

Swift's first *Examiner*, for 2 November 1710, struck a familiar pose of impartiality. 'It is a practice I have generally followed, to converse in equal freedom with the deserving men of both parties.'[34] Swift presents himself as an independent commentator on political events, assigning praise and blame disinterestedly, without reference to party labels. Two kinds of impartiality are implied here. One is the moderation or 'neutrality' of Harley's ministry itself, which sought to avoid the extremes of Whig and Tory. The other is an independence of ministerial influence. Swift claims to offer 'nothing more than the common observations of a private man', and in *Examiner* 26 asserts that the ministry have been 'so far from rewarding me suitable to my deserts, that to this day they never so much as sent to the printer to enquire who I was'.[35] This completely spurious claim is supported by the familiar tactic of identifying 'faults on both sides'. In No. 28 Swift fabricates two angry letters, one from a Whig and one from a Tory, both accusing him of bias. But whereas the 'Whig' letter is full of wild abuse and threats, the 'Tory' letter is more temperate and precise. 'It is plain', it declares, 'you know a great deal more than you write.'[36] This Tory correspondent demands that the *Examiner* should name the guilty men, which Swift very moderately refuses to do. When the *Examiner* offers criticisms of the present ministry it is merely a form of veiled praise. Thus in No. 26 we read that for all Harley's virtues 'his greatest admirers must confess his skill at cards and dice to be very low and superficial'. Yet the *Examiner* consistently maintains the fiction of independence, which allows Swift to offer this judgemental appraisal of the ministry as he signs off in *Examiner* 41.

I have with the utmost rigour examined all the actions of the present ministry . . . without being able to accuse them of one ill or mistaken step.[37]

[34] *PW* iii. 3. See also R. I. Cook, *Jonathan Swift, Tory Pamphleteer*, Seattle, 1967; and B. A. Goldgar, *The Curse of Party*, 1961.

[35] *PW* iii. 78.

[36] Ibid., p. 89.

[37] Ibid., p. 153.

The cool hyperbole of this claim is clearly preposterous, and 'arguments' such as these were hardly meant to convince the unconverted. But the *Examiner* did a valuable job in encouraging the country members and their constituents to contain their impatience at Harley's softly-softly policies. It is less easy to decide how far Swift disbelieved his own claims to independence. It was important to him that he should not appear as a mere hireling scribbler, like the Whigs whom he derided, and for the most part Harley and St John were careful to sustain this illusion. It was uncharacteristically gauche of Harley to send Swift £50 in July 1711 in return for his services. Swift refused the money and was deeply upset by the gesture. Yet Harley's inadvertency may give us an insight into his real opinion of Swift's role. On at least one other occasion in December 1712, the ministers either by carelessness or design let slip the pretence of consulting and deferring to Swift's political advice. He was furious when St John 'told me I must walk away when dinner was done, because Lord Treasurer and he and another were to enter on business: but I said it was as fit I should know their business as anybody; for I was to justify'.[38] Naturally the ministers apologized, and Swift having won the point, tried to shrug it off by describing their business as 'so important I was like to sleep over it'. But the flash of anger is a clear sign of Swift's feelings of insecurity with the ministers. On the whole both Harley and St John were remarkably careful to play a game of deferring to Swift, as if he were the great man, and they his courtiers. It was a game which gave Swift endless pleasure. But the merest hint of neglect was enough to trigger off an anxious recrimination. In April 1711 he warned St John 'never to appear cold to me, for I would not be treated like a schoolboy; that I had felt too much of that in my life already (meaning from Sir William Temple)'.[39] In his *History of the Four Last Years of the Queen,* Swift presents himself simultaneously as both a trusted insider and an impartial historian of events.

> I never received one shilling from the minister, or any other present, except that of a few books; nor did I want their assistance to support me. I very often indeed dined with the Treasurer and Secretary; but in those days that was not reckoned a bribe.[40]

[38] *Journal,* ii. 589. [39] Ibid. i. 230. [40] *PW* vii. pp. xxxiv-v.

Yet although he turned down offers of money with indignation, there is no doubt that he hoped to be rewarded in a more substantial manner, with an English bishopric or Deanery.

The *Examiner*'s most urgent task was to justify the change of ministry in wartime, and directly against the wishes of the supreme commander of the forces. Wartime expenditure had risen to unprecedented levels and Swift was soon reporting that the government was 'in a terrible loss for money'.[41] City institutions were dominated by Whigs who viewed the new Tory ministry with grave disquiet, and a considerable crisis of confidence developed. This gave Swift the central theme for his propaganda, namely a conspiracy theory. First formulated by Harley himself,[42] the theory described a network of placemen in Parliament and 'moneyed men' in the City who had intervened in the traditional relationship between the Sovereign and her people. At the centre of this network was the 'family', the Marlborough-Godolphin alliance, held together by ties of marriage and self-interest. The aim of the conspirators was not peace 'but to aggrandise themselves and to prolong the war by which they got such vast wealth'. Conspiracy theories have a simple but deadly appeal, and Swift developed this theme through all his *Examiner* papers and as the central argument of his *Conduct of the Allies*. Marlborough's behaviour, in asking to be made Captain-General of the forces for life, and in threatening to resign if Queen Anne went against his advice, seemed to offer strong confirmation of the theory, as did the refusal of Whig financiers to lend money to the new administration.

Since most readers of the *Examiner* were country squires, Swift exploited to the full their prejudices against this small conspiratorial metropolitan élite. 'I shall take occasion to hint at some peculiarities for the sake of those at a distance', he says in *Examiner* 32, and in the *Conduct* he observes that 'it is the folly of too many to mistake the echo of a London coffee-house for the voice of the kingdom'.[43] He contrasts the traditional patriotic authority of the landed gentry with the unconstitutional power exerted by a few moneyed men.

[41] *Journal*, i. 76.

[42] See 'plain English to all who are honest', ed. W. A. Speck and J. A. Downie in *Literature and History*, no. 3 (1976), pp. 100-10.

[43] *PW* vi. 53.

What people then, are these in a corner, to whom the constitution must truckle? If the whole nation's credit cannot supply funds for the war, without humble application from the entire legislature to a few retailers of money; it is high time we should sue for a peace. . . . By the narrowness of their thoughts, one would imagine they conceived the world to be no wider than Exchange-Alley.[44]

The cunning stock-jobbers of Exchange Alley are established in Swift's first *Examiner* (13) as key figures in the conspiracy that has the nation in thrall. They are the new 'dextrous men' who have grown rich lending money to the Whig ministry for the war effort, which Tory landowners had to pay off through the land tax. 'By this means, the wealth of the nation, that used to be reckoned by the value of land, is now computed by the rise and fall of stocks.'[45] A second and related prejudice that Swift plays upon is xenophobia. Not only have these 'retailers of money' no essential stake in the country; many are not even native-born. For the Whigs who 'come with the spirit of shopkeepers to frame rules for the administration of kingdoms' and who imagine that 'trade can never flourish unless the country becomes a common receptacle for all nations, religions and languages', are prepared to sacrifice Church, country, and Constitution in their pursuit of gain. In his last *Examiner* Swift denounces a proposed Bill for the naturalization of immigrants. Under the guise of refugees many foreigners had entered the country, 'who understood no trade or handicraft; yet rather chose to beg than labour; who besides infesting our streets, bred contagious disease, by which we lost in natives thrice the number of what we gained in foreigners.'[46] It is from this miasma of stock-jobbers, foreign profiteers, and corrupt Whig officials that Harley has rescued the nation.

In *Examiner* 17 Swift declared his resolution 'to concern myself only with things, and not with persons'. This conventional disclaimer is a considerable exaggeration, since probably the most effective of all his *Examiner* papers were those in which he used historical parallels to attack Wharton and Marlborough. It is true though that the 'official' tone of the *Examiner* was rather different from the style he adopted when unleashing his own personal attacks on these men. Thus

[44] *PW* iii. 134. [45] Ibid., p. 6. [46] Ibid., p. 169.

Examiner 17, an attack upon Wharton's period of rule in Ireland, is based upon Cicero's orations against Verres, the Governor of Sicily. This Ciceronian model gives a generalized gravamen and classical authority to the terms of Swift's indictment. By contrast his *Short Character of Thomas Wharton,* written at the same time, is a frankly personal attack, based on long remembered grievances that erupt in a lava of resentment.

He sweareth solemnly he loveth, and will serve you; and your back is no sooner turned, but he tells those about him you are a dog and a rascal.[47]

Swift reduces Wharton to a sub-human species, a kind of proto-Yahoo whose total insensitivity makes him impervious to more subtle thrusts.

He is without the sense of shame or glory, as some men are without the sense of smelling; and therefore, a good name to him is no more than a precious ointment would be to these. Whoever, for the sake of others were to describe the nature of a serpent, a wolf, a crocodile or a fox, must be understood to do it without any personal love or hatred for the animals themselves.[48]

Swift therefore claims to present Wharton's actions with the detached fascination of a natural scientist observing the depredations of a cannibalistic species. He was more than usually anxious to preserve his anonymity with respect to this pamphlet. 'Who the author is must remain uncertain,' he told Stella. 'Do you pretend to know, impudence?'[49] If she did, then luckily she did not tell Archbishop King who cried out in the name of Christianity against the pamphlet as a wound in the dark, an appeal to the mob.[50]

Although excluded from the *Examiner,* such outbursts of revenge were a familiar feature of Swift's personal satires, though he paid a high price for them. He may well have sunk his last remaining hopes of a bishopric with his savage attack on the Duchess of Somerset in *The Windsor Prophecy* (1711) and his ill-judged remarks on the Scottish peers in *The Public Spirit of the Whigs* (1714) led to prosecution, and a price of £300 being put on his head.

Swift's pamphleteering at this time has sometimes been

[47] Ibid., p. 179. [48] Ibid., p. 178.
[49] *Journal,* i. 148. [50] *Corr.* i. 207.

presented as a gladiatorial struggle between himself and the Duke of Marlborough, and although this is an over-simplification, it was Marlborough's power and reputation which lay at the heart of the Whig 'conspiracy'. *Examiner* 27 contains a 'Letter to Crassus' (alias Marlborough) in which Swift poses as a friend, full of praise for the Duke's wisdom and courage, but who takes the privilege of a friend to warn him frankly against his one great fault, 'that odious and ignoble vice of covetousness.'[51] Swift had already seized upon covetousness as Marlborough's weak spot in his devastating *Examiner* 16. The Whigs were loud in their protests at Tory ingratitude to the Duke, so Swift presents two contrasting bills, the first of British 'ingratitude', the other of Roman gratitude. In the British balance sheet we find Woodstock, Blenheim, gifts of jewels, pensions, and sinecures totalling 'a good deal above half a million of money' which has already been granted to the Duke, and 'all this is but a trifle in comparison of what is untold'.[52] On the Roman side we find a triumphal arch, a statue, a sacrificial bull, a crown of laurel, and some frankincense. In a meticulous display of fairness the *Examiner* costs every item, from the laurel crown, 2*d*., to 'casual charges at the triumph,' £150. Yet still the Roman bill only amounts to £944 11*s*. 10*d*. This is one of Swift's most succinct uses of statistics to 'prove' a satiric point. The 'impartial' message is that figures cannot lie, while the playful inclusion of such details as the twopenny laurel crown emphasizes the gross discrepancy between the austere Roman ideal and the vulgar British imitation.

At times Swift claimed to suffer pangs of conscience at the ferocity of some Tory attacks on Marlborough. 'I think our friends press a little too hard on the Duke,'[53] he told Stella in January 1711, only a month before he published his 'Letter to Crassus'. Two years later he again protested that 'those things that have been hardest against him were not written by me', and claimed that 'I have often scratched out passages from papers and pamphlets . . . because I thought them too severe'.[54] These were differences of tactics, rather than of intentions, however. Marlborough was a successful general with a number of victor-

[51] *PW* iii. 84.
[52] Ibid., p. 22.
[53] *Journal,* i. 159.
[54] Ibid. ii. 460, 472.

ies to his credit, and it did the Tory cause no good to treat him with the kind of abuse that Swift reserved for Wharton. Privately however, Swift believed that Marlborough was 'certainly a vile man, and has no sort of merit beside the military'. After the Duke's dismissal in January 1712 it gave him much delight to tell Stella that 'the Duke of Marlborough says, there is nothing he now desires so much as to contrive some way how to soften Dr Swift'.[55] And in his *History of the Four Last years of the Queen* he even went so far as to hint that Marlborough may have harboured an intention 'of settling the crown in his family'.[56]

The circulation of the *Examiner* varied between 500 and 1,000, but its influence spread far wider. The Tory network which it helped to sustain is clearly indicated in a letter to Harley from a contact in Scarborough, which describes how the local parson used the copy of the *Examiner* which he obtained each Sunday from the town's MP.

> After evening service the parson usually invites a good number of his friends to his house, where he first reads over the paper, and then comments upon the text; and all the week after he carries it about with him to read to such of his parishioners as are weak in the faith, and have not yet the eyes of their understanding opened; so that it is not to be doubted but that he will in time make as many converts to the true interest of the state, as ever he did to the church.[57]

Swift rarely risked disconcerting this loyal following by varying his style or approach. Sometimes his *Examiner* papers are deft and amusing in their parody of Whig themes, but often they are hectoring and narrow. They seldom strike us as works in which his full imaginative ingenuity has been engaged.

However, it was not Swift's propaganda, but a purely fortuitous event which consolidated Harley's hold on power in early 1711. On 8 March Antoine de Guiscard, a French refugee who had been put in charge of a regiment of expatriates was arrested on a charge of high treason and brought before a committee which included Harley and St John to be interrogated. There Guiscard suddenly drew a knife and stabbed Harley in the chest, 'but it pleased God, that the point stopped at one of the ribs, and broke

[55] Ibid., pp. 472, 460.
[56] *PW* vii. 7.
[57] Letter from J. Durden, 5 December 1710. Portland MSS, HMC (London 1897) iv. 641.

short half an inch. Immediately Mr St John rose, drew his sword, and ran it into Guiscard's breast. Five or six more of the Council drew and stabbed Guiscard in several places'.[58] Swift wrote this account to King on the day of the assassination attempt, in 'violent pain of mind . . . greater than ever I felt in my life' and full of fears for Harley's life. To Stella he wrote, 'my heart is almost broken . . . pity me'.[59] He told King that Harley 'hath always treated me with the tenderness of a parent', and although this was an exaggeration, it indicates that Harley had come to replace Temple in his affections as a surrogate father.

When he came to write his *Examiner* account of the incident a week later, Harley's life was out of danger, and Swift was able to exploit the attack for its full propaganda value. He did not scruple to link Guiscard's action with the political attacks of the Whigs on the Queen's ministers. Though he stopped short of suggesting any direct complicity between Guiscard and the Whigs, he did remind his readers that it was 'a Great Man of the late ministry' who had raised Guiscard from 'a profligate popish priest to a Lieutenant-General and Colonel of a regiment of horse'.[60] He also noted how the Whigs had tried to use Greg to implicate Harley himself in a charge of high treason.

> It may be worth observing how unanimous a concurrence there is between some persons once in high power, and a French papist; both agreeing in the great end of taking away Mr Harley's life, although differing in their methods.[61]

The assassination attempt was Harley's Reichstag fire, and he made the most of it. Though out of danger, he stayed away from Parliament for several weeks, while an effusion of odes in his honour poured from the presses. Guiscard meanwhile died of his wounds in prison, and his body was pickled and put on exhibition in a trough for twopence a time. When Harley finally returned to Parliament in April, Speaker Bromley delivered a fulsome eulogy in which he gave thanks 'that Providence hath wonderfully preserved him from some unparalleled attempts' made by the 'inveterate malice' of his enemies. In this emphatic use of the plural Bromley continued to associate Guiscard's attack with the indirect attempt of the Whig lords to destroy

[58] *Corr.* i. 213-14.
[59] *Journal,* i. 210-12.
[60] *PW* iii. 108.
[61] Ibid., p. 108.

Harley through Greg. The attack had another important effect in deepening the rift between Harley and St John. Jealous of the martyr's glory which Harley had gained from the incident, St John allowed a rumour to circulate that he, not Harley, had been Guiscard's intended victim. Unfortunately it was Swift's *Examiner*, in a classic *faux pas*, that did most to promote this version of events. Whether St John deceived Swift, or whether it was Swift's clumsy attempt to reconcile the two men, this was his report.

The murderer confessed in Newgate that his chief design was against Mr Secretary St John, who happened to change seats with Mr Harley, for more convenience of examining the criminal; and being asked what provoked him to stab the chancellor, he said that not being able to come at the secretary, as he intended, it was some satisfaction to murder the person whom he thought Mr St John loved best.[62]

This is pretty unconvincing stuff, particularly the make-believe that Harley was the person that St John 'loved best'. Harley let his displeasure at this account be known, and consequently Swift left the task of producing a full official version of the incident to his assistant, Mrs Manley, who cooked up his few hints into a sixpenny pamphlet of her own. 'I was afraid of disobliging Mr Harley or Mr St John in one critical point about it, and so would not do it myself,' he explained to Stella.[63] Indeed, his self-imposed task of mediating between the two men was 'a plaguy ticklish piece of work, and a man hazards losing both sides'. Both politically and personally he felt far closer to Harley than to his ambitious young rival. Yet there was a charm and excitement about St John that attracted Swift, and on the issue of press censorship at least they agreed in adopting a more hawkish line than Harley. Swift rated his efforts to reconcile the two ministers as of equal importance with his journalism. 'I have ventured all my credit with these great ministers to clear some misunderstandings between them; and if there be no breach, I ought to have the merit of it.'

Following his return to Parliament Harley was quickly promoted to be Earl of Oxford and Lord Treasurer. At the same time he had a long private session with Swift at which they talked through 'a great deal of matters I had a mind to settle with

[62] Ibid., p. 109. [63] *Journal*, ii. 244-5.

him'.[64] As a result we read in the *Examiner* for that week, 'I know not whether I shall have any appetite to continue this work much longer,'[65] and three weeks later Swift ceased authorship of the paper. However, he had not been sacked for speaking his mind. On the contrary, Harley had other plans for Swift, and advised him to take a short summer holiday before the new political campaign in the autumn.

II

It was a warm summer, and Swift enjoyed a well-earned rest. He had recently moved from St. Alban's Street, behind the Haymarket, to the relative seclusion of the riverside village of Chelsea. A lifelong believer in vigorous exercise, he hoped the daily walk to and from London would be good for his health. 'I never felt so hot a day as this since I was born,' he wrote in early June. He took to swimming in the Thames to cool himself. He must have made a comical figure, splashing about with his landlady's napkin on his head, while his servant Patrick stood on the bank, holding his nightgown, shirt and slippers.

> I have been swimming this half-hour and more; and when I was coming out I dived, to make my head and all through wet, like a cold bath; but as I dived, the napkin fell off and is lost, and I have that to pay for. O faith, the great stones were so sharp, I could hardly set my feet on them as I came out.[66]

Later that month he went for a country ramble to Lord Shelburn's estate at Wycombe, and when he returned he was delighted to discover that he had been elected to membership of an exclusive society of twelve 'brothers' whose aim was 'to advance conversation and friendship, and to reward deserving persons with our interest and recommendation'. 'We take in none but men of wit, or men of interest,' he boasted, 'and if we go on as we begin, no other club in this town will be worth talking of.'[67] Prime mover of the society was St John. When Oxford and Lord Keeper Harcourt were proposed for membership, Swift found nothing suspicious in St John's opposition to them,

[64] Ibid., p. 273.
[65] *PW* iii. p. 156.
[66] *Journal,* i. 286.
[67] Ibid., p. 294.

since he shared a disinclination for the society to become simply a social extension of the ministry.

Three months of concentrated Tory polemic had taken their toll of the former pattern of Swift's social life, which had been largely among Whigs. 'I don't go to a coffee-house twice a month,' he told Stella in March. Never a great lover of late hours, he was now regularly in bed and asleep by eleven. He observed to Stella how his companions had changed.

Prithee, dont you observe how strangely I have changed my company and manner of living? I never go to a coffee-house; you hear no more of Addison, Steele, Henley, Lady Lucy, Mrs Finch, Lord Somers, Lord Halifax etc. I think I have altered for the better.[68]

Swift affects to carry off the change with a kind of flourish, but this alteration to his social life was not all gain. Though he was happy enough to break with Somers and Halifax, the loss of Addison and Steele caused him real distress. As early as October 1710 he was reporting a new coldness in their relations. By December he and Addison met barely once a fortnight. There seemed something inevitable about the way they drifted apart, though neither man wished it. Politics lay like a sword between them, and the more they sought to avoid the topic, the more awkwardly it kept sliding back into their discussions. It was with sadness and exasperation that Swift wrote on 14 December:

Mr Addison and I are different as black and white, and I believe our friendship will go off, by this damned business of party; he cannot bear seeing me fall in so with the ministry; but I love him still as well as ever, though we seldom meet.[69]

A month later relations were even chillier; 'all our friendship and dearness are off; we are civil acquaintance, talk words of course, of when we shall meet, and that's all'. Swift was greatly saddened by this estrangement, but he would not deny his own political instincts, nor accept the silent imputation of betrayal that came from Addison's glacial aloofness.

His relations with Steele suffered a much sharper reverse. During the autumn he continued to contribute 'hints' to the *Tatler,* and his poetic 'Description of a City Shower' appeared in the issue for 17 October. A fortnight later, when his first

[68] Ibid., p. 282. [69] Ibid., p. 127.

Examiner was published, Swift described the recent *Tatlers* as 'scurvy'. 'Pray do not suspect me,' he warned Stella.[70] To mitigate the guilt of his 'defection', Swift sought assiduously to promote the interests of his former Whig friends with the new ministers. But the evident delight which he took in the exercise of such patronage rather discouraged his old friends from seeking it. Congreve, half-blind and gouty, was grateful for Swift's aid, as was Harrison, a young writer from Oxford. But Steele and Philips were more independent. 'I will do nothing for Philips,' he wrote at last, in exasperation. 'I find he is more a puppy than ever.' Had Philips, like Harrison, been prepared to be patronized as a 'little understrapper' he might have ascended the scale of animal similes to become 'an agreeable toad'. As it was, his presumption reduced him to an 'impudent puppy' or, worst of all, 'a Tisdall fellow'. The description of Swift's behaviour at court, as it appeared to a Whig bishop, explains some of the contempt which Steele and Philips felt for him in his new role as self-appointed master of requests.

> He was promising Mr Thorold to undertake with my Lord Treasurer that, according to his petition, he should obtain a salary of £220 per ann. to be settled for the minister of the English church at Rotterdam. He stopped Francis Gwyn, Esq., going in with his red bag to the Queen, and told him aloud he had something to say to him from my Lord Treasurer. He talked with the son of Dr Davenant to be sent abroad, and took out his pocket-book and wrote down several things, as memoranda, to do for him. He turned to the fire, and took his gold watch, and telling the time of the day, complained it was very late. A gentleman said 'he was too fast'. 'How can I help it!, says the doctor, if the courtiers give me a watch that won't go right.'[71]

This busy, self-important *habitué* of the ante-chamber is an indication that Swift was no more impervious to the corrupting influence of power than the Great Men whom he so often attacked.

During Harley's absence from the Commons, St John made a bid for leadership of the Tories. However, his main supporters were October Club men whose backwoods extremism made

[70] Ibid., p. 79.

[71] See *Corr.* v. 228-9. Bishop White Kennett's description of Swift at Windsor in 1713 was first printed in part by Johnson, then published in full by John Nichols in Swift's *Works,* 1801, 19 vols., vol. xix, pp. 21-2.

them an unreliable power base. With little grasp of, or interest in the technical measures essential to an efficient administration, they indulged in a brief spell of irresponsible idealism, throwing out a vital tax bill and seeking to impeach the former Paymaster-General, James Brydges. Their behaviour infuriated and embarrassed St John, who, as a personal friend of Brydges, was forced to speak against his potential 'allies' on both issues. Worse was to come. St John dispatched an expeditionary fleet to capture Quebec from the French, but eight of the ships broke up in the St Lawrence, and 900 lives were lost. Meanwhile in the City Tories failed to secure a single place in the elections to the board of the Bank of England. This conspicuous record of failures was comforting confirmation to Harley of his own indispensability to the administration.

Harley now hit upon a scheme to break the traditional alliance between 'moneyed men' and the Whigs by establishing a new financial institution that would be dominated by Tories. The South Sea Company would take on the accumulated war debt of £10,000,000, and would pay interest at the rate of 6 per cent, generating its profits from a monopoly of the slave trade with Spanish America, which Britain would demand as a condition of any peace treaty. Peace was central to all Tory policies; peace with honour, naturally, but above all, peace. Party attitudes to a treaty were quickly reduced to handy slogans. 'No Peace without Spain' was the Whig cry–that is, no peace that would leave Spain in Bourbon hands. This was a commitment that was reduced to near absurdity by the death of the Emperor Joseph of Austria in April. His heir, Charles, was the Whig candidate for the Spanish throne, yet it made little sense to fight to prevent Spain and France being ruled by the same family if the alternative was to allow Spain, Austria, the Spanish Netherlands, and half America to be ruled by the same man. By destroying much of the credibility of the 'balance of powers' argument, Swift was confident that Joseph's death would 'hasten a peace' considerably.

In July Matthew Prior was sent to Versailles to begin secret negotiations, but on his return, accompanied by two French agents, he was arrested by a zealous official at Dover on suspicion of being a spy. High-level orders for his release were immediately issued, but not before awkward rumours of an imminent sell-out began to circulate. Swift met Prior during a visit to Windsor in August, when together with Oxford and St John they concocted a plan to confuse if not confute these Whig rumours. A few weeks later *A New Journey to Paris* appeared, in which Swift gave a fictional account of Prior's embassy to France, and presented him as making ruthless demands of Louis XIV himself. The pamphlet sold well, and a thousand copies went on the day of publication. It was a 'pure bite' said Swift,[72] but it served its purpose 'by way with furnishing fools with something to talk of'.

Refreshed after his summer break, it was now Swift's task to justify the treaty that Prior was negotiating. He accepted the responsibility happily, while continuing to assert his independence of the ministry.

> There is now but one business the ministry wants me for; and when that is done, I will take my leave of them. I never got a penny from them, nor expect it.'[73]

He began work in early September, long before the negotiations had arrived at any conclusions, and stayed on alone at Windsor to get on with the job which 'will take up a great deal of time'.[74] It was a wrench to tear himself away from the diversions of the court; dinners at the Green Cloth; riding parties with the maids of honour; evenings spent playing piquet or exchanging puns with his new friend, the Queen's physician Dr John Arbuthnot. But the negotiations seemed to be progressing quickly and by the end of the month he was chiding himself for having idled so much time away. 'I have a plaguy deal of business upon my hands, and very little time to do it.' Meanwhile he begged St John to 'make examples' of one or two of the Whig pamphleteers who were taking advantage of his silence to muster their own offensive against the ministry. 'They are very bold and abusive,' he complained.

A week later two draft treaties, one public and one secret, were signed between the English and French diplomats. That night St John invited Swift to dine with himself, Prior, and the French envoys–'good rational men' according to Swift. Together they discussed how the package they had agreed was to be sold to the public at home and abroad.

[72] *Journal,* i. 358. [73] Ibid., p. 343. [74] Ibid., p. 361.

We have already settled all things with France, and very much to the honour and advantage of England; and the Queen is in mighty good humour. All this news is a mighty secret . . . there will be the devil and all to pay; but we'll make them swallow it with a pox.[75]

It is noticeable that Swift's pretence of independence has disappeared. Flattered by St John's mock-deference, he has identified himself utterly with the ministry's case for the peace. Thereafter the work piled up even faster. A fortnight later he was 'so busy now, I have hardly time to spare for our little MD'. Deadline for completion of the new pamphlet was mid-November, in time for the opening of the new session of Parliament, and the inevitable debates on the peace negotiations.

Composition of this new pamphlet was complicated by the fact that it came under continual ministerial scrutiny. This time there was no latitude for independence whatsoever. Throughout the second half of October, and well into November, Swift was in almost daily consultation with St John and Prior, with Under-Secretary of State Erasmus Lewis, or with John Barber the printer. On 30 October Swift was so full of the pamphlet that he slipped unconsciously into lecturing the ladies.

Few of this generation can remember anything but war and taxes, and they think it is as it should be; whereas 'tis certain we are the most undone people in Europe, as I am afraid I shall make appear beyond all contradiction.[76]

Despite his complete identification with the ministry's case, Swift still found the constant supervision of his work irksome. With the opening of Parliament getting ever closer, he wrote: 'Something is to be published of great moment, and three or four great people are to see there are no mistakes in point of fact: and tis so troublesome to send it among them, and get their corrections, that I am weary as a dog.'[77] *The Conduct of the Allies* was finally completed on 24 November and published three days later. It had a colossal sale for a shilling pamphlet. The first edition of a thousand copies sold out in two days, and the printer set to work 'day and night' to get out a second edition, with one or two small revisions by Oxford, within another two days. When Swift did not receive a copy of this

[75] Ibid. ii. 372. [76] Ibid., p. 397. [77] Ibid., p. 408.

second edition he assumed that sales must have dropped off. On the contrary, the edition had sold out in five hours, and the printer worked all through Sunday to prepare a third. Swift wrote to the ladies, elated with this success.

They are now printing the fourth edition, which is reckoned very extraordinary, considering tis a dear twelvepenny book, and not bought up in numbers by the party to give away, as the Whigs do, but purely upon its own strength.[78]

This last assertion was directly challenged by the Whig author of *Remarks upon Remarks,* who had a rather sharper insight into Oxford's efficient distribution service.

His party find money somewhere or other to buy his libels by dozens, and disperse them about the country to poison the minds of the people too easily impos'd upon by the plausible pretence of a concern for the public interest. I have been told they have bundled 'em up with briefs and fast-prayers, and distributed 'em by apparators gratis to the poorer vicars and curates.[79]

The argument of the *Conduct* is a straightforward extension of the conspiracy theory already presented in the *Examiner.*

No nation was ever so long or so scandalously abused by the folly, the temerity, the corruption, the ambition of its domestic enemies; or treated with so much insolence, injustice and ingratitude by its foreign friends.[80]

While never relinquishing a tone of grievance against those who have misled and cheated the nation, Swift contrives to give the impression of objectivity, as though he were simply stating matters of fact, and leaving the reader to draw his own conclusions. Actually this is far from the truth. All Swift's facts are highly selective. The ministry had decided to abandon not only Austria, but also the Catalans, the Portuguese, and the Dutch. In return for deserting these allies, the British negotiators had secured a number of important, though as yet secret gains for themselves, including Newfoundland, Gibraltar, and a monopoly of the Assiento (the privilege of supplying Negro slaves to the Spanish colonies in America). Such naked chauvinism would not

[78] Ibid., p. 430.
[79] Anonymous. Quoted by Davis, *PW* vi, p. xv.
[80] Ibid., p. 15.

be difficult to sell to the House of Commons. A more serious problem was the reaction of the heir to the throne, the Elector of Hanover who, like the Dutch and Austrian courts, had received confidential copies of the peace preliminaries. It was no doubt an enraged representative of the Austrian court who leaked the text of these preliminaries to the Whig *Daily Courant,* and sparked off a violent campaign against the proposed terms.

To justify Britain's actions, Swift fell back on the extended version of the conspiracy theory, asserting the selfishness and inefficiency of the Austrians and Dutch.

> It plainly appears that there was a conspiracy on all sides to go on with those measures, which must perpetuate the war; and a conspiracy founded upon the interest and ambition of each party; which begat so firm a union, that instead of wondering why it lasted so long, I am astonished to think, how it came to be broken.[81]

As usual, the only ones to benefit from the war were the 'monied men', 'such as had raised vast sums by trading with stocks and funds, and lending upon great interest and premiums; whose perpetual harvest is war, and whose beneficial way of traffic must very much decline by a peace'.[82]

While the Commons could be relied upon to support the peace proposals, the Lords might prove more awkward, and the ministry needed some unmistakable sign from the Queen that she endorsed the policy. Ideally too they hoped that Marlborough would maintain a soldierly silence on the matter. As Christmas approached Tory hopes slumped, and touched bottom when the Whig leaders formed an alliance with 'Dismal', the Earl of Nottingham who, as leader of the High Church extremists, had been excluded from Oxford's government. Nottingham announced his intention of voting with the Whigs in return for their support for a Bill to prohibit Occasional Conformity. It was an opportunistic alliance which reflected little credit on either side, but it posed a serious threat to the ministry. At Harley's request Swift composed a ballad on Dismal's defection, but haste and anger are both rather too evident in his 'Excellent New Song', which is not one of Swift's happiest

[81] Ibid., p. 44. [82] Ibid., p. 41.

productions. The next day Nottingham made his speech, asserting that 'No peace could be safe or honourable . . . if Spain and the West Indies were allowed to any branch of the House of Bourbon'. When the House divided the Whigs carried the vote by 62 to 54. Swift reacted with something like panic. 'It is a mighty blow . . . I am horribly down,' he wrote,[83] forseeing the ministry's collapse. He took Oxford to task for having failed to manage or pack the House to avert this crisis, urging him to sack some half a dozen unreliable office-holders, including Marlborough and Somerset. Swift's deepest fear was that the Queen herself could no longer be trusted, since when she left the House after listening to the debate 'she gave her hand to the Duke of Somerset who was louder than any in the House for the clause against peace'.[84] Two days later a conversation with St John hardened Swift's suspicions. 'We are both of opinion that the Queen is false,' he wrote, gloomily.[85] Together they indulged a despairing fantasy in which Somers and the Whigs secured a majority in Parliament 'by managing elections'. Swift also began to fear for his own safety and begged to be sent abroad on some diplomatic mission, 'then I will be sick for five or six months till the storm has spent itelf', and afterwards 'steal over to Laracor' and obscurity. As he thought of retirement he had 'in my mouth a thousand times' Wolsey's lines in *Henry VIII*:

> A weak old man, battered with storms of state
> Is come to lay his weary bones among you.[86]

Oxford's repeated assurances that all would be well did nothing to reassure him. On 13 December Morphew, the publisher of the *Conduct*, was summoned to appear before the Lord Chief Justice, reprimanded, and bound over. To Swift this was a further indication of things to come. 'He would not have the impudence to do this, if he did not foresee what was coming at court.'[87] A few weeks later it was even reported that Swift himself had been arrested for writing the *Examiner*, and Swift began to fear that the severe penalties against pamphleteers for which he had so

[83] *Journal*, ii. 432.
[84] Ibid., p. 433.
[85] Ibid., p. 435.
[86] Ibid., p. 459. See *King Henry VIII*, Act IV, sc. ii.
[87] *Journal*, ii. 438.

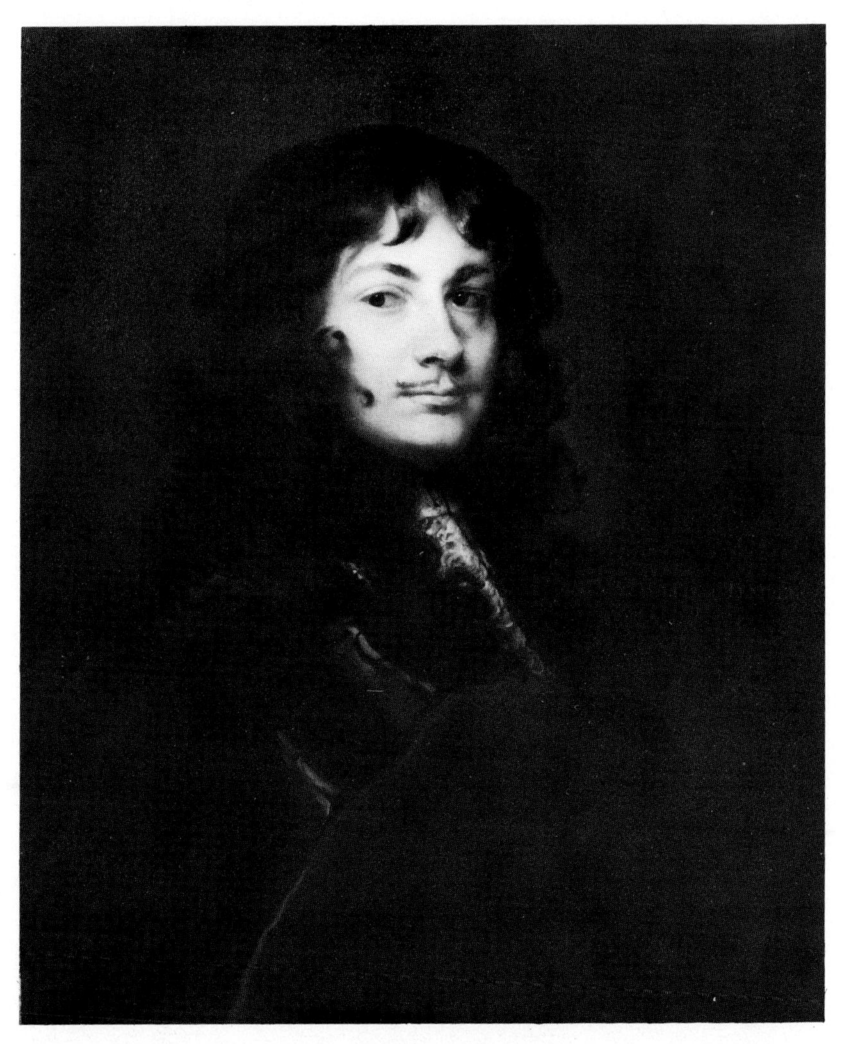

1 Sir William Temple by Lely

2 Robert Harley, 1st Earl of Oxford by Kneller

often called, would now be enforced against him. 'They lay all things on me, even some I have never read.' Such was the price of notoriety. Meanwhile the Whigs grew bolder. Prince Eugene was expected to pay a visit to England to strengthen the war party, and it was rumoured that the Whigs intended to greet him with a parade of 40,000 horsemen. In the Lords, Nottingham complained of the ballad attacking him, and the House agreed to take up the printer. Fearful of prosecution, low in spirits, and suffering from both a lingering cold and pains in the back, Swift anticipated Christmas as 'a languishing death'. Even his election as president of the Society of Brothers for the following week did not cheer him up. 'It will cost me five or six pounds,' he grumbled.[88]

However, since outward confidence and cheerfulness were still required, Swift's fears had to be kept concealed. Privately he was convinced that the malign influence at work upon the Queen could be identified with the Duchess of Somerset, and in his frustration he channelled his rage against her into a savage little satire. *The Windsor Prophecy*. It is a crude and ill-judged work, motivated by desperation, which seems almost to invite prosecution. Written on 23 December and printed on Christmas Eve, Swift declared that he liked the *Prophecy* 'mightily'. He was the only one who did. The poem is a piece of gnomic doggerel, full of riddling allusions to the current political situation, and including a punning summary of the much-married Duchess's eventful early career of murder and elopement.[89] Mrs Masham, the Queen's Tory Lady of the Bedchamber, advised strongly against publishing the *Prophecy* 'for fear of angering the Queen',[90] and Swift made a show of writing to the printer to stop production. But the following evening, as Swift presided over a meeting of his 'Brothers' at the Thatched-house tavern, he made no attempt to stop the printer from distributing the poem as an after-dinner entertainment. He continued to boast that the poem was 'an admirable good one, and people are mad for it'. He knew that Mrs Masham was right, but there was a familiar self-destructive bravado about this Christmas gesture of ill-will.

The gloom of that Christmas was increased by some sad news that he heard on Christmas Day, while dining with the Vanhom-

[88] Ibid., p. 443. [89] *Poems*, i. 145-8. [90] *Journal*, ii. 446.

righs. Anne Long, whom he had first met at their house some years earlier, had died. She had contracted rather too many debts in the expectation of a legacy from an 'odious grandmother' (Swift's phrase) who was such an unconscionable time a-dying that her granddaughter was forced to go into hiding from her creditors in King's Lynn, where she lived under an assumed name. She kept up a lively correspondence with Swift, describing the rituals of provincial life, and the mysterious figure that she cut there. Swift had written only a week before her death, full of affection and good humour, but including a phrase that becomes more piquant with hindsight: 'Health is worth preserving, though life is not'.[91] He told Stella that he was 'never more afflicted at any death', and wrote this eloquent obituary.

> She was the most beautiful person of the age she lived in, of great virtue, infinite sweetness and generosity of temper and true good sense.[92]

If the Duchess of Somerset represented the type of feminine evil to him that Christmas, it was perhaps natural that he should describe Anne Long in terms of exaggerated idealism, as if to compensate for the injustice of life which had made such disparity between their fortunes.

On 29 December the period of suspense was over. The Queen finally acceded to the advice of her ministers and agreed to the creation of twelve new Tory peers. Delighted, Swift broke open the seals of the letter he had just completed to the ladies to give them the good news. 'We are all safe; the Queen has made no less than twelve lords to have a majority . . . and has turned out the Duke of Somerset. She is awaked at last.'[93] The reference to Somerset was premature; he was not removed until the following month, but Swift's information was otherwise correct. It was, he admitted, 'a strange unhappy necessity' to pack the House in this way, but 'the Queen has drawn it upon herself by her confounded trimming'. Only one further measure remained to be taken, and on the last day of the year the ministry grasped that nettle. The *London Gazette* for 31 December announced

[91] *Corr.* i. 277.
[92] Account Book for 1711-12. Forster (48 D 34/4).
[93] *Journal,* ii. 449-50.

that the Duke of Marlborough had been dismissed, and replaced as Captain-General of the forces by Ormonde

Relieved as he was, Swift experienced some embarrassment at this desperate strategy, and a chill of apprehension at the removal of a commander who had had 'constant success in arms. . . . If the ministry be not sure of a peace, I shall wonder at this step,' he confided, fearing that the move might encourage French resistance. 'I do not love to see personal resentment mix with public affairs,' he concluded with ringing disingenuousness.[94] Not only had he just vilified the Duchess of Somerset, but only two days later he was collaborating with Arbuthnot on yet another attack on Marlborough in a ballad entitled 'The Widow and her Cat'.

Victories bring their own problems, and the loyal Tories who had supported the ministry's peace proposals by a thumping majority of 232 to 106 were determined to enjoy the fruits of success. Surely now Oxford must realize the folly of trimming and must finally expel all Whigs and waverers from office. St John did nothing to lessen the clamour for more extreme men and measures, and as the Tory zealots grew more vociferous it was Swift who came to Oxford's aim with *Some Advice humbly offered to the Members of the October Club*.[95] This is one of Swift's more artful productions, specifically contrived for a known audience. He professes himself entirely sympathetic to the club's grievances, who 'expected to see a thorough change with respect to employment'. He does not challenge or refute their views; rather he assures his readers that 'this letter is sent you, gentlemen, from no mean hand, nor from a person uninformed',[96] but hints at several weighty reasons of state, which may not be divulged, that make it impossible for changes to be made as quickly as they all would wish. Trust us, is the message of the pamphlet. Don't ask for explanations that might expose our weaknesses to our common enemies. 'A minister . . . is sometimes forced to preserve his credit, by forbearing what *is* in his power, for fear of discovering how far the limits extend of what *is not*.'[97] Swift maintains his confidential tone by making his pamphlet allusively enigmatic. Not a single person or event is

[94] Ibid., pp. 452-3.
[95] *PW* vi. 67-80.
[96] Ibid., p. 74.
[97] Ibid., p. 79.

identified by name. Instead there are veiled references to 'an unlucky proceeding some months ago', a 'person on whom we much depend', a 'tender point', a 'certain great minister', and so on. Actually the matters referred to, the machinations of the Somersets against the peace, were common knowledge. By these oblique references however, Swift gives the impression of confiding sensitive information to his political allies in a conspiratorial manner which confirms the general need for secrecy. No doubt he hoped to convince the ale-fuddled squires with this battery of nods and winks, that they had been entrusted with some great state secrets, when in fact they had been told nothing new at all. In addition he makes an unashamed plea to the patriotism of the October men. The old divisions of court and country have been superseded, he argues, since 'the Queen and Ministry are at this time fully in the true interest of the kingdom'. True patriots, such as they, should be motivated by higher considerations than the Whigs who, as they all knew, were driven by mercenary motives of self-interest. But the case was very different now, when everyone who adhered to the ministry served 'God, his Prince and his country'. Though sometimes sharing St John's sense of impatience and frustration, Swift fundamentally believed that Oxford's policy of moderation had the best chance of success. But it was a straight and very narrow path between Whig and Tory extremists who both seemed 'fully resolved to have his head'. Like Gulliver, Swift could admire Flimnap's agility on the high wire, but he lacked the head to imitate his prolonged balancing act. 'I am sick of politics,' he wrote, after completing *Some Advice.* 'I have not dined with Lord Treasurer these three weeks; he chides me, but I don't care; I don't.'[98] A fortnight later he made another of his sporadic resolves to return to Ireland, with an authentic note of weariness. 'I don't much care; I shall not be with them above two months; for I resolve to set out for Ireland the beginning of April next . . . and see my willows.'

In an effort to press home their political advantage, the Tories charged and convicted both the Duke of Marlborough and Robert Walpole, the former Secretary at War, for corruption in their handling of war supplies. Meanwhile Swift began work on

[98] *Journal,* ii. 466.

his *Remarks on the Barrier Treaty.* This treaty allowed the Dutch to garrison a defensive line of fortresses on their borders with France and the Spanish Netherlands. It had been concluded in 1709, at a time when Dutch support for the war had been lukewarm, and therefore offered massive advantages to the Dutch, and relatively few to the British. In fact Marlborough, sensing that this might become a source of later friction, wisely refused to sign the final document. It was therefore relatively easy for Swift to use the treaty as proof of Dutch avarice and British innocence.

When he had completed these *Remarks*, Swift could look back on a press campaign of considerable success. On 4 February his excitement bubbled over his weariness as his months of effort were finally rewarded.

> The house of commons have this day made many severe votes about our being abused by our allies. Those who spoke drew all their arguments from my book, and their votes confirm all I writ; the court had a majority of a hundred and fifty; all agree, that it was my book that spirited them to these resolutions.[99]

Those Commons' votes were sweet music to his ears–reward and justification for his months of effort. This glow of satisfaction was to be his only reward however. The familiar ironies of his life reasserted themselves the following day when, fired with a sense of his just deserts, he sent a brief note to Oxford containing the heavy hint that 'the Dean of Wells died this morning at one o-clock; I entirely submit my good fortune to your lordship'.[100]

He sent this note after playing cards in a joyful mood with Oxford and Mrs (now Lady) Masham. That morning St John had shown him fifty guineas which were to pay off a French spy. It reminded Swift of the time when Oxford had sought to reward him the same way, and the coincidence strengthened his resolve to seek a more appropriate recompense. But he was not granted the Deanery of Wells, nor of Ely, nor of Lichfield, which likewise fell vacant that year; though he had to endure the mortification of being frequently congratulated on each. 'No, if you will have it, I am not Dean of Wells, nor know anything of being so,' he

[99] *Journal,* ii. 480. [100] *Corr.* i. 288.

wrote in exasperation to Stella in March.[101] 'I can serve everybody but myself,' he commented ruefully.

Of all the disappointments in Swift's life, his failure to secure a place in the Church of England was the most severe. Not only did it condemn him to a life of 'exile' in Ireland; it also cast doubts, as he recognized, on his real value to the men he served so faithfully and with whom he was in almost daily contact. Despite Oxford's promises, he had received neither acknowledgement nor reward for his part in negotiating the First Fruits; now it seemed he was to receive only the taunts and vilification of the Whig press for his efforts to promote and protect the government's peace policy. Swift was only too well aware that in the absence of any such visible symbol of his status, many would treat his claims to an intimacy with great men with scepticism, if not outright disbelief. He himself believed that it was the Queen who blocked his preferment, just as she refused to hear him preach before her. The picture he gives of her in 1714, with his sense of rejection fresh in his mind, and a Royal Proclamation of a price of £300 on his head, is very different indeed from the sustained idolatry of the loyal *Examiner*.

> . . . a royal prude
> The Queen insensed, his services forgot,
> Leaves him a victim to the vengeful Scot.[102]

There is little doubt that Oxford would have promoted Swift had he run no risk of incurring Royal displeasure. It cannot have been easy to maintain an optimistic composure in the company of a vicar of Laracor stony-faced with disappointment. But Oxford was a supremely accomplished politician, whose facility in handling such awkward situations was most memorably displayed in his treatment of the dramatist Nicholas Rowe. Rowe, a Whig, sought Oxford's patronage and was advised to learn Spanish. Anticipating some diplomatic appointment, Rowe applied himself to the task, only to be told by the foxy Lord Treasurer when he returned, 'Sir, I envy you the pleasure of reading *Don Quixote* in the original'.[103]

[101] *Journal,* ii. 518.
[102] 'The Author upon himself', *Poems,* i. 193-5.
[103] Spence, i. 96.

Apart from his own writings, Swift also operated as an unofficial press chief for the ministry, co-ordinating a team of literary 'under-strappers' and printers. He referred to Abel Roper, editor of the *Postboy* as a 'humble slave' always prepared to accept pro-government copy. It was Swift who obtained the editorship of the *Gazette* for Dr King, formerly in charge of the *Examiner*, and when King proved unequal to the task, the position went to another of Swift's nominees, Charles Ford. Mrs Manley, who took over the *Examiner* from Swift, was another loyal assistant, though Swift described her work to Stella as 'trash'. As an ex-mistress of Steele, her attacks on him had a special piquancy which Swift did not fail to exploit. William Oldisworth took over the *Examiner* from her with orders to take the paper further down-market. 'We would have it a little upon the Grub street,' said Swift in December 1711, 'to be a match for their writers.'[104] It was Oldisworth's attacks on Marlborough that sometimes caused Swift to wince with embarrassment. 'I would fain have hindered the severity of the two or three last, but could not.'[105]

On a more congenial level than these Grub Street acquaintances was his new friendship with the Queen's physician, John Arbuthnot. Swift first met the doctor on a visit to Windsor in the summer of 1711. The two men soon discovered that they shared a common sense of humour, which they quickly put to use entertaining and teasing the ladies of the court. A Scot, a scientist, and a family man, Arbuthnot's deep humanity and genial humour were to be constant sources of reassurance for Swift over the years ahead. In the first months of their acquaintance they scribbled a number of satirical squibs together, and in the spring of 1712, Arbuthnot's lively series of *John Bull* pamphlets developed the arguments of Swift's *Conduct of the Allies* in a witty allegory. But it was as a trusted friend, an honest, kindly man of unswerving integrity and unpretentious erudition, that Arbuthnot was of most importance to Swift. Years later, just before the publication of *Gulliver's Travels,* Swift declared to Pope, 'O, if the world had but a dozen Arbuthnots in it, I would burn my *Travels*.[106] From the first, the doctor represented to him all that was best in human nature.

The frosts of winter dragged on well into March that year, and

[104] *Journal,* ii. 430. [105] Ibid., p. 472. [106] *Corr.* iii, 104.

Swift suffered a series of colds, never able to shake off the last before a new one assailed him. No doubt his idleness and unsatisfied hopes of preferment made him an easy prey to illness. 'I have no longer work on my hands,' he wrote in early March. Prior too was ill, and Oxford was taking physic for a 'swimming in the head'. Even dinner with the Arbuthnots was a lenten occasion 'not in point of vittals, but spleen, for his wife and a child or two were sick in the house, and that was full as mortifying as fish'.[107] March also brought a new anxiety for Swift. The Mohocks, a gang of aristocratic youths, reputedly the sons of Whig grandees, had taken to terrorizing Tories in the streets of London at night. The newspapers were soon filled with stories of their atrocities, of how they slashed faces and cut off noses. On 17 March a proclamation offering £100 for the discovery of any of the offenders was issued. As tales of their savagery multiplied, some sceptics claimed to doubt the very existence of the Mohocks. Swift himself began 'almost to think there is no truth, or very little, in the whole story'. Nevertheless he took no chances, since the word was that the Mohocks had 'malicious intentions against the ministers and their friends'[108] and Swift's servant overheard a rumour in a coffee-house 'that one design of the Mohocks was upon me'.[109] In consequence he would take a chair when returning late from dining at Lady Masham's or the Lord Treasurer's, though he grumbled at the cost. 'The dogs will cost me at least a crown a week in chairs.' Even this was no security though, for Oxford, who took the threats seriously, warned him that they 'insult chairs more than they do those on foot'. For a fortnight or so leading Tories were all rather panicked by the scare, and the worst of it was, that none could be sure whether they were merely the victims of a hoax. As Swift cowered in his rooms in Panton Street, counting the crowns the scare had cost him, it was the thought that he might merely be the target of a Bickerstaffian imposture that annoyed him most.

Before he could be sure that the streets were safe for him to walk in once more, his health finally broke down under the combined pressures of frustration, anxiety, and continual colds. At the end of March he began complaining of rheumatic pains.

[107] *Journal*, ii. 505. [108] Ibid., p. 509. [109] Ibid., p. 511.

These were quickly followed by a rash of great red spots. He had shingles, and with a despair that bordered closely on relief, he abandoned visits, ambitions, and even his *Journal* for the whole of April. For a month it was as though he had disappeared from the real world, into a private zone of clysters, blisters, plasters, pimples, and purges.

III

Even in illness Swift had to be different. 'I can never be sick like other people, but always something out of the common way.'[110] By the beginning of May he was starting to emerge from his sick-room, still itching terribly and sweating profusely. He was a good deal thinner too, having lost so much weight that his breeches needed to be taken in two inches. It was not until the autumn that he was fully recovered, but after a month's solitude he had grown weary of his own company and ventured back into the social world that meant rather more to him than he liked to admit. It was to be another year of suspense. In March he expected news of the peace 'in a few days' and put off his return to Ireland until the treaty was signed. 'Yet I would fain be at the beginning of my willows growing,' he wrote, with evident regret. By June he was still hoping that the delays would be over 'in a month or two'. By the end of October it was almost with despair that he wrote 'as for my stay in England, it cannot be long now'. Archbishop King was neither surprised nor disappointed at these delays. In November he had wagered that peace would not be signed before the following August, and having won that bet, he was prepared to double it. 'I durst venture on the same terms that we shall have no universal peace in Europe till this time next year.[111] The daily expectation of an announcement of peace had a catastrophic effect on the prosecution of the war. Since no treaty was signed before the opening of the campaigning season, Ormonde was sent to the field with orders, so the allies were assured, to pursue the war 'with all possible vigour'. Actually this was quite untrue. Ormonde had been given secret orders by St John to avoid any engagement with the enemy. The French had been informed of these secret

[110] *Journal,* ii. 531-2. [111] *Corr.* i. 303.

orders and had agreed to co-operate in the charade of mock-hostilities. This dangerous ministerial gamble rested on the assumption that the treaty would be signed before Ormonde was required to act upon these humiliating directives. However, a remarkable series of deaths in the Bourbon family put the whole peace policy at risk. In 1711 the Dauphin had died of smallpox. In February 1712 his eldest son, the Duke of Burgundy died; on 8 March *his* eldest son, the Duke of Brittany died, leaving only the sickly two-year-old Louis to bar the accession of Philip, the allied candidate for the throne of Spain, to the throne of France as well. After ten long years of war, it seemed that the Bourbons might still achieve the empire they had always sought. So, while Ormonde marched his men aimlessly hither and thither, St John redoubled his diplomatic efforts to extract from Philip a renunciation of any claim to the French throne. Philip agreed, but not before Ormonde had been forced to retire awkwardly from an attack planned by Prince Eugene. When the nature of his restraining orders leaked out in England there was uproar among the Whigs. There were censure debates in both Houses of Parliament, and although, by a combination of truculence and mendacity, the ministry won both votes, they were hardly inspiring performances. Swift came to the aid of the ministry with his *Letter to a Whig Lord.* Adopting his familiar role as a moderate, he argues that the only real political questions are, 'Whether the Queen shall choose her own servants? And whether she shall keep her prerogative of making peace?' The true and genuine cause of Whig animosity, he asserts, is simple ambition.

> Those who are out of place would fain be in; and . . . the bulk of your party are the dupes of half a dozen, who are impatient at their loss of power.[112]

The reduction of all political arguments to the cynical formulas of La Rochefoucauld seems tired and somewhat glib; but the example that he cites, of Nottingham's alliance with the Whigs, was certainly appropriate. Swift passes over in silence the equally opportunistic compromises by which the ministry were fighting to retain their hold on power. His apparent cynicism

[112] *PW* vi. 127.

may reflect a growing unease about the measures which he was called upon to defend, and an increasing despondency about his own hopes of preferment. Unwell and idle he chafed away his time, tortured by congratulations on appointments which he had failed to obtain. Stella's thirty-second letter infuriated him so much that he sealed it up immediately 'and shall read it no more this twelvemonth . . . you talk as glibly of a thing as if it were done, which for ought I know, is further from being done than ever, and the court always giving me joy and vexation'.[113] It was unlike Stella to make such an error, and his rebuke is sharp. A week later, however, he ventured so far as to make a tentative disposal of the living of Laracor to Raymond, 'if ever the court should give me anything'[114] but the mood of his letter was that of a man barely surfacing above bitter depression. 'If I had not a spirit naturally cheerful, I should be very much discontented at a thousand things.'[115] One wonders how Stella received Swift's description of himself as such a natural optimist.

Archbishop King continued to torment Swift with platitudinous advice. 'Pray make hay whilst the sun shines,' he counselled in November. Swift's unease made it impossible for him to settle to anything–even to his *Journal.* From May till December 1712 the ladies received only a few brief laconic letters, full of weariness and pain. Yet at least Swift had become a familiar figure at the court 'where dinners are to be found'. In June, when the court removed to Kensington he quickly followed, taking lodgings nearby. Yet although the court supplied him with meals, card-partners to divert him, and maids of honour to tease him, he found little pleasure there. The insecurity of his position made him awkward and defensive. He soon found that Kensington shared the same drawback as Chelsea–an absence of transportation to and from town. After visiting the Duchess of Ormonde at Sheen one day, he walked along the river as far as Kew.

> But no boat; then to Mortlake, but no boat; & it was 9 o-clock; at last a little sculler called, full of nasty people; I made him set me down at Hammersmith; so walked 2 miles to this place.[116]

[113] *Journal,* ii. 552.
[114] Ibid., p. 559.
[115] Ibid., p. 560.
[116] Ibid., p. 538.

On another occasion, at night, he was forced to walk all the way home in the rain from the city because no coach could be found. These 'shabby difficulties' depressed him mainly because 'they arise from not having a thousand pounds a year' and marked him out as a poor relation at court. The one great boon of living out of town was that he was no longer 'teased with solicitors'. But even that freedom palled after a while. It was, after all, rather flattering to be sought out so constantly for his advice, and his repeated complaints at these 'solicitors' come very close to boasts. By mid-July he was heartily sick of Kensington and looked forward to the court's next move, to Windsor.

To divert himself he 'plied it pretty close' in the fortnight before Grub Street 'died' on 1 August, when Oxford's Stamp Duty of 1*d.* a sheet on newspapers came into force. Swift applauded the measure as a step towards the suppression of Whig libels, but he was as premature in announcing the death of Grub Street, as Bickerstaff had been in predicting that of Partridge. 'No more ghosts or murders now, for love or money,'[117] he declared, but in fact the circulation of papers dropped very little, and by the end of October he was again complaining that 'these devils of Grub Street rogues that write the *Flying Post* and the *Medley* . . . will not be quiet'.[118]

Swift set himself to work that summer compiling an 'official' Tory history of recent events for the benefit of the young Prince Posterity. He sensed no contradiction in attempting to combine the roles of propagandist and historian, and at the end of 1713 in fact made a determined effort to secure the post of Historiographer Royal. He was dismayed when Oxford, who had willingly provided him with all the documents necessary for writing the *Conduct*, seemed reluctant to show him the papers he needed for this new work. 'They delay me as if it were a favour I asked of them,' he complained. It was not until the end of October that Swift could get seriously to work, toiling 'like a horse' with hundreds of letters, in order to 'squeeze a line perhaps out of each, or at least the seeds of a line'.[119] By December he was able to report that he had '130 pages in folio to be printed', but again the ministers prevaricated and delayed. Oxford had a strictly limited view of Swift's role, as a producer

[117] Ibid., p. 553. [118] Ibid., p. 568. [119] Ibid., p. 569.

of propaganda timed exactly for specific political situations; he was far less concerned about the judgements of posterity. A similar conflict of interests had arisen in 1711, when Swift had sought to engage Oxford's support for the creation of 'a society or academy for correcting and settling our language'. Swift's fears of the decay of the language were a common anxiety at the time, and usually related to fears of corruption in society as a whole.

> During the usurpation, such an infusion of enthusiastic jargon prevailed in every writing, as was not shaken off in many years after. To this succeeded that licentiousness which entered with the Restoration, and from infecting our religion and morals, fell to corrupt our language.[120]

The seriousness with which he approached his scheme for reforming the language can be seen in the unusual care he took preparing his *Proposals for correcting . . . the English Tongue* (1712), a pamphlet which has the almost unique distinction of bearing his own signature. The idea of an élite institution working to preserve the nation's culture, and free of party-political prejudices, appealed to him strongly, and he made strenuous efforts to enlist support for such an Academy from Addison and Steele. But as usual Oxford procrastinated, promising general support for the scheme, but always 'too busy' to do anything specific. Archbishop King liked the idea too, all except for the name, 'for I find the generality so prejudiced against all French precedents that as parties are more formed I apprehend the name will be of great disadvantage'.[121] Swift might have done well to ponder further on the limitations of the French Académie in its attempts to fix their own language, but he waved all such objections aside with an offhand chauvinistic jibe at the 'natural inconstancy of that people'. Still, despite his high hopes and careful planning, Oxford effectively sabotaged Swift's scheme by simply doing nothing. Such differences of emphasis and intention often placed strains on the relationship between Swift and Oxford, but in November the quasi-filial bond between them was strengthened when Swift was on hand to save the Lord Treasurer from another apparent attempt on his life. One morn-

[120] *PW* iv. 10. [121] *Corr.* i. 304.

ing Oxford received a bandbox through the post. He was beginning to open it when something aroused his suspicions. Swift took hold of the box, carried it to the window and there, very gingerly, undid it. Inside was a booby-trap device made up of two pistols with hair-sprung triggers. As before, Swift exploited the incident for its propaganda, while the Whigs protested that the whole thing was a figment of his imagination. 'The Whig papers have abused me about the bandbox,' he remarked innocently to Stella, 'God help me, what could I do? I fairly ventured my life.'[122] By December the full force of his frustration was evident. 'I dislike a million of things in the course of public affairs; & if I were to stay here much longer, I am sure I should ruin myself with endeavouring to mend them.'[123] The Countess of Orkney, a former mistress of William III, and owner of a beautiful house at Cliveden, was a new friend who attempted to lift his low spirits. She gave him presents of a writing-table and a nightgown of her own making and was, he told Stella, 'the wisest woman I ever saw . . . perfectly kind, like a mother'.[124] But it was another woman who did most to comfort him during these long months of depression, not as a mother figure, but rather as a surrogate daughter. For it was at this time that Hester Vanhomrigh–Vanessa–came to assume great importance in his life.

Swift and Andrew Fountaine had been regular visitors at the Vanhomrigh household since Swift's return in September 1710. That autumn Mrs Van suffered the death of her youngest son, Ginkel, who was buried in St James's church the day after Swift's first interview with Harley. Swift's visits were a great comfort to both mother and daughter during that winter when he was their neighbour in Bury Street. In February he and Ford spent a convivial evening drinking punch to celebrate Vanessa's twenty-third birthday. Later in the spring, after barking his shin, Swift who had now moved his lodgings round the corner to St Albans Street, dined three days out of four with the Vanhomrigh family. On Good Friday he called there for both breakfast and supper. Soon it became a regular habit for him to share a meal with his obliging neighbours, and pay for it in riddles and puns, anecdotes and advice. When Stella remarked upon his new par-

[122] *Journal*, ii. 579. [123] Ibid., p. 580. [124] Ibid., p. 557.

tiality for the Dublin family he replied with a facetious evasiveness, turning all to sport and epigram.

> You say they are of no consequence: why, they keep as good female company as I do male; I see all the drabs of quality at this end of the town with them; I saw two lady Bettys there this afternoon, the beauty of one, the good breeding and nature of t'other, and the wit of neither, would have made a fine woman.[125]

Stella was shrewd enough to recognize that females 'of no consequence' socially must have other attractions. Swift silently acknowledged the justice of her suspicions by omission. From this point on, although visits to the Vanhomrighs remained frequent, they are mentioned less often in the *Journal*, and when they are noted, he is usually careful to add that he was accompanied by Sir Andrew, Ford, or Lewis. There is a revealing discrepancy between what is described in the *Journal*, and what is recorded in his private account books. Time and again where the account books tell of a visit to the Vans, his letters back to Ireland merely mention a meeting with 'a friend'. One Tuesday in November he writes 'I dined privately with a friend today in the neighbourhood', where the account book reads 'Wine; Van's 1s 6d.'[126] The following Monday dinner with 'a friend in St James's street' disguises the fact that he spent the evening playing piquet at Mrs Van's. Stella was quite used to these evasive formulas. Swift's characteristic code of hints and nods usually had political motives, but his evasiveness concerning Mrs Van and her two daughters is of another, though equally familiar kind.

When in the summer of 1711 Swift decided to move to Chelsea, Mrs Van's was a useful resting place on journeys to and from town. He left some belongings there, including a chest of wine, and his best wig and gown, to change into before making an appearance at court. Mrs Van went so far as to put a small room aside for him, where he could read and write, and when the court was dispersed he would often idle a day away there. 'An ugly rainy day,' he remarked once in July; 'I was to visit Mrs Barton, then called at Mrs Vanhomrigh's, where Sir Andrew Fountaine and the rain kept me to dinner; and there did I loiter

125 *Journal*, i. 202.
126 Ibid. ii. 411; Account Book 1711-12.

all the afternoon like a fool, out of perfect laziness, and the weather not permitting me to walk; but I'll do so no more.'[127] Despite this resolution, he spent much of the following day at Mrs Van's. 'I struck into Mrs Vanhomrigh's, and dined, and stayed till night.' This time he did not even have the excuse of Sir Andrew's company.

During his attack of shingles in the spring of 1712, Mrs Van's was the only place that he felt fit enough to visit. It was Vanessa who nursed him, talked to him, and helped to ease the pains of his illness and frustration. This is a further reason for the perfunctoriness of the *Journal* at this period. He was now confiding more of his private feelings to this new young pupil, and could not bring himself either to acknowledge, or to conceal the fact from Stella. Between 7 August and 15 September there is a long awkward gap in the *Journal,* to which Swift himself calls guilty attention. 'I never was so long without writing to MD as now, since I left them, or ever will again while I am able to write.'[128] It is a silence which coincided with the first of the many major crises in his relationship with Vanessa.

Weary of court life at Windsor, and missing Vanessa's company, Swift wrote to her on 1 August:

> I am resolved to leave [this place] in two days, and not return in three weeks. I will come as early on Monday as I can find opportunity; and will take a little Grub street lodging; pretty near where I did before; and dine with you thrice a week; and will tell you a thousand secrets provided you will have no quarrels to me.[129]

To Stella he merely remarked that he was now 'in a hedge lodging, very busy'. However the Grub Street holiday was less delightful than he had hoped, for his account book records he was 'sick with giddiness much', and on the 10th he was required to return to Windsor. He was accompanied by Vanessa's brother Bartholomew, on vacation from Oxford, and there was a half-serious suggestion that the rest of the family should pay them a surprise visit. In a light-hearted letter Swift enters fully into the spirit of this adventure.

[127] *Journal,* i. 309.
[128] Ibid. ii. 556.
[129] *Corr.* i. 305.

How came you to make it a secret to me that you all design to come up for 3 or 4 days? Five pounds will maintain you and pay for your coach backwards and forwards.[130]

With a lover-like intimacy he seeks to know how Vanessa spends the hours of her day.

I cannot imagine how you pass your time in our absence, unless by lying a-bed till twelve, and then having your followers about you till dinner. . . . What do you do all the afternoon? . . . I will steal to town one of these days and catch you napping. I desire you and Moll will walk as often as you can in the park, and do not sit moping at home . . . I long to drink a dish of coffee in the sluttery, and hear you dun me for secrets, and–drink your coffee–why don't you drink your coffee . . .[131]

Vanessa promptly seized on the idea of a visit, taking it out of the realm of fantasy. Swift became afraid, and, sensing an awkward scene, said nothing. Vanessa's response was a flurry of letters, half-teasing, half-chiding. On 1 September she took him to task for his silence.

Had I a correspondent in China I might have had an answer by this time. I never could think till now that London was so far off in your thoughts and that 20 miles were by your computation equal to some thousands . . .[132]

Next day she wrote again full of mockery, having heard of his embarrassment at the impending visit.

Mr Lewis tells me you have made a solemn resolution to leave Windsor the moment we come there. Tis a noble res: pray keep it, now that I may be no ways accessory to your breaking it. I design to send Mr Lewis word to a minute when we shall leave London, that he may tell you . . .[133]

'You railly very well,'[134] replied Swift shamefacedly, amused and embarrassed in equal degrees by this exposure of his inhibitions. 'You may come,' he concedes at last, but it's a grudging admission, teased out of him against his better judgement. He sent a haunch of venison to Mrs Van and best wishes for Moll's recovery from illness, though he may secretly have hoped that this illness might yet prevent a visit. He added that he was 'full of business and ill humour', but Vanessa was deaf to all dissuasion.

[130] Ibid., p. 308. [131] Ibid., p. 308. [132] Ibid., p. 309.
[133] Ibid., p. 310. [134] Ibid., p. 311.

Towards the end of September she came to Windsor alone, having left the ailing Molkin behind. Swift, shocked by this unchaperoned appearance, refused to see her, giving the excuse of an indisposition. It was the first of several such stand-offs between them, when she would force a situation, and he would back away. She had foreseen just such an evasion in her teasing letter of 2 September.

> If there be a by-way you had better take it, for I very much apprehend that seeing us will make you break through all.

Swift took the 'by-way' of illness, not concealing from her that it was a deliberate excuse to punish her for placing him in a false position.

> I would not see you for a thousand pounds if I could . . . you should not have come, and I knew that as well as you.[125]

Annoyed, Vanessa went on to Oxford and Swift left for London feeling ill and cowardly, 'deep in pills with *assa fetida* and a steel bitter drink'.

Swift had earlier boasted to Vanessa of his commitment to a principle of personal integrity. It was a boast which she never forgot, and used to taunt him with when she observed signs of cowardice or evasion in his behaviour.

> You once had a maxim, which was to act what was right and not mind what the world said.[136]

Discussion of this principle figured prominently in their conversations, and Swift refers to it again in his versified account of their relationship. 'Cadenus and Vanessa'.

> That virtue, pleased by being shown,
> Knows nothing which it dare not own;
> Can make us without fear disclose
> Our inmost secrets to our foes;
> That common forms were not designed
> Directors to a noble mind.[137]
>
> (ll. 608-13)

The conscious moral superiority upon which Swift had prided himself in his contempt for the nasty gossips of Leicester,

[135] Ibid., pp. 313-14. [136] *Corr.* ii. 148. [137] *Poems,* ii. 706.

became something rather different in Vanessa's mouth. It was a kind of bravado that rather frightened a churchman-politician who had grown more diplomatic with age.

> Cadenus felt within him rise
> Shame, disappointment, guilt, surprise.
> (ll. 624-5)

The 'indisposition at Windsor' became one of the mental events in the calendar of their relationship which each remembered as a crucial indication of the differences between them.

Swift's relationship with Vanessa had begun as practically an exact counterpart of that with Stella. What could be more agreeable than, having established the precise kind of domicile that he favoured with MD in Dublin, to find its mirror-image here in London? Like Esther Johnson, Hester Vanhomrigh was a sickly fatherless girl; like her she was surrounded by other females. In Swift's early dealings with the Vanhomrighs, he wrote to mother and daughters, Mrs Van, Hessy, and Moll with equal regularity, maintaining the fiction that his interest was engaged by all three equally. As with Stella, he treated Hester Vanhomrigh to a variety of nicknames; Mishessy, Misheskinage, Governor Huff, and, naturally, Vanessa. With both girls he employed a 'little language' of private allusions; with both he sought to supervise their reading, and correct their spelling; he would scold or praise them both in terms that were specifically masculine, 'sirrahs, boys, rogues'. 'Cadenus and Vanessa' makes insistent mention of Vanessa's exemplary *masculine* qualities.

> Wisdom's above suspecting wiles
> The queen of learning gravely smiles;
> Down from Olympus comes with joy
> Mistakes Vanessa for a boy;
> Then sows within her tender mind
> Seeds long unknown to womankind.[138]
> (ll. 198-203)

All the mechanisms for damping down sexual recognition and desire which he had successfully employed with Stella were present in his new relationship. He allowed himself to enjoy the flattering, exciting company of both young women as tutor,

[138] Ibid., pp. 692-3.

father, and mock-gallant, without ever committing himself to any one of these roles.

He had first offered advice to the Vanhomrighs in connection with the prolonged disputes over their estate following Mr Vanhomrigh's death in 1703. The vivacious Mrs Van, who had an incorrigible tendency to live beyond her means, talked to Swift of sending Vanessa to Ireland 'to look after her fortune, and get it into her own hands'.[139] He soon learnt that it was better to offer his counsel to Vanessa than to her mother, and persuaded her, instead of going to Ireland, to submit a private bill to Parliament to transfer the family's property to England. He promised to do all in his power to assist the bill in its progress. On another occasion he acted a fatherly part, warning Mrs Van that he suspected her new landlady of being a bawd. It was, apparently, her eyebrows that gave her away. A month later, again following his advice, the family moved to lodgings which were just around the corner from his own new rooms in St Martin's Street. The Vanhomrighs were now boarders, however, and it seems clear that despite maintaining a brave exterior, they were gradually slipping down the social scale in response to the same unforgiving economic forces that had banished Anne Long to King's Lynn. In addition to such protective help, Swift also brought them the excitement of a vicarious contact with the court. Vanessa loved to tease him about his 'secrets', and Mrs Van no doubt encouraged his visits in the hope of seeing her daughter well married. Swift was well able to deflect any such designs, but found Vanessa herself more difficult to control. By the time that he ceased mentioning her name in the *Journal* he had realized that she was an altogether more combative woman than Jane Waring or Stella. The subterfuges required for corresponding with Anne Long afforded Swift a clandestine means of flattering his new young admirer. His last letter to Mrs Long was enclosed with a letter to Vanessa, in a third, false letter; 'see what art people must use, though they mean ever so well'. This complicated intrigue allowed Swift to allow Vanessa to 'eavesdrop' on his praise of her to Anne Long.

> I think there is not a better girl upon earth. I have a mighty friendship for her; she had good principles, and I have corrected all her faults; but

[139] *Journal,* i. 333.

I cannot persuade her to read, though she has an understanding, memory and taste that would bear great improvement. But she is incorrigibly idle and lazy; thinks the world made for nothing but perpetual pleasure.[140]

He refers also to Vanessa's characteristic lack of tact, which both excited and intimidated him: 'She will bid her sister go downstairs, before my face, for she has some private business with the doctor. ' His praise has an uneasy patronizing quality that seems to recognize its own inability to control those passions–moody, tempestuous, provocative–that it affects to mock. The letter to Vanessa herself concludes with a show of intimacy.

Now are you and puppy lying at your ease, without dreaming anything of all this; Adieu till we meet over a pot of coffee, or an orange and sugar in the sluttery, which I have so often found to be the most agreeable chamber in the world.[141]

The playfully sensuous language here, like much of the language of the *Journal,* combines idioms from the bedchamber and the nursery, and the letter itself is addressed to 'little Misshessy'. But Vanessa would not be satisfied with the false intimacy of baby-talk. She forced him to take their relationship seriously, and sometime in 1712 she made it clear to Swift that she loved him.

The only account we have of their relationship is that provided by Swift in 'Cadenus and Vanessa'. This was written a year later, in Windsor, when the first shock of her declaration had subsided and when, although there was still no satisfactory equilibrium to their relations, he had at least managed to formulate for himself a set of attitudes to serve as a cover for feelings. It is a curious poem with uncertain, embarrassing dips of tone. He seeks on the one hand to praise Vanessa as the perfection of her sex, while at the same time justifying his refusal of her advances. She appears at times as a paragon of wisdom and prudence, at others as impulsively predatory. He himself appears sometimes as the flattering author of the poem, sometimes as a patronizing avuncular figure, and sometimes as a timid, passive male, cowering away from *Vénus toute entière à sa proie attachée.* It is a justification, a defence, a celebration, but most of all an assertion of control. However, all too often the pleasure of finding

[140] *Corr.* i. 278. [141] Ibid., p. 276.

himself loved by a young woman, and the panic that her impulsiveness caused him, break through the coy ironic tone. Having set his face and pen, twelve years earlier, against the possibility of being loved by a young woman, it caused Swift some consternation to find that the impossible had happened. The poem is top-heavy with compliments, beginning with an elaborate series of mock-heroic compliments that detach Vanessa from any realistic setting.[142] Modern love, it is alleged 'is no such thing/ As what those ancient poets sing' since the passions of modern women 'move in lower spheres' of 'gross desire'. Venus, to refute these charges, determines to produce a woman capable of raising 'a flame that will endure/For ever, uncorrupt and pure'. The result is Vanessa. At her birth, Venus summoned the Graces to give her skin that essential Swiftian requirement 'a cleanliness . . . incapable of outward stains' and her mind that equally necessary 'Decency . . . where not one careless thought intrudes'. Pallas, believing the 'lovely infant' to be a boy, sows within the baby Vanessa's mind

> Seeds long unknown to womankind
> For manly bosoms chiefly fit,
> The seeds of knowledge, judgement, wit . . .
> (ll. 203-5)

Vanessa emerges as the de-sexed, hygienic Swiftian ideal, a woman whose cleanliness, decency, and manly temperament make her a truly rational companion. But Pallas, once undeceived, tells Venus that her plan is doomed to failure, since love grows by 'want of sense'. In this way Swift introduces the paradox that such an exemplary woman should find no more dashing admirer than an elderly clergyman. Vanessa is shown in conversation with a crowd of fops 'just issued from perfumers' shops' whose talk is all gossip and innuendo. She gives them a prim little lecture.

> That present times have no pretence
> To virtue, in the noblest sense,
> By Greeks and Romans understood.
> (ll. 346-8)

Swift deliberately presents her as a slightly comic bluestocking, as a way of further explaining her 'crush' on an old man with

[142] *Poems,* ii. 683-714.

whom she can discuss the antique virtues. Among the ladies too she exhibits an unbending earnestness. While they enjoy a little salacious tattle Vanessa prefers to read Montaigne. With her male mind in a female body, she is a ridiculous anomaly, a *lusus naturae.* Yet while the books that she reads act as shields against most of Cupid's darts, it is finally from a book–one of Swift's–that an arrow goes home.

> Some lines, more moving than the rest
> Struck to the point that pierced her breast.
> (ll. 520-1)

The lines that follow blend comedy with tenderness as Vanessa, blinded by love, and Cadenus, almost blind from study, stumble towards each other. Swift deliberately emphasizes disparity, incongruity, and sexlessness.

> Vanessa, not in years a score
> Dreams of a gown of forty-four;
> Imaginary charms can find
> In eyes with reading almost blind . . .
> (ll. 524-7)

The portraits here take on the tone of a defensive statement. He was a father-figure, a tutor, and 'knew not if she were young or old'.

> His conduct might have made him styled
> A father, and the nymph his child.
> That innocent delight he took
> To see the virgin mind her book,
> Was but the master's secret joy
> In school to hear the finest boy.
> (ll. 543-53)

This is where elements of disingenuousness, and even panic enter, as the mock-heroic form breaks down, and something more like experience shows through. He describes 'tutorial' sessions in which the student becomes listless and inattentive. In a fit of pique, Cadenus offers to release her from her 'dull studies' in order to shine in the *beau monde.* But Vanessa replies with a boldness that takes his breath away.

> Now, said the nymph, I'll let you see
> My actions with your rules agree,

That I can vulgar forms despise
And have no secrets to disguise . . .
Your lessons found the weakest part,
Aimed at the head, but reached the heart.
(ll. 614-17, 622-3)

Suddenly Swift found himself the victim of just such a trick of role-reversal as he loved to perform on the men in power.

The nymph will have her turn to be
The tutor: and the pupil, he.
(ll. 806-7)

Nothing like this had ever happened to him before. No other woman had ever, even momentarily, taken him over. This poem is an attempt to reassert control, to put back a conventional frame on their relationship–but he cannot quite decide which frame. Hence the failure of tone. The coyness of these lines rebounds against him.

But what success Vanessa met,
Is to the world a secret yet:
Whether the nymph, to please her swain
Talks in a high romantic strain;
Or whether he at last descends
To like with less seraphic ends;
Or, to compound the business, whether
They temper love and books together;
Must never to the world be told,
Nor shall the conscious muse unfold.
(ll. 818-27)

Early in 1712, ill with shingles, disappointed in his career, and fearing the imminent collapse of the ministry, he found himself loved by a woman half his age.

Letters to Stella dwindled to a guilty few. Those that he did write dwell at length on his intentions of returning speedily to Dublin and Laracor to tend his beloved willows. He told himself that this love which had come so late in life could not be accepted. He had no intention of marrying Stella, and by this time she understood that. But it was equally understood between them that their relationship was of paramount importance to them both, though in a way which conventional minds would not comprehend. Inevitably the new relationship raised awkward comparisons in his mind. Vanessa was bold, witty, and assertive. He had no need to assemble a collection of her *bon mots* to persuade posterity of the sharpness of her mind. It strikes us, independently, from her letters. In contrast, all that remains of Stella's writing are some dutiful housekeeping accounts in an unformed childlike hand that is obviously modelled on Swift's. Since it was he who destroyed all her replies to his *Journal,* the comparison is hardly fair. Yet the fact that he had it in his power to destroy all the independent evidence of Stella's mind, while Vanessa kept careful copies of all her letters to him, is a further indication of the difference of attitude of the two women in their behaviour to him. Vanessa was never afraid of confronting Swift's moods, whereas Stella would always seek to comply with them, and, if necessary, modify his ideas by soothing, rather than ruffling him.

Despite the strait-jacket that Swift had sought to impose on his emotions in 1699, he discovered that he could still love passionately. In his writings there is a lacuna where healthy sexual love is concerned. There is false gallantry, naïve idealism, cynicism, and physical revulsion in his presentation of women. Vanessa could not be assigned to any such category. But he realized that, although he did not love Stella, he had built up a life in which he had grown dependent upon her unchallenging submissiveness. She kept his house, entertained his friends, and repeated his stories throughout Dublin. She was his ideal audience, her mind having been modelled on his own. The lack of a formal tie between them paradoxically made his sense of duty towards her, in the ambiguous position she had accepted at his request, even greater. It was with feelings of guilt and loss that he rejected Vanessa. These inhibitions, some conscious, others not, account for the various subterfuges of revelation and concealment, apology and recrimination that hover in the lines of this poem. But Vanessa would not be shaken off, nor would she allow their relationship to be transmuted into a form that he could more easily control.

> . . . Friendship in its greatest height
> A constant, rational delight.
> (ll. 780-1)

They continued to meet, with Vanessa becoming less and less

restrained in the expression of her feelings. He preferred to meet in her mother's house, where he could maintain the pretence that his visits were to the whole family, but sometimes at her request, they would meet alone at Barber's house in the city, where they could drink their coffee undisturbed.

IV

BY THE summer of 1712 the rift between Oxford and St John was a matter of common knowledge. In his self-appointed role as go-between Swift prided himself on the frankness with which he treated both men, making a virtue of his indifference to the consequences of such plain dealing. 'This proceeding of mine was the surest way to send me back to my willows . . . but I regarded it not, provided I could do the kingdom service.' They did not send him back to his willows, however, but merely listened to his chiding with deaf ears. The atmosphere was particularly bad in October. 'I have helped to patch up these people together once more,' he wrote at the end of the month. 'God knows how long it may last.'[143] He looked forward to Ormonde's return from his season of mock campaigning. 'I design to make him join with me in settling all right among our people.' St John's jealousy of Oxford had been seriously exacerbated by the matter of their respective titles. Harley's self-elevation to the titles of Earl of Oxford and Mortimer was particularly audacious, since he had no descent from either of the families to which these titles traditionally belonged. Meanwhile political necessities required that St John must stay in the Commons to marshal the Tory members there, and when the twelve new peers were created at Christmas, he was not among them. He was assured, however, that he would lose nothing in precedence or rank to these latest creations. When in the summer his time finally came for a peerage, he expected to receive a title of equivalent prestige with Oxford's. He settled on the earldom of Bolingbroke, lately extinct in the elder branch of his family. Oxford, however, would allow him only to become Viscount Bolingbroke. St John was furious, considering this lesser honour as hardly adequate to compensate him for relinquishing his leadership of the

[143] *Journal*, ii. 568.

Commons. 'I was dragged into the House of Lords in such a manner as to make my promotion a punishment, not a reward,'[144] he later recalled. He told Swift that from now on he would 'never depend upon the earl's friendship as long as he lived, nor have any further commerce with him than what was necessary for carrying on the public service'. Swift tried to believe that Oxford was 'wholly innocent' of malice towards the Secretary, but admitted that 'the appearances were so strong' that he could understand St John's suspicions. Earlier that year he had said of St John, 'if I were a dozen years younger, I would cultivate his favour, and trust my fortune with his'.[145] This assertion of a generation gap is really a cover for an instinctive suspicion of a streak of opportunism in the younger man's personality. He admired St John's dynamism, but felt more at home in the family atmosphere that Oxford was careful to provide. As the schism deepened, Swift spent more time with Oxford, who would be angry 'if I don't dine with him every second day'. This was highly flattering, but involved him in late hours which he did not really relish. Still, despite private protests, he acceded to Oxford's invitations with some regularity. 'I have now dined 6 days successively with Ld Tr,' he boasted in March, and the dinners took on an increasingly intimate and familial aspect. 'Of a dozen at table, they were all Harley family but myself.' In October, Oxford's behaviour reminded him of Temple in a way that was comically reassuring. 'I was playing at one-and-thirty with him and his family t'other night. He gave us all 12 pence apiece to begin with; it put me in mind of Sr W.T.'[146] Both Oxford and Bolingbroke continued to annoy him by their indifference to his proposed history of recent events, but at least they had given a convincing excuse for their hesitation. 'Some think it is too dangerous to publish,' he wrote, not entirely unhappy at this reaction. He had deliberately written a work whose controversial nature would, he hoped, force the ministry to protect and provide for him.

I forbear printing what I have in hand, till the court decides something about me: I will contract no more enemies, at least I will not embitter worse those I have already, till I have got under shelter.[147]

[144] Bolingbroke, 'A Letter to Sir William Windham', *Works,* 4 vols., 1967 reprint of the 1844 edition, i. 117.

[145] Ibid., p. 495. [146] Ibid., p. 561. [147] Ibid., p. 590.

His miscalculation was in believing that the ministry really wanted his work published. On the contrary, Oxford had reasons, which he preferred not to admit to Swift, for wishing to leave some of the details of recent history conveniently vague. Swift tried to encourage Oxford by showing him 'that part of my book in manuscript to read, where his character was'. Even this ploy failed, for the Bishop of St David's arrived to interrupt them, and Oxford put the manuscript aside. 'Delay is rooted in Eltee's heart,' muttered Swift gloomily.[148]

One reason for Oxford's delay was that the peace, which had been expected weekly for over a year, had still not been signed. Parliament sat on 3 February, and was immediately prorogued for a fortnight, causing grumbles and 'cloudy faces'. On the 17th it was prorogued for a further fortnight; on 3 March for a week, likewise on the 10th. By this time Swift was beginning to share the general misgivings.

I hope they are sure of the peace by next week, and then they are right in my opinion, otherwise I think they have done wrong & might have sat 3 weeks ago.

Oxford maintained his usual calm in the face of criticism; 'he cares not a rush,' said Swift. On St Patrick's Day the prorogation was continued for a further week, disappointing the Irish folks who swarmed down the Mall with their crosses. By the 24th the strain was beginning to tell on close ministerial friends. Swift dined at Sir Thomas Hanmer's with Ormonde and Lord and Lady Orkney; 'I left them at 6, everybody is as sour as vinegar.'[149] At last on 3 April Swift reported 'the great work is in effect done'.[150] That saving 'in effect' is a nervous tic occasioned by so many disappointments. Later that week he received the heartening confirmation 'pax sit' from Prior in Paris. There were deep sighs of relief in the ministry.

During all these months of delay Swift had to suffer sniping criticisms from Archbishop King who strongly suspected a sell-out to the French. Swift's irritation at having to defer to opinions with which he totally disagreed, is evident in his squirming letters that are by turns obsequious and contemptuous. 'Your

[148] Ibid., pp. 613, 644. (Eltee = L.T., Lord Treasurer.)
[149] Ibid., p. 644.
[150] Ibid., p. 652.

Grace hath certainly hit upon the weak side of our peace,' he writes in disingenuous concession to King in January, but immediately counters, 'but I do not find you have prescribed any remedies.'[151] His general approach to King was to assure him, confidentially, that all was really well, despite appearances. 'Your Grace, who are at a distance . . . is likely to be more impartial than I.' In this case, for 'impartial' read ignorant. He makes a show of considering King's arguments, but in fact his position rests wholly on authority. 'Some accidents and occasions have put it in my way to know every step of this treaty better, I think, than any man in England.'[152] King would not be easily quashed, and replied in a long self-righteous letter that laboriously exposed Swift's evasions. Yes, he *is* impartial. 'If a set of men fall under the displeasure of the government, shall I immediately look on them as abandoned wretches, and avoid their conversation as infected?'[153] This was in response to Swift's ill-advised confession that perhaps, as an insider, he tended 'to converse only with one side of the world'. It must have made him groan to read King's specious boasts of Christian magnanimity. 'I have two notions,' King declared:

> that to prefer the public to private interest is virtue and . . . that to prefer the future good of myself and posterity to the present is wisdom. Perhaps I had this from my own practice and other children's with their butter cakes. I remember we would thrust off the butter from one part of the cake and eat it without any, that we might have the more on the last, and that it might be the more pleasing and relishing.[154]

King's butter-cake sermon left Swift feeling sick, and he could not bring himself to reply until the peace was agreed. King then sent his grudging hopes 'that this may prove a firm and lasting peace that may answer the expectations of good people and be for the honour of those that have negotiated it'.[155] The unmistakable implication here is 'but I doubt it'. However there were pressing reasons why a reconciliation between Swift and King was necessary. Only a week earlier Swift had enclosed a brief note 'not above two lines' to Stearne, the Dean of St Patrick's, in a letter to Stella. It was a note so secret that he told Stella only to give it to him 'under conditions of burning it immediately after

[151] *Corr.* i. 327. [152] Ibid., p. 328. [153] Ibid., p. 331.
[154] ibid., p. 334. [155] Ibid., p. 343.

reading it, & that before your eyes'.[156] The note was to sound out Stearne's attitude to being offered a bishopric, a promotion that would leave the Deanery of St Patrick's vacant.

Increasingly Swift confined himself not merely to 'one side of the world' but to a rather small group, including Oxford, Lady Masham, Lady Orkney, and Dr Arbuthnot, within that hemisphere. Lady Jersey would often press him to take chocolate with her, and play a hand at threepenny ombre. But her delight in crowding her house with foreign dignitaries who played at 'gacco' and 'jump' was not to his taste. How often he dined with the Vanhomrighs we cannot tell, since this information was excluded from the *Journal*. On 7 January he admitted to having spent an evening there playing cards, but 'merely for amusement' he added defensively.

By now the Society of Brothers had resumed its regular Thursday meetings, but Swift's enthusiasm for them had declined. The original group of twelve brothers had swollen to over twenty, and the society's high ideals had degenerated into a mere ritual of banqueting. In December Swift was among those suggesting that 'our meetings should be only once a fortnight, for betwixt you and me,' he told Stella, 'we do no good.'[157] He had neither the stomach, nor the pocket for the society as it had changed. Ever since his election as president for the dinner at the Thatched House tavern a year earlier, his references to the society had been edged with criticism and disappointment. His presidential dinner cost him 'seven good guineas' which was nothing compared with Ormonde's feast in March, but still more than Swift could happily afford. Such extravagant meals only emphasized the differences between himself and his artistocratic 'brothers', undermining the family spirit that the meetings were supposed to foster. He had, anyway, as he confessed, 'a sad vulgar appetite . . . I cannot endure above one dish, nor ever could since I was a boy and loved stuffing.'[158] His childhood of relative poverty left a taste for simple dishes that all the acquired tastes of the town could not alter.

On 26 February, when Ormonde chid him for avoiding the society meeting that day, Swift told him, 'I never knew 16 people good company in my life; no (faith) nor 8 neither.'[159] Swift

[156] *Journal,* ii. 655. [157] Ibid., p. 583.
[158] Ibid., p. 637. [159] Ibid., p. 628

always preferred small groups to large ones, which was another reason for attending Oxford's Saturday meetings rather than the Brothers' Thursday feasts. He was already complaining that these intimate Saturday meetings of Oxford's kitchen cabinet were getting overcrowded.

> I was of the original club when only poor lord Rivers, lord Keeper, & lord Bolingbroke came. But now Ormonde, Anglesea, lord Steward, Darmouth & other rabble intrude, and I scold at it, but now they pretend as good a title as I, & indeed many Saturdays I am not there: the company being too many, I don't love it.[160]

Though irritated at having to share Oxford's 'whipping day' with such newcomers, he continued to enjoy these meetings. On 11 April they were ten at table, 'all lords but myself & the Chancellor of the Exchequer'. The intimate contact with this 'rabble' of aristocrats compensated in some measure for having to share Oxford's company. By now it was clear that the Society of Brothers had largely become a power base for Bolingbroke, and Swift found himself torn between the two clubs, the Thursday and the Saturday, both of which had become political lobbies as much as social gatherings. At the end of January 1713 fourteen of the Brothers met, and at Swift's suggestion agreed to hold full meetings only once a fortnight, alternating these with smaller 'committee' meetings of some half a dozen members 'about doing some good', that is, about arranging patronage for some Tory writers. With his usual facility for modest computation, Swift drew up a sliding-scale for 'taxing' the Brothers according to the value of their estates, charging Ormonde ten guineas, and himself a third of a guinea, to support starving writers. At the Duke of Beaufort's 'prodigious fine dinner' in March, he prevailed on the Chancellor Robert Benson to send twenty guineas to Oldisworth, and made similar efforts on behalf of Diaper, Parnell, and the 'young brat' Harrison. At a meeting in January he extracted a 'tax' of sixty guineas from his Brothers, and persuaded three of them to urge Oxford to give a hundred guineas to William Harrison, whose short unhappy career is a poignant illustration of the insecurity of the life of a writer at the time.

Harrison, a mild-mannered Fellow of New College with

[160] Ibid., p. 599.

modest literary abilities, first met Swift in 1710, when aged twenty-five. He quickly became a favourite 'hedge companion' and Swift set himself about making the career of this young fellow who could be trusted to spread his benefactor's virtues wherever he went. Harrison's modesty, deference, and ill health made him the perfect protégé, a kind of male Stella, a man so obviously in need of advice and assistance that he brought out all Swift's fatherly instincts. He recommended Harrison not only to Addison and Steele, but also to Harley and St John. As a result, when Steele laid down the *Tatler* in January 1711, Harrison took up a continuation of the periodical. Nervously he approached his mentor for an opinion of the first issue. 'I am tired with correcting his trash,' Swift reported back to the ladies.[161] In fact he was sad and rather embarrassed at what he read. Much as he liked the 'little brat' he could see that he lacked the 'true vein' for journalism. He gave what help he could, and when in mid-March the poor frantic youth came to beg him to compose a whole paper, he did so. But the incident proved to Swift that the paper would not survive much longer. However, when after the fifty-second issue the enterprise sank, Swift was on hand to suggest a new employment. In his conversations with St John he had persuaded the Secretary to offer Harrison 'the prettiest employment in Europe', namely secretary to Lord Raby, ambassador extraordinary at The Hague. Swift took delight in breaking the good news to him. 'I will send Harrison tomorrow morning to thank the secretary,'[162] he told Stella, showing her that he schooled this young man's manners as carefully as he had done hers. But the employment proved considerably less pretty than anticipated. St John despatched Harrison in April with fifty guineas in his pocket. 'Aren't I a good friend?' crowed Swift to Stella. 'Why are not you a young fellow that I might prefer you?'[163] He received dutiful reports back from his young friend whose earnestly grateful tone makes it embarrassingly obvious why Swift liked him so much. 'I long to see the little brat; my own creature,' he boasted.

Whitehall was notoriously dilatory in the payment of salaries, and although Harrison's post was worth a thousand pounds a year, Swift knew that they would never pay him 'a groat' till he

[161] *Journal*, i. 162. [162] Ibid., p. 217. [163] Ibid., p. 246.

3 Jonathan Swift by Charles Jervas

4 Hester Van Homrigh ('Vanessa') by Philip Hussey

returned. Still the problem was so common that it warranted only a gentle teasing by Swift of the Lord Treasurer. 'He must be three or four hundred pounds in debt,' Swift guessed, looking forward to marching the lad round, with self-righteous indignation, from one paymaster to the next. When Harrison actually returned Swift's tone changed abruptly. A depressing little letter lay on his table one night in February, after he had indulged himself at the society with 'the greatest dinner I have ever seen'.[164] It told him that the brat had returned, but was lying ill and wanted to see him. It was too late for a visit that night, and Swift fretted away the hours till morning in anxious foreboding. He had obtained Oxford's promise of an immediate payment of a hundred guineas, but he knew Oxford's promises of old. The next day he found Harrison cooped up in squalid airless lodgings, seriously ill with fever and inflammation. He immediately had him moved to the fresh country air of Knightsbridge, and gave him £30 from Bolingbroke. Much affected and shocked by the poor youth's condition he spent the following day on a tour of mercy distributing money to other starving writers. He found the poet Diaper 'in a nasty garret, very sick"[165] and gave him another £20 from Bolingbroke. He divided a further £60 between two other unnamed authors. All night he worried over Harrison's condition, and, not daring to visit him himself, sent out a servant to see how he was. 'Extremely ill' was the reply. He was, as he said 'very much afflicted for him, for he is my own creature'. Guilt is obvious in Swift's melancholy. 'His mother and sister attend him, & he wants nothing,' he added. The surrogate father who had betrayed the boy had been displaced. First thing next morning he went with Parnell–for he dared not go alone–to see Harrison. He had at last a Treasury note for £100 in his pocket. 'I told Parnell I was afraid to knock at the door; my mind misgave me. I knocked & his man in tears told me his master was dead an hour before.'[166]

It was a terrible blow that struck at the heart of all the paternal feelings he had done so much to nurture. 'I went to his mother, & have been ordering things for his funeral with as little cost as possible.'[167] He subsequently referred to Harrison's

[164] *Journal,* ii. 618.
[165] Ibid., p. 619.
[166] Ibid., pp. 619-20.
[167] Ibid., pp. 620-1.

mother as an 'old bawd' and to his sister as 'no better' suggesting that sharp words were exchanged between them. It was part of his normal pattern to seek to detach his protégés from their natural families, but now the family had returned to claim the corpse. As if in spite Swift did all he could to stop Harrison's £100 from falling into their hands. 'No loss ever grieved me so much, poor creature–Pray God almighty bless poor MD–adieu!' he wrote. 'I send this away tonight and am sorry it must go while I am in so much grief.' As with the death of Anne Long, Swift's sorrow was sharpened by a deep sense of the injustice of the world. Six weeks earlier the death of Lady Ashburnham had led him to articulate that sense.

> I hate life when I think it exposed to such accidents. And to see so many thousand wretches burdening the earth while such as her die, makes me think God did never intend life for a blessing.[168]

Harrison's death was the most glaring instance of the emptiness of court promises. The funeral was a sad little affair, held at ten o-clock on a wet and windy night, attended by four silent people in one inexpensive coach in the pouring rain. On the way back from the graveyard the braces of the coach snapped, and they were forced to sit in angry silence till Swift's man could run and fetch sedan chairs to take them home.

Another poet whose career Swift sought to advance, with motives which were as much therapeutic as financial, was Thomas Parnell, Archdeacon of Clogher. Parnell had recently suffered the deaths of his wife and child, which left him subject to fits of depression, alleviated only by alcohol and the composition of morbid allegories. On Christmas Day 1712 Swift took Parnell to dine with Bolingbroke, having previously shown the Secretary Parnell's poem celebrating the peace. On New Year's Day the three men met again to correct the poem, which now included a fulsome euology of Bolingbroke. Soon Swift was congratulating himself on having launched Parnell 'who hardly passed for anything in Ireland' into fashionable London society. The friendship between Parnell and Bolingbroke received a macabre consolidation when the Archdeacon met the Secretary's wife, who took them by surprise one day. Looking up, 'Parnell stared at her as if she were a goddess.' Or a ghost, Swift

[168] Ibid., p. 595.

might have said, for, as he explained, 'I thought she was like Parnell's wife, & he thought so too.'[169]

Swift's desire to enlist Parnell's pen for the Tory cause was part of his consistent campaign to build up a nucleus of talented Tory wits to challenge the more numerous Whig writers. Two days after Christmas he met Addison on the Mall. It was their first meeting for several months and, as he indicates, party divisions had now grown so bitter that it was almost impossible for them to converse at all. 'I met Mr Addison and pastoral Philips on the Mall today, & took a turn with them; but they both looked terrible dry and cold; a curse of party!'[170]

After doing what he could for his fellow writers, Swift made a New Year resolution to be more assertive on his own behalf. At the very least he determined to 'contract no more enemies . . . till I have got under shelter; and the ministers know my resolutions'. On 20 January there are two entries in the *Journal* that have no causal or syntactial link, but indicate clearly Swift's association of ideas. Oxford 'gives us nothing but promises', he writes, and later, 'our English bishopric [Hereford] is not yet disposed of'.[171] Four days later a new Bishop of Hereford was appointed, and that particular train of thoughts shunted into a siding. The irony was not lost on either Swift or Oxford of Swift's officiousness in soliciting for others, while nothing was done for himself.

> I never go to his levee unless to present somebody . . . I expected they would have decided about me long ago; & as hope saved, as soon as ever things are given away, & I not provided for, I will be gone with the very first opportunity, & put up bag and baggage.[172]

He wrote that in March when the 'things' to be given away were the three vacant deaneries of Wells, Ely, and Lichfield. He had finally let it be known that he expected one of these in recompense for his services. *'Ubi nunc?'* he wrote in his account book for the beginning of April as the suspense became intolerable. Where now? Dr Arbuthnot was a constant solace during these spring weeks when he felt his whole future hung in the balance. On April Fool's Day, with Lady Masham's help, they concocted a hoax in the old Bickerstaffian manner. 'We hope it will spread,'

[169] Ibid., p. 623.
[170] *Journal,* ii. 589.
[171] Ibid., p. 605.
[172] Ibid., p. 633.

said Swift, gleefully.[173] But it didn't. People had more serious matters on their minds these days.

A new rumour began to circulate at this time which caused Swift deep misgivings. It was said that Oxford was planning to form a new alliance with the Whigs as soon as the peace was signed. On 21 March Swift learnt that Oxford had been at a meeting with some of the leading Whigs at the house of Lord Halifax. It was even rumoured that Swift himself had written privately to friends in Ireland, advising them to 'keep firm' to the Whig cause. In a letter to King, Swift was at pains to quash this story, explaining that Oxford's meeting with the Whigs was merely an expression of the impartiality that King so much admired. 'His lordship is somewhat of your grace's mind in not refusing to converse with his greatest enemies; and therefore he is censured, as you say you are, upon the same account.'[174] King insisted that he was innocent of spreading the rumour in Ireland, but Swift continued to suspect him. In fact Swift was deeply disturbed by the story. His trust in Oxford was always more a matter of faith than reason, and large blinkers were required to trust a man so consistently devious and secretive. Swift's instinct was simply to deny the rumours and plan a new offensive to coincide with the opening of Parliament. However, the signing of the peace treaty at the beginning of April did lead to a general proffering of olive branches. Oxford was not the only one to reopen conversations with old adversaries, and Swift himself made an attempt to be reconciled with Addison. On 28 March, most unusually for him, Swift held a 'mighty levee' when half a dozen people came to drink chocolate, despite his own aversion to the stuff. Among them was Addison, their first meeting apparently since their chilly encounter on the Mall. A few days later he prevailed on Bolingbroke to invite Addison to dinner on Good Friday, though he anticipated a certain formality; 'I suppose we shall be mighty mannerly',[175] and so it proved. That very day Bolingbroke's brother arrived from Utrecht with final confirmation that the treaty was signed. Addison's presence naturally put something of a damper on Tory celebrations. Polite congratulations were offered, but then 'in a friendly manner of party, Addison raised his objections, and Lord Bolingbroke

[173] Ibid., p. 650. [174] *Corr. i. 338-9.* [175] *Journal,* ii. 651.

answered them with great complaisance.'[176] They found they were able to argue without offensiveness, and Addison, slily entering into the spirit of the occasion, proposed Lord Somers's health, 'which went about' in the mood of reconciliation. But Swift bid him 'not name Lord Wharton, for I would not pledge it, and I told Lord Bolingbroke frankly that Addison loved Lord Wharton as little as I did'. 'We all laughed,' said Swift. Both men retired from the encounter pleased that mutual respect, if not friendship, had been re-established. Addison had his own reasons for making peace. He was chiefly occupied at the time with his tragedy *Cato*, that 'splendid exhibition of artificial and fictitious manners', as Johnson called it. Fearing a hostile reception to the play, Addison sought to placate potential critics in advance, and invited Swift to a rehearsal. However Swift did not enter into the spirit of impromptu, and was unable to suspend his disbelief.

We stood on the stage & it was foolish enough to see the actors prompted every moment, & the poet directing them, & the drab that acts Cato's daughter out in the midst of a passionate part, & then calling out, what's next?[177]

Mrs Oldfield (the drab) was enough to strain anyone's credulity as Cato's virgin daughter, being in an advanced state of pregnancy. Despite Addison's best endeavours, scenes of political riot greeted the play on its first night, and were observed with particular apprehension by the young Alexander Pope who had composed the prologue, and found himself 'clapped into a staunch Whig, sore against his will, at almost every two lines'.[178] Only a month earlier Swift had detected Tory leanings in Pope's georgic poem 'Windsor Forest', and was soon doing his best to recruit the young poet to the Tory cause.

On 8 March Oxford showed Swift a draft of the speech the Queen would deliver at the opening of Parliament, and asked him to improve the style. 'I corrected [it] in several places,' he noted.[179] Swift's anxieties about his letters miscarrying, or being intercepted, increased at this time, and he adopted the nick-

[176] Ibid., p. 652.
[177] Ibid., p. 654.
[178] Pope's *Correspondence,* ed. George Sherburn, 5 vols., Oxford, 1956, i. 175.
[179] *Journal,* ii. 635.

name of Eltee (L.T.–Lord Treasurer) for Oxford, but explained it so often, that it rather lost its value as a code. On 9 April the much prorogued parliament finally met, and the Queen delivered her speech, including Swift's corrections, from the throne, 'very well but a little weaker in her voice'.[180] Two days before Swift had sent his secret note to Stearne at St Patrick's. This was a consequence of a frank discussion between himself, Oxford, and Ormonde at which, finally acknowledging the problem of the Queen's stubbornness, the expedient of finding Swift an Irish post was openly suggested. For some weeks the prospect of Ireland had been looming ominously larger in Swift's mind. There was a horrible inevitability about it that made him sick. He squirmed to think what a ridiculous figure he would cut there, after all his talk of Eltee, and the court. How they would exult, the people who had envied him the excitement of his life of court and political intrigue! And when he thought of the exulters, he thought of Tisdall.

When I come back to Ireland with nothing, he will condole with me with abundance of secret pleasure. I believe I told you what he wrote to me, that I have saved England, & he Ireland. But I can bear that. I have learnt to hear and see and say nothing.[181]

On the unlucky 13th of April he knew the worst.

My friend Mr Lewis came to me, and showed me an order for a warrant for the 3 vacant deaneries, but none of them to me.[182]

It was a heavy blow. 'This was what I always foresaw', he wrote. It would be truer to say it was what he always feared. To Lewis he put on a brave face, taking the news 'better I believe than he expected'. He told Lewis to assure Oxford 'that I took nothing ill of him but his not giving me timely notice, as he promised to do'. Actually for all his apparent calm, there was a deep bitterness in his heart at this disappointment. The situation was made still worse by the fact that even now there were people advising him to hope. There was still a last rope for him to dance or dangle on. The fortnight which followed was the most excruciating of his life, and is best told in his own words.

April 13: At noon, Lord Tr hearing I was in Mr Lewis's office, came to me, & said many things too long to repeat. I told him I had nothing to

[180] Ibid., p. 657. [181] Ibid., p. 632. [182] Ibid., p. 660.

do but go to Ireland immediately, for I could not with any reputation stay longer here, unless I had something honourable immediately given to me; we dined together at D. Ormonde's, he there told me, he had stopped the warrants for the Deans, that what was done for me, might be at the same time, & he hoped to compass it tonight; but I believe him not.

14th: . . . I will leave this end of the town as soon as ever the warrants for the Deaneries are out, which are yet stopped; Lord Tr told Mr Lewis, that it should be determined tonight; & so he will for a hundred nights, so he said yesterday, but I value it not . . .

15th . . . Lord Bolingbroke made me dine with him today. I was as good company as ever; & told me the Queen would determine something for me tonight; the dispute is Windsor or St Patrick's; I told him I would not stay for their disputes, & he thought I was in the right . . .

16th . . . I was this noon at Lady Mashams . . . she said much to me of what she had talked to Queen and Lord Treasurer. The poor lady fell a crying, shedding tears openly: She could never bear to think of my having St Patrick's &c. I was never more moved than to see so much friendship . . . Mr Lewis tells me, that D. Ormonde has been today with Queen & she was content that Dr Stearne should be Bishop of Dromore and I Dean of St Patrick's, but then out came Lord Treasurer, & said he would not be satisfied, but that I must be prebend of Windsor, thus he perplexes things–I expect neither: but I confess, as much as I love England, I am so angry at this treatment, that if I had my choice I would rather have St Patrick's. Lady Masham says she will speak to Qu-- tomorrow . . .

17th . . . Qu-- says she will determine tomorrow with Ld Tr. The warrants for the deaneries are still stopped, for fear I should be gone. Do you think anything will be done? I don't care whether it is or no . . . Ld Tr told Mr Lewis it should be done tonight, so he said 5 nights ago . . .

18th . . . At 3 Ld Tr sent to me to come to his lodgings at St James's, and told me the Qu was at last resolved, that Dr Stearne should be Bishop Dromore, and I Dean of St Patrick's . . . I do not know whether it will yet be done; some unlucky accident may yet come; neither can I feel joy at passing my days in Ireland: and I confess I thought the ministry would not let me go; but perhaps they can't help it.

19th . . . I went this evening and found D. Ormonde at the cockpit, & told him, and desired he would go to Qu-- and approve of Stearne. He made objections, desired I would name any other deanery, for he did not like Stearne, that Stearne never went to see him, that he was influenced by Archbishop Dublin & c; so all is now broken again . . . This suspense vexes me worse than anything else.

20th : I went today by appointment to the Cockpit, to talk with D.

Ormonde; he repeated the same proposal of any other deanery etc. I desired he would put me out of the case, & do as he pleased; then with great kindness he said he would consent, but would do it for no man alive but me & c, and he will speak to the Qu-- today or tomorrow. So perhaps something will come of it. I can't tell.

21st: D. Ormonde has told Qu-- he is satisfied that Stearne should be Bishop, & she consents I shall be Dean, and I suppose the warrants will be drawn in a day or two.

22nd: Qu says warrant shall be drawn, but she will dispose of all in England and Ireland at once, to be teased no more, this will delay it sometime; & while it is delayed I am not sure of the Qu-- my enemies being busy; I hate their suspense.

23rd: This night the Qu-- has signed all the warrants, among which Stearne is bishop of Dromore, & D. Ormonde is sent over an order for making me Dean of St. Patrick's . . . I think tis now past: and I suppose Md is malicious enough to be glad & rather have it than Wells. But you see what a condition I am in. I thought I was to pay but £600 for the house, but Bp Clogher says £800 . . . so that I shall not be the better for this deanery these three years.

25th: I know not whether my warrant be yet ready from the D. Ormonde . . .

26th: I was at court today; & a thousand people gave me joy, so I ran out.[183]

Throughout this painful narrative one senses the bitterness of his disillusionment with Oxford. Rather than openly confess his inability to influence the Queen, as Lady Masham's tears so eloquently did for her, the Lord Treasurer continually deceived Swift with false hopes. Some instinct of natural justice, deeper than his tirades against the world's unfairness, made Swift feel that he could not go unrewarded. 'I confess I thought the ministry would not let me go.' That is an unusually unguarded comment, almost shocking in the deep sense of disappointment it betrays. Little wonder that he preferred the anonymity of ale-houses, where he could dine with Arbuthnot, Berkeley, or Parnell, to the company of ministers and courtiers with their false promises and insincere congratulations. On the 19th he told Stella that he dined 'with a private friend', almost certainly Vanessa. She at least would share his sense of desolation. If Stella

[183] Ibid., pp. 660-7. The Cockpit was an office in Whitehall used for various Government purposes.

was secretly pleased to have him back in Dublin, Vanessa was openly inconsolable.

Swift soon worked out that the appointment to St Patrick's was not only a snub, it was also probably a financial disaster.

I shall be ruined or at least sadly cramped unless the Queen will give me £1,000. I am sure she owes me a great deal more. Ld Tr rallys me upon it, and I believe intends it; but quando?[184]

He was back on the same treadmill. He never received his £1,000 of course, and for years afterwards it rankled with him as the insult added to the injury of exile.

As if all this were not enough, his life was further soured by an unpleasant row with Steele. At the beginning of March Steele began a new periodical, the *Guardian*. 'Good for nothing' was Swift's comment, although, with contributions from Pope and Berkeley, the early issues were both entertaining and non-partisan. However on 12 May, *Guardian* 53 contained a letter, signed by Steele, attacking the author of the *Examiner* as either 'an estranged friend or exasperated mistress'. The references were obviously to Swift and Mrs Manley, and Swift appealed in protest to the fair-minded mediation of Addison. He 'could hardly believe' that Steele would attack him in this 'grossest manner' and was particularly distressed that a former friend should have attacked him in print without having the decency to check his facts. 'Should not Mr Steele have first expostulated with me, as a friend?' he asked Addison. The sense of another betrayal stung him deeply. 'Have I deserved this usage from Mr Steele, who knows very well that my Lord Treasurer has kept him in his employment upon my entreaty and intercession?'[185] In contrast with the Whig lords who had promised so much and performed so little, Swift always pictured himself as a loyal friend, who had striven to help his old associates, despite their political differences. After his chilly Christmas meeting with Addison he ran through the roll-call of those he had helped in an anxious reflex of self-justification.

I have taken more pains to recommend the Whig wits to the favour & mercy of the ministers than any other people. Steele I have kept in his place; Congreve I have got to be used kindly and secured. Rowe I have recommended, and got the promise of a place.[186]

[184] Ibid., p. 669. [185] *Corr.* i. 347-8. [186] *Journal,* ii. 589.

One remembers what that promise of a place to Rowe was worth–but that was hardly Swift's fault. It baffled Swift that his Whig friends should be so stiff-necked about accepting his patronage. Steele was particularly ungrateful. He had been deprived of his position as Gazetter on the change of ministry, but retained his post as Commissioner of the Stamp Office. In December 1710 Swift made a great to-do about arranging an interview between Harley and Steele to ensure that he kept this place, but Steele failed to attend. Swift was furious at this discourtesy. 'Whether it was blundering, sullenness, insolence, or rancour of party I cannot tell; but I shall trouble myself no more about him.[187] Actually Steele's non-appearance was merely a device to avoid any indebtedness to Swift. He made his own arrangements to see Harley, and the two men had a short conversation which resulted in Steele retaining his post at the Stamp office, much to Swift's mystification. This secret agreement with Harley was a trump card which Steele had kept well hidden until now, when he played it with devastating finesse. Instead of apologizing, he replied with a calculated piece of provocation. 'They laugh at you, if they make you believe your interposition has kept me thus long in office.'[188] He could not have delivered a more insidious thrust, for his phrase invited Swift to contemplate an abyss of ministerial duplicity. The Whigs laughed at the absurd posturing, the self-importance of this dedicated ministry-man, this official solicitor of places, who had so conspicuously failed to provide a place for himself. The courtiers laughed at him as they 'wished him joy' of his Irish appointment, which amounted to no more than being bundled unceremoniously back to oblivion. Steele's taunt threw into sharp relief all Swift's half-suppressed anxieties concerning Oxford. Had he been in touch with Steele? Was he, even now, negotiating an alliance with the Whig lords? Were there whole areas of policy to which Swift had never been allowed access? Steele pushed his point home with an ironic twist worthy of Swift himself.

You . . . make it an argument of your innocence that the *Examiner* has declared you have nothing to do with him. I believe I could prevail upon the *Guardian* to say there was a mistake in putting my name in

[187] Ibid. i. 128-9. [188] *Corr.* i. 351.

his paper: But the English would laugh at us, should we argue in so Irish a manner. I am heartily glad of your being made Dean of St Patrick's.[189]

Steele deflates Swift's political objections by comparing their argument to the brawling of a couple of teagues. 'The English would laugh at us,' Steele presents Swift as nothing but an Irish clown, and 'congratulates' him on returning where he belongs, to Dublin. Swift replied in a furious letter, raw with self-justification, returning obsessively to that humiliating taunt 'they laugh at you'. Though the letter is now torn, the phrase is clearly repeated at least six times.

Suppose they did laugh at me, I ask whether my inclinations to serve you merit to be rewarded by the vilest treatment, whether they succeeded or no? If your interpretation were true, I was laughed at only for your sake; which I think is going pretty far to serve a friend.[190]

A tone of self-pity shows through the rage, as Swift's confidence shows signs of breaking down under Steele's goading. He genuinely felt this to be an unmerited attack, a betrayal, and he utters a cry of pain, a shout of 'foul'. Yet in his pain he exaggerates his innocence.

I have several times assured Mr Addison, and fifty others, that I had not the least hand in writing any of those papers; and that I had never exchanged one syllable with the supposed author in my life that I can remember, nor ever seen him above twice . . .

This stretches the truth a good deal, and the phrase 'that I can remember' is a transparent evasion. Steele replied with ironic condescension.

I am obliged to you for any kind things said in my behalf to the Treasurer; and assure you, when you were in Ireland, you were the constant subject of my talk to men in power at that time.[191]

Touché. It's a neat point for Steele to remind Swift of his ambiguous relationship with the Whig lords only four years earlier, when their respective roles were reversed. There is little doubt that throughout their exchange, Swift comes off worse. Steele is deft and precise, constantly wrong-footing his opponent. Swift is awkward with self-justification, blundering with rage and con-

[189] Ibid. [190] Ibid., p. 355. [191] Ibid., p. 358.

scious innocence. His writing is repetitious, self-righteous, and without either shape or sharpness. The exchange shows Swift feeling raw with a sense of betrayal, not just by Steele, but by Oxford, the court, the Queen, and the whole of London society.

It was a sad and angry man who left London the last night of May. Before setting out for Chester he sent a brief note to Vanessa, in fulfilment of a promise she had extracted from him. 'It is impossible for anybody to have more acknowledgements at heart for all your kindness and generosity to me.'[192] Despite the friendly tone, these are cold words for a lover to use. The appointment to St Patrick's was a defeat and he was in no mood to dress it up as anything else. He did not know when, or whether, he would return. In such circumstances the additional heartache that Vanessa might cause was more than he could bear. 'I will write a common letter to you all' is the most that he will promise, in response to her entreaties for a regular correspondence, thereby clearly ruling out any private intimacies. 'Adieu brat' the letter ends, an affectionate flourish, but one with a distinct whiff of finality about it.

V

SWIFT WAS as good as his word and sent common letters addressed to Vanessa from St Albans, to her sister Moll from Dunstable, and to Mrs Van from Chester. As soon as the family received the first two letters, Vanessa rushed to reply, though her apologetic admission, 'I believe you little thought to be teased by me so soon', suggests she was infringing a prohibition. 'Pray why did not you remember me at Dunstable as well as Moll,' she complained.[193] In reply, Swift told Mrs Van that he could find no marks 'in the chimney at Dunstable of the coffee Hessy spilt there' at their first meeting.[194] He hoped that this little private allusion would pacify Vanessa, though he was scrupulously careful, beneath his bantering tone to express an equal interest in both the Vanhomrigh girls. 'Who will Hessy get now to chide, or Moll to tell her stories, and bring her sugar-plumbs?' But Vanessa was now in the clutches of the green-eyed

[192] Ibid., p. 360. [193] *Corr.* i. 364. [194] Ibid., p. 366.

monster. Obsessively she counted up the references to herself and Moll in his last letter.

tis true the number is equal, but you talk to Moll, and only say now Hessy grumbles. How can you possibly be so ill-natured to make me either quarrel or grumble when you are at so great a distance, that tis impossible for me to gain by doing so: beside you promised the letter should be directed to me.[195]

He did not reply and a week later she wrote again, using the stale excuse of a concern for his health.

Oh what would I give to know how you do at this instant. . . . I have done all that was possible to hinder myself from writing to you till I heard you were better for fear of breaking my promise but twas all in vain . . . I beg you won't be angry with me.[196]

Her pathetic desire to be acknowledged only embarrassed Swift further and still he did not reply. A week passed and she wrote again. If he were too busy to write, could he perhaps get Parvisol, his steward, to write to her, to assure her he was well?

If you think I write too often your only way is to tell me so or at least to write to me again that I may know you don't forget me for I very much fear that I never employ a thought of yours now except when you are reading my letters which makes me ply you with them.[197]

At last he did reply, but in cold unemotional terms, designed to chill rather than to cheer.

I had your last splenetic letter; I told you when I left England, I would endeavour to forget everything there, and would write as seldom as I could.[198]

Ill health and ill temper meant that he had written to very few people. 'I stayed but a fortnight in Dublin, very sick, and returned not one of a hundred visits that were made me. . . . I did indeed design one general round of letters to my friends, but my health has not yet suffered me.' Rather than stay in the 'great house' that Stearne had built in Dublin, Swift preferred to confine himself hermit-like to his little 'cabin' at Laracor, with its field bed and earthen floor. In unspoken reaction to her tears and entreaties he paints a deliberately bleak and down-to-earth

[195] Ibid., p. 368.
[196] Ibid., pp. 369-70.
[197] Ibid., p. 372.
[198] Ibid., p. 373.

portrait of himself in his new surroundings. He is facing reality, and by implication he requires her to do the same.

At my first coming I thought I should have died with discontent, and was horribly melancholy while they were installing me, but it begins to wear off, and change to dullness.

Was she really so keen to hear what he was doing? Very well then, he treats her to some of the details of his new life.

Mr Warburton has but a thin school; Mr Percival has built up the other side of his house, but people whisper that it is but scurvily built. Mr Steers is come to live in Mr Melthorpe's house, and tis thought the widow Melthorpe will remove to Dublin–Nay if you do not like this sort of news, I have no better. So go to your dukes and duchesses and leave me to Goodman Bumford and Patrick Dolan of Clanduggan.[199]

This is the voice he was to perfect in later years, the voice of stoic resignation without stoic calm; the immersion in a common life that he both cherished and despised.

After writing to Mrs Van from Chester, he wrote to Stella the same night. It was the final letter of his long *Journal,* and in it he began to imagine the people who would replace Oxford, Bolingbroke and the Queen in his thoughts. 'I am sorry for Raymond's fistula,' he wrote; and to personify all that was ugly and unappealing about Dublin, he thought of Tisdall. 'Do his feet stink still?' he asked.[200]

VI

ON HIS return to Ireland Swift suffered an immediate attack of his old disorder of giddiness and pains in the head. He was besieged by visitors who came with chores associated with his new office 'all to the Dean, and none to the doctor'.[201] His official reception was decidedly cool. One Oxford Tory asserted that 'the clergy [of Ireland] detest Dr Swift because they think him an enemy to the order'.[202] Ironically, it was the Whigs who applauded his arrival for vexing 'the godly party beyond expression'.[203] Archbishop King a few years later was injudicious enough to express relief that Swift had not gained higher preferment since 'a Dean

[199] Ibid., p. 374. [200] *Journal,* ii. 671. [201] *Corr.* i. 372-3.
[202] Landa, p. 73. [203] Ibid., pp. xiv, xv.

could do less mischief than a bishop'. King was not in Dublin to welcome Swift, having just set sail for Bath. Instead, according to legend, Swift found these lines affixed to the door of St Patrick's Cathedral on the day of his installation.

Look down, St Patrick, look we pray
On thine own church and steeple;
Convert thy Dean on this great day;
Or else, God help the people.[204]

Swift was left in little doubt of the lack of enthusiasm for his return. For three years, in his letters back to the ladies, he had drawn an idealized picture of the cosy intimacy of Dublin life, with its old friends and card parties. Now that he was back he found a tiny capital city full of petty rivalries, the hypocrisy and back-biting of party politics, that could imitate London only in its vices. Even the company of Stella, to which he had so much looked forward, could not lift his melancholy or dispel his giddiness. At the earliest possible moment he escaped to the seclusion of Laracor, where he rode and rode 'for life' and tried to lead a simple, ascetic life. It was a solace to see his canal 'in great beauty'[205] and the trout playing in it. Gradually the melancholy left him, and his mood changed to a more neutral numbness under the therapeutic influence of his country domain. Yet even here he was nagged by a sense of guilt, for he had promised to pass on the living at Laracor to his friend Dr Raymond, the minister at Trim, 'if ever the court should give me anything'. But now Swift found that he could not bear to part with it. Not only did he value the recuperative atmosphere of the spot; he also needed its revenues, for he continued to fear that the costs of taking over the Deanery would be ruinous. Stearne was now trying to sell him his furniture at top prices. 'I shall buy Bishop Stearne's hair as soon as his household goods,' he wrote tartly.[206] He was particularly annoyed that Stearne had disposed of two livings before departing, in defiance of Swift's explicit wishes. Almost certainly King had encouraged him in this, to prevent Swift from filling the posts with his own nominees.

[204] Smedley, *Gulliveriana: or a Fourth Volume of Miscellanies,* 1728, p. 78.
[205] *Corr.* i. 373.
[206] *Journal,* ii. 669.

To assuage his guilt at not vacating Laracor, Swift made strenuous efforts to obtain the modest benefice of Moymet for Raymond, worth a mere £40 a year. It was so small that Swift told the Secretary for Ireland that it was 'hardly worth while for anybody else to pass patent for it'. He was wrong. A great many people were anxious to make a claim for £40 per annum, and in May he wrote in exasperation to Stella.

> Tell Raymond I cannot succeed for him to get that living of Moymet; it is represented here as a great sinecure, several chaplains have solicited for it, & it has vexed me so, that if I live, I will make it my business to serve him better in something else.[207]

Among the applicants was a blind clergyman, Dunbar, who described himself as being 'without bread or subsistence'.[208] On the contrary, the unsentimental Raymond quickly produced evidence that Dunbar already held a satisfactory living, and Stella, in an aside worthy of her mentor, suggested hanging the fraudulent blind man. It took two more months, and several emphatic letters for Swift to have his way in this tiny matter. His other attempts at patronage met with complete failure, and Swift soon discovered that there were very strict limits to his power in his new position. He resented these restrictions acutely, and particularly disliked being under King's authority once more. The gulf between them is indicated by an offhand remark in a letter from King at Bath that August. Having heard that Swift had spent only a fortnight in Dublin before retiring to the country, he wrote,

> An odd thought came into my mind on reading that you were among willows imagining that perhaps your mistress had forsaken you and that was the cause of your malady. If that be the case, cheer up, the loss may be repaired, and I hope the remedy easy.[209]

It is difficult to know whether King is merely being facetious here, or whether he has Stella in mind. Swift's friendship with 'the ladies' was obviously common knowledge in Dublin, but one would hardly expect King to refer to Stella as Swift's 'mistress'. Possibly the influence of Bath's fashionable society led King to affect this rakish worldliness. Since King had failed to oblige him in the matter of preferments, Swift was less than

[207] Ibid., p. 668. [208] *Corr.* i. 370–1. [209] Ibid., p. 382.

enthusiastic about pursuing King's pet scheme for the erection of a 120 foot spire on top of the cathedral. King had put aside £200 for this purpose, but Swift feared that the spire would cost far more than that, and that the local brick would not endure 'being exposed so much to the air'.[210] On the contrary, replied King, 'Our Irish brick will do very well.'[211] And so it went on, an uncomfortable exchange of bricks and begging letters. But Swift had learnt the art of delay by observing a master at close quarters. No spire was erected over St Patrick's in his lifetime.

As he lived his hermit's life of riding, hedge-cutting, and walking by his canal, he found letters from England an unwelcome reminder of the world he had lost. Vanessa's letters especially, which arrived every week, were excruciating with their mixture of pleading and railing. Charles Ford, now Gazetteer, was another regular London correspondent. He sent Swift six letters in as many weeks till Swift begged him to stop. He found Ford's account of political events in England, depressing. 'I do not at all like the general face of them.'[212] The worst feature was the ever deepening division between Oxford and Bolingbroke. It was a 'shuddering question' which of them would succeed in the struggle for power. Like Vanessa, Ford called on Swift to return. The idea was tempting but he was not prepared to be humiliated again. The bait that was held out to him was the £1,000 which he needed to pay his costs at the Deanery, but he was tired of hints and promises. They would not 'draw me over to solicit, except I am commanded to come'. He insisted that the great men should explicitly acknowledge their need for him before he would return. Erasmus Lewis was in no doubt of that need. On 18 June the commercial clauses of the Peace treaty were defeated by 194 to 185, the worst ministerial defeat since 1710. 'We are all running headlong into the greatest confusion imaginable,' he wrote. 'I heartily wish you were here, for you might certainly be of great use to us by your endeavours to reconcile.'[213] Vanessa agreed. 'I wish you had been here last Thursday, I am sure you could have prevented the bills being lost.'[214] Actually they were lost because, being entirely Bolingbroke's concern, Oxford had done little to support them. But

[210] *Corr.* i. 354. [211] Ibid., p. 357. [212] Ibid., p. 375.
[213] Ibid., p. 374. [214] Ibid., p. 368.

Bolingbroke fought back, and gradually persuaded the Queen to promote more High Tories to key positions. This led Lewis to become still more pressing in his demands for Swift to return. 'My Lord Treasurer desires you will make all possible haste over, for we want you extremely.'[215] But Oxford's desires were no longer enough; Swift would only return if ordered to do so. This was not just his usual coyness, but reflects his real pain at his treatment during April. Lewis wrote again with patient entreaties.

I have so often and in so pressing a manner desired you to come over, that if what I have already said has no effect I shall despair. . . . The *desires* of great men are *commands*, at least, the only ones I hope, they ever will be able to use.[216]

At last Swift relented and agreed to return. He grew happier at once. He wrote a merry letter to Archdeacon Walls, agreeing to act as godfather to his new-born daughter. The ladies moved up to Trim, where they stayed with Raymond, and together they all enjoyed a happy rural interlude in the Irish countryside that summer. Swift could now enjoy the simple life with no overtones of self-mortifying asceticism, secure in the knowledge that a more complex life awaited him. He set sail from Dublin on 29 August and landed at Parkgate two days later. King arrived back in Dublin simultaneously and was not the only person to find something suspicious about the way that he and Swift kept crossing on the Irish sea.

VII

SWIFT WAS no sooner back in London than he was writing back to Walls, 'I protest I am less fond of England than ever.'[217] It was a familiar cry, a defence against his own excitement; but this time there were powerful reasons for disenchantment. He returned with few of his former illusions, and a clear task before him, to reconcile the two men who seemed determined to tear the ministry apart. He set about the task immediately, contriving various schemes to bring them together. One day he was due to travel to Windsor with Oxford.

[215] Ibid., p. 378. [216] Ibid., p. 383. [217] *Corr.* i. 387.

I pretended business that prevented me; and so I sent them to Windsor . . . in the same coach; expecting they would come to some éclaircissement, but I followed them to Windsor; where my Lord Bolingbroke told me . . . that my scheme had come to nothing.[218]

After a month of such efforts the rift was as wide as ever. 'I am heartily weary of courts, and ministers and politics,' wrote Swift.[219] He was exasperated to despair by the streaks of wilfulness which often led both men to act with almost childish spite. Oxford particularly took delight in wrecking the plans of his younger rival, apparently careless of the larger effect of such behaviour on Government policy. In the elections that summer he took few measures to ensure a Tory victory, with the result that the Tory majority was reduced. Swift drew up a private memorandum of points which he intended to urge upon the Lord Treasurer, all of them designed to make him act more decisively in public matters. 'No orders of any kind whatsoever given till the last extremity' is a typical note from Swift's list of criticisms.[220] He also had a personal case to argue with Oxford, but for this he adopted a more deferential approach, composing a verse epistle in imitation of Horace's 'Seventh Epistle of the First Book'. Written only weeks after 'Cadenus and Vanessa' this poem gives the 'authorized version' of another ambiguous relationship. But whereas 'Cadenus and Vanessa' sought to set limits to a relationship, the 'Epistle' to Harley stresses an intimacy deeper than any political alliance.[221] At the start of the 'Epistle' Swift presents himself as a bookish parson 'cheapening old authors on a stall'. By the end, he has become a harassed and penniless Dean, 'so dirty, pale and thin/Old Read would scarcely let him in'. Swift celebrates his friendship with Oxford in a back-handed fashion that makes clear the pains it has caused him. The bookish parson that Harley first spies from his coach is described (quite falsely) as 'a perfect stranger to the spleen'. At first the cheerful parson Swift plays hard-to-get, but as Harley persists with his invitations, he is finally persuaded to become a regular visitor.

[218] Ibid. v. 45-6.
[219] Ibid., i. 389.
[220] Rylands English MS 659, No. 13. Printed as Appendix E in Ehrenpreis, ii. 773-4.
[221] *Poems,* i. 169-75.

> Soon grows domestic, seldom fails
> Either at morning, or at meals;
> Came early and departed late:
> In short, the gudgeon took the bait.
> (ll. 77-80)

Hooked by Harley's hospitality, Swift admits a desire for a canon's place at Windsor.

> A canon! that's a place too mean,
> No doctor, you shall be a Dean.
> (ll. 87-8)

Passing quickly over the painful memories of the previous year, Swift simplifies several months of Harleian evasions into this:

> You need but cross the Irish seas
> To live in plenty, power and ease.
> (ii. 91-2)

But in place of plenty, Swift in Dublin lives on 'borrowed money': in place of 'power and ease' the new post brings endless vexations.

> Patents, instalments, abjurations,
> First-Fruits and Tenths, and Chapter-treats,
> Dues, payments, fees, demands and cheats.
> (ll. 102-4)

The jargon of terms fits neatly into the hectic rhythms of the verse. What the poor exhausted Dean needs above all is–a thousand pounds. In fact the poem is a begging letter, witty, courteous, polite, and good-natured, but for all that a serious attempt to goad or shame Oxford into fulfilling this promise at least. Swift does not wish to accuse Oxford directly of neglect and hopes that humour may succeed where cajolery had failed. The poem is a fascinating glimpse into the teasing intimacy between them, and into that never perfect ease between superior and inferior, patron and client. It was soon afterwards that Swift wrote, in a different context: 'Nothing is so hard for those, who abound in riches, as to conceive how others can be in want. How can the neighbouring vicar feel cold or hunger, while my lord is seated by a good fire in the warmest room of his palace, with a dozen dishes before him?'[222] The feeling of

[222] *PW* iv. 65.

injustice evoked here may well derive from his observation of Oxford's failure to conceive how his thoughtlessly unfulfilled promises could cause real distress to others.

However, Oxford's complacency did not go long unpunished. In November his daughter Betty gave birth to a son amid family rejoicing. A fortnight later she was dead. It was a severe blow and affected Oxford deeply. Acting as both friend and spiritual adviser, Swift wrote a moving letter of condolence which nevertheless betrays some social awkwardness. His eulogy of Lady Betty sounds artificially fulsome, when compared with his simple summary of the virtues of Anne Long.

> My lord, I have sat down to think of every amiable quality that could enter into the composition of a lady, and could not single out one, which she did not possess in as high a perfection as human nature is capable of.[223]

Swift is still paying court, and is conscious of addressing a superior. This is evident too in the way he attempts to turn Oxford away from his grief.

> To say the truth, my Lord, you began to be too happy for a mortal. . . . Perhaps your lordship has felt too much complacency within yourself, upon this universal success: and God Almighty, who would not disappoint your endeavours for the public, thought fit to punish you with a domestic loss, where he knew your heart was most exposed.

There is a hint of self-righteousness here. God has seen, like Swift, the dangers of Oxford's complacency and has done what Swift could not do, punished him for it. Swift could not decide whether to send this letter or not. He did not wish to intrude upon a private grief, but his hesitation was less one of feeling than of etiquette. Much as he wished to be, he was not one of the family, and felt the anxieties of an outsider who might one day have the door shut in his face by Read.

After his daughter's death, Oxford fell into a deep melancholy which left public affairs in a state of paralysis. Rumours spread that he was about to resign, and Bolingbroke took new heart. At the last moment, however, Oxford heeded the admonitions of his friends and returned to his duties. It was not a moment too soon, for on 23 December the Queen suffered a violent ague.

[223] *Corr.* i. 405.

Over Christmas there were strong rumours of her imminent death. The Tories were in despair, while the Whigs found it hard to disguise their jubilation with a decent show of grief.

Swift's main interest in returning to politics was to repay the attacks of Steele who had finally resigned from the Stamp Office in order to stand for Parliament. The *Guardian* for 7 August delivered a blistering attack on the ministry for failing to force the French to demolish the fortifications at Dunkirk. The British nation, according to Steele 'expects the demolition of Dunkirk'. This phrase quickly became the new Whig slogan and the *Flying Post* printed at the end of every issue that summer 'Dunkirk is not yet demolished. The Pretender is not yet removed from Lorraine.' At the end of September Steele expanded his attack into a full-length pamphlet, *The Importance of Dunkirk, Considered.* It is a tedious, pedestrian work, padded out with lengthy quotations. The tone lurches unsteadily from synthetic rage to whining self-righteousness as Steele defies the 'prostitute pens' of his opponents and glories in the joy of serving his fellow men. Swift prudently waited until the demolition of the Dunkirk fortifications had begun before replying. He could therefore brush that issue aside and concentrate upon his real aim, the demolition of Richard Steele. His reply is called, with appropriate irony *The Importance of the Guardian Considered* and begins with a character-sketch of Steele, a long-pondered revenge which is rich in half-truths and remembered details.

> To take the height of his learning, you are to suppose a lad just fit for the university, and sent early from thence into the wide world, where he followed every way of life that might least improve or preserve the rudiments he had got. He hath no invention, nor is master of a tolerable style; his chief talent is humour, which he sometimes discovers both in writing and discourse; for after the first bottle he is no disagreeable companion.[224]

Swift takes great fun in ridiculing Steele's style, but the main charges against him are those of insolence and ingratitude. Swift presents a fictitious exchange of letters between Steele and the Queen in which Steele lectures the monarch in a bombastic, self-important manner. Yet, reading through Swift's attack one feels the familiar sense of unease that often arises from his

[224] *PW* viii. 5-6.

defences of authority. Take, for example, his argument that it is right for the *Examiner* to be polemical in defence of the ministry, but wrong for the *Guardian* to use the same techniques in opposition.

It is more impudent, immoral and criminal to reflect on a majority in power, than a minority out of power.

One blinks at Swift's logic here. Only governments may write polemics, not oppositions; to be out of power is to be denied the free use of the press as well. This is a chilling defence of Big Brother, and Swift's analogies indicate the inegalitarianism of his argument.

Mr Bailiff; suppose your worship, during your annual administration, should happen to be kicked and cuffed by a parcel of Tories, would not the circumstance of your being a magistrate, make the crime the greater than if the like insults were committed on an ordinary Tory shopkeeper, by a company of honest Whigs?[225]

Elected officials, it seems, deserve greater protection than ordinary shopkeepers. But why? It would not be many years before Swift, in the character of a shopkeeper, would be putting an exactly contrary argument, and would rigorously resist the attempts of ministers to have a monopoly of the press.

By now Steele had decided that the *Guardian* was no longer an appropriate vehicle for his political writings, and in October he began the *Englishman,* a frankly partisan Whig paper. Addison attempted, unsuccessfully, to restrain him from descending further into the Grubean depths of political invective. 'I am in a thousand troubles for poor Dick,' he wrote, 'and wish that his zeal for the public may not be ruinous to him.'[226] In addition to his new periodical, Steele was preparing a major new pamphlet to be called the *Crisis,* offered for subscription at a shilling a copy. An expensive publicity campaign heralded the appearance of this new work, but as delays grew, so did the rumours. It was said that not only Steele, but also Addison and a team of Whigs were working on the pamphlet. At Christmas, Steele sought to disguise his difficulties by announcing that publication of the *Crisis* had been postponed to allow the ladies

[225] Ibid., p. 14.
[226] *Letters,* ed. Graham, Oxford, 1941, p. 280.

a last chance to join the subscription list. When at last it appeared in January, after so many fanfares, it was, almost inevitably, an anticlimax.

Inspired both by personal rivalry and public expectations to produce a work comparable with Swift's *Conduct*, Steele was forced back on the hack's last resort–padding. The *Crisis* is stuffed full of documents. Its argument, buried under a weight of legal texts, is that the Hanoverian Succession is still in danger. To prove this, Steele offers his own version of recent history, including extensive eulogies of his hero, Marlborough. He loudly laments the undiminished strength of France, the undemolished Dunkirk, and has a special sympathy for the sad plight of the Catalans: 'I mention the Catalonians, but who can name the Catalonians without a tear? Brave unhappy people . . .' Yet the unhappiness of the Catalans is as nothing to the miseries to be feared in Britain if ever the Pretender should return. The pamphlet, repetitious, rambling, and frequently self-righteous, has all the deftness and dexterity of an elephant on a trapeze. Clearly it was through the eyes of faith that Abel Boyer read his copy of the work, which he found 'so excellent that there can be no abridging of the least part, without maiming the whole'.[227]

Swift began an immediate reply, and within four weeks *The Public Spirit of the Whigs* was on sale. It was his last major Tory pamphlet, and one of his happiest efforts in the style of an 'answerer'. At first he treats Steele like a schoolboy who has produced a rather shoddy essay. 'He hath a confused remembrance of words since he left the university, but hath lost half their meaning.'[228] It is when he comes to Steele's suggestion that the clergy should read from the *Crisis* in their pulpits, that he demonstrates real anger. 'This is the right Whig scheme of directing the clergy what to preach,' he thunders, linking Steele's suggestion with the arguments of Bishop Burnet of Salisbury: 'This is the spittle of the bishop of Sarum, which our author licks up, and swallows, and then coughs out again, with an addition of his own phlegm.'[229] There is a sureness of touch and confidence about this pamphlet, which comes from having identified the glaring contradiction in Steele's position. For on

[227] Boyer. *The Political State of Great Britain,* 8 vols., (2nd edn. 1719), vii. 2.
[228] *PW* viii. 36.
[229] Ibid., p. 38.

the one hand Steele expressed complete confidence in the Queen, the British people, and the Constitution, while on the other he argued that the Church and the succession were in deadly danger from within and without. But if the nation were not secure when both parties were pledged to uphold the Act of Settlement, then what possible security could there be? Swift becomes playful, parodying Steele's alarmism. 'Must we send to stab or poison all the Popish princes, who have any pretended title to our crown?' he asks. He is unashamedly tearless over the plight of the Catalans, and crows with triumph that the fortifications of Dunkirk have been demolished. Steele was reduced to complaining that they had not been demolished properly–a sad retreat from the heady rhetoric of 'Britain expects . . .'.

However, at one point in the pamphlet Swift's political instinct failed him. In response to Steele's complaint that the Scots, with 'as numerous a nobility as England', were restricted to sending only sixteen peers to Westminster, Swift's animosity against the Scots bursts out.

> Their nobility is indeed so numerous, that the whole revenues of their country would be hardly able to maintain them according to the dignity of their titles.[230]

As a result the Scottish peers demanded the prosecution of the pamphlet. When Swift heard this, he called in the second edition of the pamphlet and cancelled the offending paragraph, but this only confirmed the Whigs in their determination to proceed. On 2 March Wharton rose in the Lords, with the pamphlet in his hand, to begin the attack. Justifying the call for prosecution, Wharton cited the Queen's speech, a speech polished and corrected by Swift himself, which had complained bitterly against seditious libels. With magisterial solemnity Wharton offered to read the offending paragraph, only to find that in his copy no such paragraph existed. Somewhat deflated, he had to call for a copy of the first edition for the attack to proceed. Oxford then gave a superb simulation of amazement, assuring the House that he knew nothing about the pamphlet at all. Nevertheless their Lordships agreed to condemn *The Public Spirit of the Whigs* as a 'false, malicious and factious libel'.

[230] Ibid., p. 50.

Swift's immediate instinct was for flight and the following night he received a cryptic little note from Oxford, written in a heavily disguised hand.

I have heard that some honest men who are very innocent, are under trouble touching a printed pamphlet. A friend of mine, an obscure person, but charitable, puts the enclosed bill in your hands to answer such exigencies as their case may immediately require, and I find he will do more, this being only for the present. If this comes safe to your hands it is enough.[231]

The enclosed bill was for £100. Two days later the printer and publisher of *The Public Spirit* were hauled before the Lords, but admitted nothing. On 9 March they were released from custody, but Wharton would not give up his prey so easily, and petitioned the Queen to offer a reward for the discovery of the author of the pamphlet. A few days later she consented, and Swift found himself a wanted man, with a price of £300 on his head.

Steele fared little better. Two days after the Whigs began their hunt for the author of *The Public Spirit,* the Tories in the Commons demanded the prosecution of Steele for his final issue of the *Englishman* in February. Undismayed, Steele made a brilliant little speech in his own defence before being expelled from the House after only a month's membership. It is perhaps worth noting however that whereas Swift's instinct, when threatened, was to hide, Steele's was to expose himself as a willing martyr.

Swift had another notable adversary in Gilbert Burnet, Bishop of Salisbury. Burnet appeared to Swift as the very type of the sanctimonious Low Church prelate, forever disguising Dissenting sympathies under a clamorous invective against Catholicism. In 1713 the Bishop published an *Introduction* to the long-delayed final volume of his *History of the Reformation.* It was a sensationalist work, full of the horrors of Counter-Reformation Popery, and launched, Swift believed, in consultation with Steele and the Whig lords. In December Swift published a *Preface* to Burnet's *Introduction,* ridiculing both the form and content of the Bishop's work. Mock-heroic and fairground analogies are used to reduce the status of Burnet's work from serious history to venal hack-work. But Swift's more

[231] *Corr.* ii. 12.

serious purpose is to decode the innuendoes of Burnet's alarmist rhetoric. He translates the Bishop's implied attack on the ordinary clergy into these 'plain terms'.

The bulk of the clergy, and one third of the bishops are stupid sons of whores, who think of nothing but getting money as soon as they can: If they may but procure enough to supply them in gluttony, drunkenness, and whoring, they are ready to turn traitors to GOD and their country, and make their fellow-subjects slaves.[232]

Burnet asserted that the nation was 'in more danger of popery than towards the end of King Charles the Second's reign' for, he argued, the Tories were so 'impiously corrupted in the point of religion, that no scene of cruelty can frighten them'. Nevertheless he did his best, describing at length the tortures of the Inquisition with 'halter, gibbet, faggots . . . slavery' and the burning of heretics at the stake in Smithfield. What angered Swift about this farrago was not simply its sensationalism, but the implication that most Tories had no doctrinal objections to Catholicism and hence could only be bribed to remain loyal Protestants by an appeal to their material self-interest. For Burnet's central argument was a warning to country squires that any new Catholic regime would seek to claw back its sequestered monastery lands and dues. One need not be a papist, Swift retorts, to feel that the impropriation of ecclesiastical dues and tithes into lay hands was unjust, often leaving the incumbent 'wholly at the mercy of his patron for his daily bread'. It is at this point, writing in defence of the poor, loyal lower clergy, that Swift's writing achieves a moral authority and eloquence in the face of injustice that is the hallmark of his major writings.

I take his lordship's bishopric to be worth near £2500 annual income; and I will engage, at half a year's warning, to find him above 100 beneficed clergymen who have not so much among them all to support themselves and their families . . . nothing is so hard for those, who abound in riches, as to conceive how others can be in want.[233]

Throughout the autumn every post from Dublin brought some fresh vexation. Swift's agent, the 'knave' Parvisol, was collecting far less in rents and tithes than Swift had anticipated, and he sent weekly reminders for him to be more vigorous in his

232 *PW* iv. 69. 233 Ibid., p. 65.

demands. Even at this distance he also did all he could to safeguard the rights, privileges, and traditions of the Dean and Chapter of St Patrick's. There was a vacancy in the choir, and Swift, though he claimed to understand music not at all, or rather 'like a Muscovite', insisted that musical considerations must take precedence over anything else in making the appointment. However, keen as he was to maintain the musical standards of the Vicars Choral, he was equally determined to keep their temporal ambitions in check. In December they attempted to renew a lease on some land near St Stephen's Green without seeking his consent. He wrote a stormy letter, threatening to sack any of them who went ahead with the scheme without his permission. He also sought to cancel the residual subordination of St Patrick's to its neighbouring cathedral, Christ Church. Traditionally the canons of St Patrick's were required to take turns at preaching in Christ Church. Swift himself had been fined for failing to comply with this expectation. Now, as Dean, he simply omitted to send any list of preachers to Bishop Wellbore Ellis. Ellis, no friend to Swift, had anticipated some such trouble as soon as he heard of his appointment. 'The new dean of St Patrick's is bringing a new broil upon us,' he groaned, having heard that Swift had 'given out he will break this usage'.[234] When the list of preachers from St Patrick's failed to appear, he wrote Swift a diplomatic note, in which he claimed the tradition could be traced as far back as the reign of Elizabeth, 'whatever opinion you may entertain of it'. Indeed, he argued, hardly any of the duties which the Dean and Chapter 'swore to observe' at their admission, was of longer continuance. Swift was not impressed. A year later he told Ellis that he considered this particular ancient usage to be now 'null and void'.

In Convocation Swift was proposed for the post of Prolocutor, and he was not disinclined to serve, yet would only accept nomination if he were sure to be successful. His humiliation of the spring was too recent for him to offer himself again for a dignity which he might not achieve. 'I should make the foolishest figure in nature', he wrote to Walls, 'to come over hawking for an employment I no wise seek or desire and then fail of it.'[235] Not surprisingly, such an attitude was rather too arrogant for the

[234] Landa, pp. 72-3. [235] *Corr.* i. 395.

bishops to tolerate, and his name did not go forward. In October, Constantine Phipps, the high Tory Lord Chancellor, gave Swift a friendly hint that 'your great neighbour at St Pulchers' [i.e. Archbishop King] is very angry with you. He accuseth you for going away without taking your leave of him and intends in a little to compel you to reside at your Deanery.'[236] This news of King's resentment came as little surprise to Swift–but the confirmation of it was peculiarly ill-timed for King himself. Only a week later Archbishop Marsh, the Primate of all Ireland, died, and Swift was immediately consulted by another Tory, Richard Nutley, Justice of the Queen's Bench in Ireland, about a successor. In fact, in Nutley's letter, the matter of finding a successor to Marsh ranks second to the opportunity for advancing Swift himself; 'if you can be tempted to part with your fine house in Dublin for an ill contrived one on a country bishopric'.[237]

Swift could not be so tempted, but thanked Nutley for his kind thoughts. He savoured the chance of deciding the fate of the man who had for so long patronized him. Yet it was less personal spite than political considerations which led Swift to pass over the claims of King to the Primacy. Three years before he had promised to 'drop in a word' for King in just such an eventuality, but King had given all too obvious indications of his Whiggish tendencies ever since. Instead Swift put forward the name of Thomas Lindsay, an undistinguished man, but a sound Tory. Nutley entirely agreed that this appointment would be 'of great advantage to the Queen's affairs at this juncture'. Swift could not resist letting King know what he had done: 'I should be thought a very vile man if I presumed to recommend to a (Minister) my own brother if he were the least disinclined to the present measures of her majesty.'[238] He also made sure that Lindsay knew of his instrumentality in the appointment. Lindsay was appropriately grateful, and offered to reciprocate whenever it might be possible. The choice of Lindsay precipitated a new outbreak of Whig belligerence in the Irish House of Commons. Instead of voting supplies for two years, they voted only for

[236] Ibid., p. 398.

[237] Ibid., p. 402. Nutley's High Tory sympathies earned him the nickname 'Nutbrain' among his enemies.

[238] Ibid., p. 419.

three months; they loudly called for the Lord Chancellor's dismissal, and proposed far-reaching changes in the legislative procedures in the country.

Writing with supreme detachment to King on New Year's Eve, after these tumults, Swift remarked, 'nothing could do more hurt to the Whig party in both kingdoms, than their manner of proceeding in your House of Commons'.[239] Yet, slyly recalling King's sanctimonious pose of universal tolerance, Swift hastened to assure the Archbishop that he acquitted him personally of such contumelious Whiggery. 'I conceive you to follow the dictates of reason and conscience,' he purrs, 'and whoever does that, will in public management, often differ as well from one side as another.' Swift was the master of the poisonous compliment. It must have given him pleasure to pen this, after having carefully shut the door on King's ambitions.

It was in the spring of 1714 that the meetings of the Scriblerus Club began. Swift's disappointment with the Society of Brothers had by now extended to include all such large and self-important gatherings. Yet he still saw it as his role and duty to foster an atmosphere in which right-minded writers might exchange 'hints' and find patrons. His hopes of establishing a literary academy had come to nothing, and the virulence of his feud with Steele meant that there could now be no hope of any co-operation across the party divide. He had already been in close consultation, if not co-operation with Arbuthnot in the production of the *History of John Bull*.[240] The Scriblerus Club came into existence when these two elder satirists joined forces with the younger writers, Pope and Gay. Both of these young poets had come to London as Whigs, and had sheltered briefly under Addison's aegis. However, in the literary-political tussles of 1713 they soon found themselves at odds with Addison's more favoured protégés, Tickell and Philips, and open rivalries quickly developed.

'The design of the Memoirs of Scriblerus', Pope subsequently recalled, 'was to have ridiculed all the false tastes in learning,

[239] Ibid., p. 425.

[240] The account of the collaboration by Swift and Arbuthnot offered by Teerink in his edition of *The History of John Bull* (Amsterdam, 1926) is highly misleading. For a more reliable discussion of this matter see *The History of John Bull*, ed. A. Bower and R. A. Erickson, Oxford, 1976.

under the character of a man of capacity enough that had dipped in every art and science, but injudiciously in each.'[241] He also added, modestly, that 'it was begun by a club of some of the greatest wits of the ages'. Pope had been nursing this idea at least since the summer of 1712, when he had submitted a suggestion to the *Spectator* for a monthly paper entitled *An Account of the Works of the Unlearned.* When Swift read Pope's *Windsor Forest* the following March, and detected its Tory leanings, he determined to win over this rising young talent. He began his campaign on an uncharacteristically gauche note, however, by offering Pope twenty guineas to change his religion from Catholicism to Anglicanism. His motive was not so much a missionary zeal as a wish to be able to offer Pope further patronage in the form of government sinecures. As a Catholic Pope was prohibited from holding any government post. It is a further insight into the extent to which Swift took 'nominal Christianity' to be the norm, that he should make such a crude attempt to buy Pope over. Pope turned aside the offer as skilfully as he could, only remarking that it was 'almost as many pieces of gold as an apostle could get of silver from the priests of old'.[242] Despite this false start the friendship developed and flourished, though Pope was still sometimes surprised by the Dean's 'odd, blunt way' as he described.

Dr Swift has an odd, blunt way that is mistaken by strangers for ill-nature. 'Tis so odd that there's no describing it but by facts. I'll tell you one that first comes into my head. One evening Gay and I went to see him; you know how intimately we are all acquainted. On our coming in, 'Hey-day, gentlemen' says the Doctor, 'what's the meaning of this visit? How come you to leave all the great lords that you are so fond of to come hither to see a poor dean?' Because we would rather see you than any of them. 'Aye, anyone that did not know you so well as I do might believe you. But since you are come I must get some supper for you, I suppose.'

No, Doctor, we have supped already.

'Supped already! that's impossible–why, 'tis not eight o'clock yet.'

Indeed we have.

'That's very strange. But if you had not supped I must have got something for you. Let me see, what should I have had? A couple of lobsters? Aye, that would have done very well–two shillings. Tarts–a shilling. But

[241] Spence, i. 56. [242] *Corr.* i. 412.

you will drink a glass of wine with me, though you supped so much before your usual time, only to spare my pocket?'

No we had rather talk with you than drink with you.

'But if you had supped with me as in all reason you ought to have done, you must then have drank with me: a bottle of wine–two shillings. Two and two is four, and one is five: just two and sixpence apiece. There, Pope, there's half a crown for you, and there's another for you, sir, for I won't save anything by you. I am determined.' This was all said and done with his usual seriousness on such occasions, and in spite of everything we could say to the contrary, he actually obliged us to take the money.[243]

Meetings of the Scriblerus Club were usually held in Arbuthnot's rooms at St James's, and two other men were regular attenders: Parnell and Lord Oxford. According to Pope's recollection, Oxford 'used to send trifling verses from court to the Scriblerus Club almost every day, and would come and talk idly with them almost every night'. Though Swift continued to enjoy Oxford's company, he could not approve such obvious neglect of duties. In May he wrote that the Queen's least disorder 'puts all in alarms, and when it is over, we act as if she were immortal. Neither is it possible to persuade people to make any preparations against an evil day'.[244] It was largely for Oxford's benefit, who liked to show off his skill in versifying in company with the professionals, that elaborate rhymed invitations to the full meetings of the club were exchanged during March and April.

> I will attend to hear your tuneful lays
> And wish your merits meet with one who pays–[245]

wrote Oxford, with unconscious irony, in reply to one such invitation. 'I really believe . . . he will prove a very good poet', remarked Arbuthnot of this and similar effusions. There were frequent meetings of the full club during the early spring, when the first 'hints' which were eventually to form the *Memoirs of Scriblerus,* and *The Art of Sinking in Poetry* were collected. The first break in the group came on 21 April, when Pope and Parnell left to work on Homer at Pope's family home near Windsor. From there they wrote deploring 'this miserable age . . . sunk between animosities of party and those of religion'.[246]

[243] Spence, i. 53-4.
[244] *Corr.* ii. 21.
[245] *The Memoirs of Martinus Scriblerus,* ed, Kerby-Miller, 1966, p. 357.
[246] Sherburn, i. 220.

Swift remained in London for another month, but then, 'weary to death of courts and ministers, and business and politics', he stole away to the Berkshire countryside, to stay with the Reverend John Geree, at his rectory at Letcombe Bassett. He reached there at the beginning of June, and, after a week's rest, wrote to Vanessa. He loved Geree very well, he wrote, 'but he is such a melancholy thoughtful man . . . that I shall soon catch the spleen from him'.[247] Meals were regular, but without wine, and life was uneventful, 'different in every circumstance from what I left'. He had come to the country to escape from the political turmoil of rumour and counter-rumour in London, but after only a week he was bored. 'I have a mind to steal to Ireland', he mused, insisting that he cared 'not threepence' to hear any news. 'Our strawberries are ruined . . . and our parson's pigs have done ten shillings worth of mischief upon goodman Dickens's corn,' he told Ford.[248] But he was not really in a mood to play the countryman. It was a time for telling home truths, after months of prevaricating. When Arbuthnot wrote in the middle of the month, Swift replied in proud and bitter tones.

> I have a mind to be very angry, and to let my anger break out in some manner that will not please them. . . . You are a set of people drawn almost to the dregs; you must try another game; this is at an end . . . writing to you much would make me stark mad . . .[249]

More than anyone else, it was Oxford who made Swift mad, and at the beginning of July he finally wrote to him. 'In your public capacity you have often angered me to the heart', he wrote, 'but as a private man, never once.'[250] He promises Oxford that 'posterity shall know that and more, which, though you and somebody that shall be nameless, seem to value less than I could wish, is all the return I can make you'. From his detached position at Letcombe, one can detect him assuming the mantle of an impartial observer, as he awaits the final catastrophe of the great historical drama of his age. His *History of the Four Last Years of the Queen* was already nearly completed. While at Letcombe he also wrote *Some Free Thoughts upon the Present State of Affairs*, and *Memoirs relating to That Change . . . in the Queen's Ministry in 1710*. After his return to Ireland, he embarked upon

[247] *Corr.* ii. 26.
[248] Ibid., p. 32.
[249] Ibid., p. 36.
[250] Ibid., p. 44.

his *Enquiry into the behaviour of the Queen's last Ministry,* but, with the exception of *Some Free Thoughts,* none of these historical writings was published during his lifetime. 'Will you give me leave to say how I would desire to stand in your memory,' he writes to Oxford, having indicated that he will take it upon himself to transmit Oxford's own memory to posterity.

As one who was truly sensible of the honour you did him, though he was too proud to be vain upon it. As one who was neither assuming, officious nor teasing, who never wilfully misrepresented persons or facts to you, nor consulted his passions when he gave a character. And lastly as one whose indiscretions proceeded altogether from a weak head, and not an ill heart.[251]

This is one of the many self-conscious attempts that Swift makes, throughout his life, to control his own reputation with future ages. He concludes with a gesture that courts obsequiousness in its parody of gallantry.

I have said enough, and like one at your levee having made my bow, I shrink back into the crowd.

Contemporary history is a difficult and dangerous genre. Bolingbroke for one resented Swift's attempt to give the finality of historical fact to a situation which, he hoped, was still fluid with possibilities. He made a 'splutter' over *Some Free Thoughts,* and wanted to 'soften a particular that seems to fall hard upon him'.[252]

Arbuthnot claimed that Oxford had been pleased by Swift's letter. Swift took leave to doubt that. 'I nipped him a little in it,' he told Ford.[253] He was soon complaining of being 'as splenetic as a cat' in his retirement. In London there were many rumours concerning his abrupt departure. Had he gone to Hanover to bring over the Elector? Had he gone to France to treat with the Pretender? Arbuthnot sought to quash all speculation by insisting that Swift's only design in leaving was 'to attend at full leisure to the life and adventures of Scriblerus'. He also did what he could to make this true, sending Swift endless notes of schemes and projects that cried out for Scriblerian ridicule.

[251] Ibid., p. 45. [252] *Corr.* ii. 71. [253] Ibid., p. 72.

'Pray remember Martin, who is an innocent fellow,' he counselled, trying to divert him from his brooding.[254]

Swift received almost daily bulletins from London, which did little to lessen his spleen. Bolingbroke (the man of mercury, the captain), was making a determined attempt to wrest power from Oxford (the Dragon, the Colonel). 'The Dragon dies hard, he is now kicking and cuffing about him like the devil,' reported Arbuthnot on 26 June.[255] A week later Ford told him that 'the Colonel' had given up the game for lost. 'I believe by next week we shall see L.B. at the head of affairs.'[256] Both the Duchess of Somerset and Lady Masham were now strongly supporting Bolingbroke's ambitions, and Lewis blamed Masham in particular for Oxford's decline. The Dragon's supporters 'call her ten thousand bitches & kitchen wenches', he wrote. Yet still Oxford clung on to power 'with a dead grip'. If only he had shown half as much determination in pursuing positive policies, Arbuthnot complained, the crisis would never have arisen.[257] But now he feared that 'the great person', that is the Queen herself, had grown tired of him. Arbuthnot himself remained on amicable terms with all parties and continued to lobby Lady Masham on Swift's behalf about the historiographer's place. Swift pretended to be angry at this. 'I am not asking a favour,' he insisted, and 'I would not accept it if offered'.[258] But these were no more than his usual defensive devices for forestalling disappointment. He was right to be sceptical. The place had in fact already been awarded to Thomas Madox. 'A worthless rogue, that nobody knows,' Swift complained with obvious bitterness. 'I am satisfied; let them take their own course; perhaps it may produce a history they will not like.'

He was not entirely without diversions at his retreat. Pope and Parnell visited him at Letcombe in the first week of July, when they amused themselves burning holes, by the use of a magnifying glass, in a House of Commons voting list. It was a schoolboyish game that suited well their hopeless sense of political impotence. Later in the month he visited Oxford with Lord Harley, and later still Vanessa paid him an unexpected visit–the 'Berkshire surprise' of their private system of references. As usual, he was not entirely pleased by her sudden

[254] Ibid., p. 42. [255] Ibid., p. 41. [256] Ibid., p. 51.
[257] Ibid., p. 57. [258] Ibid., pp. 62-3.

arrival. 'You should not have come by Wantage for a thousand pound. You used to brag you were very discreet; where is it gone?'[259] He endeavoured to satisfy his own creature comforts by supplementing his host's homely fare with some wine. In early July he had a hamper, filled with two dozen bottles of French red and a dozen of strong Arazina white, sent to him; and a fortnight later Ormonde added some bottles of burgundy. Suitably fortified, he watched with horrified fascination as events in London unfolded. He was now receiving several, often contradictory, reports every day. On the 17th Ford admitted that he couldn't imagine 'why we have no mischief yet. Sure we are not to be disappointed, at last, after the bustle that has been made.'[260] 'The Dragon' and 'his antagonist' continued to meet every day at cabinet, 'they often eat, and drink, and walk together' but on parting 'give one another such names, as nobody but ministers of state could bear without cutting throats'. A week earlier Swift had received a brief, ingratiating note from Bolingbroke himself, in which the man of mercury insisted that 'no alteration in my fortune or circumstances' could ever alter their friendship, and declared that it was 'amongst the eternal scandals of the government' that more had not been done for Swift.[261] Swift was unimpressed. 'I cannot rely on his love to me,' he told Vanessa candidly.

At last, on the 24th, Swift heard from Ford that 'everybody agrees . . . the great affair . . . will be tonight.'[262] He wrote back by return, telling Ford that he would abandon his plan to visit Pope. 'Entre nous, I will set out for Ireland tomorrow se'nnight. Say nothing of it yet.'[263] However, he also sent a brief note to Oxford, offering to attend him, in retirement, in Herefordshire. Oxford replied at once, and in apparent good spirits. Tomorrow morning he would be a private man again, and would like nothing better than Swift's company, for he believed that 'in the mass of souls, ours were placed near each other'. Swift, who was 'almost in rags', having already sent on his trunk, with all his clothes and linen in it, to Chester, for embarkation to Ireland, immediately changed his plans, and wrote to Walls to obtain a renewal of his licence of absence in England.

[259] *Corr.* ii. 123. [260] Ibid., p. 66. [261] Ibid., p. 61.
[262] Ibid., p. 78. [263] Ibid., p. 83.

Three days later, on 1 August, occurred an event which again changed his, and everyone else's plans. The Queen died. The Whigs rejoiced and, according to Ford, 'what is most infamous, stocks rose three per cent'. Lewis urged Swift to return to town, 'to see how the world goes, for all old schemes, designs, projects, journeys etc. are broke by the great event'.[264] The blow was most crushing to Bolingbroke who, after only four days of power, found himself replaced by a Whig-dominated council of Lords Justices. He wrote to Swift with a plangent complaint against fortune's wheel.

The earl of Oxford was removed on Tuesday, the Queen died on Sunday . . . what a world is this, & how does fortune banter us? . . . I have lost all by the death of the queen, but my spirit.[265]

He appealed to Swift to attack the Whigs as 'a pack of Jacobites', but it was a despairing gesture. 'It is true my lord', Swift agreed, 'the events of five days last week might furnish morals for another volume of Seneca.' As with Oxford, he determined to make his own–impartial–position clear. 'All I pretended was, to speak my thoughts freely, to represent persons and things without any mingle of my interests or passions.'[266] Yet he is not without hope. 'We are certainly more heads and hands than our adversaries.' It is clear from this and other letters that although Swift was preparing to go back to Ireland to take his oath of allegiance to the new monarch, he had high hopes of returning to London the following year. There is a strong valedictory note in all his exchanges from Letcombe, and, with retrospect, one is tempted to identify his renewed interest in history with a sense that a particular chapter in his life was closing. But he was still not aware of the finality that Queen Anne's death was to represent for himself and his fellow Tories. When he left Letcombe for Ireland on 16 August, he did not suspect that he was going into exile for the rest of his life.

He spent his last week in England settling his accounts for handkerchiefs and hats. In the City, stocks continued to rise as never before. He also wrote Vanessa a letter that mingles endearments with reproofs. Her mother's death had left her with several awkward debts to settle, and he had recently arranged a

[264] Ibid., p. 98. [266] Ibid., pp. 109-10. [265] Ibid., p. 101.

loan from Benjamin Tooke to help her. He promised to continue his care of her affairs, and hoped his stay in Ireland would not be long. While away he would write, but secretly, 'always under a cover'. He begged her to be similarly circumspect. 'I beg you will write nothing that is particular, but what may be seen, for I apprehend letters will be opened and inconveniences will happen.'[267]

Though he had sought to keep his departure a secret from his fellow Scriblerians, Arbuthnot at least came to hear of it. By way of a farewell he wrote this deeply moving testimony to the friendship that Swift's many private qualities could inspire in those who knew him well, a friendship that was to survive intact despite twelve years of separation.

> Dear friend, the last sentence of your letter quite kills me. Never repeat that melancholy tender word, that you will endeavour to forget me. I am sure I never can forget you, till I meet with, (what is impossible) another whose conversation I can so much delight in as Dr Swift's & yet that is the smallest thing I ought to value you for. That hearty sincere friendship that plain & open ingenuity in all your commerce, is what I am sure I never can find in another, alas. I shall want often, a faithful monitor, one that would vindicate me behind my back & tell me my faults to my face. God knows, I write this with tears in my eyes . . .[268]

[267] Ibid., p. 123. [268] Ibid., p. 122.

Part Four
THE DEAN AND THE DRAPIER

I

BACK IN Dublin, Swift felt himself in the grip of a deadening listlessness. 'I cannot think nor write in this country,' he complained. 'Being in England only renders this place more hateful to me, which habitude would make tolerable.'[1] Even his customary retreat of Laracor was denied to him, for his cabin there had gone to ruin. 'The wall . . . is fallen down, and I want mud to rebuild it, and straw to thatch it.' Worse, his neighbour Percival had seized on some of his land, and chopped down several trees. 'I have not fortitude enough to go and see those devastations,' he lamented.[2] As it was, conditions in the Deanery were quite bad enough. It took several weeks for him to make Stearne remove his cat 'who by her perpetual noise and stink must be certainly a Whig'.[3] He had no success at all in attempting to remove Stearne's elderly uncooperative housekeeper. His horses had all developed ailments, and he could find no hay for them. When finally he got some, his groom stacked it on a rainy day 'so that it is now smoking like a chimney'.[4] The groom promptly ran off, having first robbed Swift and several of his neighbours. A carpenter was hammering all day putting up library shelves, 'who has employed a fortnight, and yet not finished what he promised in six days'. At least Swift was not alone in his sufferings. Arbuthnot wrote that all the Queen's former servants 'are like so many poor orphans, exposed in the very streets, and those whose past obligations of gratitude and honour ought to have engaged them to have represented their case, pass by them like so many abandoned creatures'.[5] As often happened, depression led to illness, and Swift complained of 'perpetual colds and twenty ailments'. It was at this time that he wrote his dismal poem *In Sickness.*

[1] *Corr.* ii, 127.
[2] Ibid., p. 130.
[3] Ibid., p. 138.
[4] Ibid., p. 135.
[5] Ibid., p. 136.

> My state of health none care to learn;
> My life is here no soul's concern.
> And, those with whom I now converse,
> Without a tear will tend my hearse . . .
> Some formal visits, looks and words,
> What mere humanity affords,
> I meet perhaps from three or four,
> From whom I once expected more;
> Which those who tend the sick for pay
> Can act as decently as they.
> But no obliging, tender friend
> To help at my approaching end.[6]
> (ll. 5-8, 13-20)

What is most disturbing about the morbid self-dramatization of this poem is the attitude to Stella which it implies. No one is exempted from censure; he insists that 'no obliging tender friend' exists to comfort him. The wilful loneliness of these lines, and the pointed omission of her name, seem to hint at a certain resentment of her as the embodiment of the life to which he has been reduced. She became, in Murry's phrase, 'the symbol and the scapegoat of his discomfiture'.[7] Our view of Stella has become sentimentalized; the adoring pupil, the secret wife–these elements are the raw material of romance. Yet a sense of distance is unmistakable in the first year after his return to Ireland. Not the least of his reasons for bitterness at this enforced exile was his reluctance to forsake the flattering passion of Vanessa for the patient devotion of Stella. His despondency was unfair, he knew, but that merely drove him to express it more sulkily. Stella was exactly as he had moulded her. Trained from girlhood to play a particular role in his life, she had left home and family, accepting the rules of a relationship that had been established by him. Vanessa refused to be bound by his rules. She shocked, appalled, and infuriated him, yet her arrogant belief that 'common forms were not designed/Directors to a noble mind' fed his own fantasies of superiority. In losing Vanessa, Swift was losing the dream of power.

Not that he did lose her immediately. Within two months of his arrival in Dublin, she followed him across the Irish Sea. Ostensibly she came to settle some complications about her

[6] *Poems*, i. 203-4. [7] Murry, p. 277.

estate; but in fact she came to see him. This was exactly what Swift had feared most. The two women who had come to represent the contrasting lifestyles of London and Dublin were now together in one city, and he was forced to choose between them. Little wonder that he fell into sickness and dejection. There could be no doubt which choice he must make, for the ladies of William Street had provided him with emotional security for fifteen years, and although familiarity may have led him to regard them with less warmth, his need for them was still great. But he had also come to need the excitement of Vanessa. It was not, as Vanessa sometimes feared, merely to please her that he would often recall in affectionate detail the various episodes of their relationship. He had just as great a need for such an illuminated chronicle of happy days, which he stored and told over to himself in his isolation:

> the chapter of the blister, the chapter of madam going to Kensington, the chapter of the colonel's going to France, the chapter of the wedding with the adventure of the lost key. Of the strain, of the joyful return, two hundred chapters of madness. The chapter of long walks. The Berkshire surprise, fifty chapters of little times.[8]

This catalogue, from a letter written in 1720, is an inventory of such happy times, recollected in their own private code. He wrote to both women in specially coded languages, but there are important differences between the two. The 'little language' of the *Journal* is a system of games, riddles, and endearments delivered in the presence of a third person. Rebecca Dingley, an inseparable part of 'MD', makes all Swift's intimacies safe by turning them into play. The language to Vanessa, being genuinely private, uses its codes to enhance a truly lover-like feeling, with pet terms for special shared memories and experiences. Yet even when they seem to enhance this sense of intimacy, Swift's codes are part of an instinctive distancing process. He dramatizes the role that he is playing, which helps to save him from experiencing it. He has a range of subterfuges which assist him to conceal his feelings, and, characteristically, his only surviving declaration of love to Vanessa is contained in a letter written entirely in bad French.

[8] *Corr.* ii. 356. The colonel is Vanessa's brother Bartholomew who made a visit to France.

As soon as Vanessa reached Ireland she sent a servant inviting Swift to visit her at her house at Celbridge, eleven miles west of Dublin. The servant arrived just as Swift was leaving to spend a fortnight with his new friend Knightley Chetwode, and he declined. He would have declined anyway, since the invitation displayed Vanessa's usual lack of tact. 'I would not have gone to [Celbridge] to see you for all the world. I ever told you, you wanted discretion.'[9] Discretion was now more important than ever, and not only on Stella's account. Swift's relationship with King, never easy at the best of times, entered a new awkward phase. It would take several years of patient endeavour for him to win control of his Chapter and the approval of his parishioners. The last thing that he wanted was any rumour of scandal. He had warned Vanessa before that Dublin was such a 'nasty tattling town' that they could not see each other privately without causing gossip. Any meetings would have to be open, formal occasions at her town lodgings in Turnstile Alley. He saw her there once a week, but she was not satisfied with such a meagre ration of his time. She pestered him with impatient notes, to which he replied with a mixture of regret and asperity. 'I will see you tomorrow if possible, you know it is not above 5 days since I saw you.'[10] His coyness puzzled and annoyed her. Though she knew of the existence of Stella, she can have had no notion of her special status in Swift's life. He was put upon inventing 'a thousand impediments' to explain his inability to see her more often, when the truth was that he shrank back from all but the most pressing business, and had only these visits to cheer him. For her part she argued that she had more need of a trusted friend now than ever. She found herself involved in endless problems caused by 'cunning executors' and 'importunate creditors' and felt her spirits sinking. Once, she complained, she had had a friend that 'would either commend what I did or advise me what to do which banished all my uneasiness'. But now, 'when my misfortunes are increased by being in a disagreeable place among strange, prying, deceitful people', her friend had deserted her for no better reason than a fear of gossip. Again she taunted him to have the courage of his convictions and defy the gossips.

[9] Ibid., p. 142. [10] Ibid., p. 147.

You once had a maxim (which was to act what was right and not mind what the world said.) I wish you would keep to it now, pray what can be wrong in seeing and advising an unhappy young woman?[11]

Swift would not be shamed into jeopardizing his position. However, when Vanessa reminded him that poor Moll was equally involved in these difficulties, and even less able to fend for herself, he convinced himself that there was nothing reprehensible about seeing *both* Vanhomrigh sisters, and his visits increased. Almost immediately the rumours began, and he dashed off this panicky little note. 'This morning a woman who does business for me, told me she heard I was in–with one–naming you.'[12] There is barely suppressed anger in this letter for having put him at risk with her emotional blackmail. She had even been confiding in Archbishop King, whose deviousness Swift had every reason to suspect. If Vanessa really cared for him, he argued, she must surely understand the delicacy of his situation. But she could not understand, for he would not tell her openly what the major obstacle was. Instead he prevaricated with excuses which it was easy for her to ridicule. She became more, not less emphatic in her demands, precipitating crisis after crisis. Her letters became morbid, whining, and accusatory.

Well, now I plainly see how great a regard you have for me. You bid me be easy, and you'd see me as often as you could get the better of your inclinations, so much. Or as often as you remembered there was such a one in the world. If you continue to treat me as you do, you will not be made uneasy by me long. Tis impossible to describe what I have suffered since I saw you last, I am sure I could have bore the rack much better than those killing, killing words of yours. Some times I have resolved to die without seeing you more, but those resolves, to your misfortune, did not last long. . . . Oh that you may but have so much regard for me left that this complaint may touch your soul with pity.[13]

For several months she hung on, hoping to win a change of heart, moping about the great empty house at Celbridge, writing and tearing up a hundred pathetic letters to him. There has been some argument about why she kept copies of all the letters she sent to Swift. Some have suspected her of compiling 'evidence' as a way of forcing him into marriage. But the suggestion is as unlikely as the attempt would have proved

[11] Ibid., pp. 148-9. [12] Ibid., p. 149. [13] Ibid., p. 150.

unsuccessful. The reasons for her care are probably much simpler. Sometimes, when anxiously awaiting a letter, she would experience a dozen different emotions in an hour: anger, resentment, excitement, despair. In such 'splenetic' states she would draft and redraft umpteen replies, and when she came at last to send one of them, it was natural that she should take special care with it. Moreover, like Stella, she never entirely shook off the pupil/tutor aspect of their relationship. However much she might tease and wheedle him, she never lost a certain girlish fear of his schoolmasterly disapproval, and, as Murry observes, 'much of the peculiar pathos of her letters comes from this'.[14]

When it became obvious that Swift would not marry her, she began to make loud noises about leaving her large, dreary house, and the endless entanglements of the lawsuit. She decided to sell up and return to England, but first she asked Archbishop King what he thought of her plan. 'I confess I did not approve their resolution of selling their estate and turning it into money in order to their living in London,' he wrote. 'That way of living succeeded so ill with their mother that in my opinion it was advisable for them to change it.'[15] King makes no mention of Swift here, though he must have known of the Dean's friendship for Vanessa and her sister Moll. Still smarting from the loss of the Primacy, for which he held Swift largely to blame, King may well have made a point of championing Swift's cast mistress. The evidence is inconclusive, but it seems probable that the Archbishop would have been reluctant to see the removal of this particular thorn in Swift's side. While she remained in Dublin, the sick, sorrowful, and solitary figure of Vanessa was a walking accusation. By comforting her, King could gain an easy moral authority over his troublesome Dean.

The pressure was not all on one side, however, for Stella also had demands to make. Swift's homecoming was very different from the lyrical anticipation so often expressed in the *Journal*. Naturally she understood his deep sense of disappointment and betrayal at the sudden curtailment of all his ambitions, but when she tried to sympathize or to suggest diversions, she found herself shunned as merely another feature of his new bleak life. As the winter months brought no alleviation of his depression,

[14] Murry, p. 283. [15] Ibid., p. 282.

her own anxieties increased. Almost certainly it was at this time that she wrote the anguished little poem 'On Jealousy' which is traditionally ascribed to her, and which begins,

> Oh! shield me from his rage, celestial powers!
> This tyrant that embitters all my hours.[16]

However, she did more than merely confide her fears to paper. Dutiful and deferential as she was, she did not lack courage. Once, as Swift recalled with admiration, she had demonstrated the 'personal courage of a hero' in repulsing an attack on her home.

> The other women and servants being half-dead with fear, she stole softly to her dining-room window, put on a black hood to prevent being seen, primed the pistol fresh, gently lifted up the sash; and taking aim with the utmost presence of mind, discharged the pistol loaden with the bullets, into the body of one villain.[17]

It took no less courage to confront Swift with her suspicions of Vanessa, but at some point in 1715 it seems certain that she did just that. The gossip surrounding Vanessa made her own ambiguous relationship with Swift even more awkward than usual. In October Chetwode wrote teasingly to Swift that he need not fear being condemned to the company of his hard-faced housekeeper, 'since the world says you may command a very agreeable one and yet defer it',[18] though on second thoughts, he tactfully struck out these words. Swift who, when it suited his purposes, had been sensitive to the theoretical slight to Stella's reputation posed by Tisdall's attentions, could not simply shrug off the world's opinion in this matter any longer.

At this point the story becomes largely a matter of anecdote and tradition. Swift's 'secret marriage' to Stella is probably the most famous episode in the considerable volume of legendary Swiftiana. As with all the best of these legends, it is supported by a combination of circumstantial details, personal reminiscences, and intuitive commentary. There have been several critics of the following account, which cannot therefore be offered as a matter of fact, but remains still the most plausible explanation of the enigmatic relationship between Swift and Stella.

[16] *Poems*, ii. 738. [17] *PW* v. 230. [18] *Corr.* ii. 139.

As her apprehensions about her character's suffering seemed to weigh the heaviest on her mind, in order to put an end to those, he was ready to go through the ceremony of marriage with her, upon two conditions. The first was, that they should continue to live separately, exactly in the same manner as before: the second, that it should be kept a profound secret from all the world, unless some urgent necessity should call for the discovery. However short of Stella's expectations these conditions might be, yet as she knew the inflexibility of Swift's resolutions, she readily embraced them. And as it is probable that her chief uneasiness arose from jealousy, and the apprehensions she was under that he might be induced to marry Miss Vanhomrigh, she would at least have the satisfaction by this measure, of rendering such a union with her rival impracticable. Accordingly the ceremony was performed without witnesses, and the connubial knot tied in the year 1716, by Dr Ash, Bishop of Clogher, to whom Swift had been a pupil in the college; and who, as I have been informed, was the common friend to both, employed in the above negotiations. But the conditions upon which this union was formed were punctually fulfilled. She still continued at her lodgings in a distant part of the town, where she received his visits as usual, and returned them at the Deanery, in company with her friend, Mrs Dingley.[19]

As a corollary of this secret arrangement, Swift sought to extinguish Vanessa's residual hopes by encouraging the suit of Archdeacon Winter who had already proposed to her once. The Archdeacon was most gratified to receive Swift's endorsement, but Vanessa was not at all happy at the substitution, and lost no time in rejecting Winter for a second and final time.

[19] Sheridan, pp. 322-3. There is no consensus among biographers concerning this important episode in Swift's life. Ehrenpreis evades the issue altogether by simply asserting that Swift 'evaded marriage' (iii. 405)- For this he has been severely criticized by Denis Donoghue in a review for *TLS*, (10 Feb., 1984). 'The notion that Swift and Stella married in 1716 has strong credentials', Donoghue observes. 'It is mentioned . . . by Bishop Evans, Orrery, Delany, Deane Swift, Sheridan and Samuel Johnson.' On the other hand, J. A. Downie in his recent book *Jonathan Swift, Political Writer,* (1984) repeats John Lyon's refutation of the story of the marriage. 'There is no authority for it, but a hearsay story, and that very ill founded' (pp. 342-3). In Downie's view the fact that Stella continued to sign herself 'Esther Johnson, of the City of Dublin, *spinster*' in legal documents, including her will, is conclusive evidence against the marriage. 'Would the God-fearing Stella have signed her name thus if she had been married?' (p. 342). These are strong arguments, and cannot be easily dismissed; yet my own view is that the balance of probabilities still favours the notion of the secret marriage in 1716. Apart from the weight of circumstantial evidence and hearsay reports, it seems to me to confirm, rather than contradict the defensive network of reticences with which Swift surrounded this most private relationship.

II

IN HIS public life too, Swift was tormented by contrary feelings which became sharper and more urgent as the winter of 1714/15 wore on. During the seven weeks before George I arrived in England to claim his throne, the country was administered by a council of regents of whom at least four were Tories. Swift was not alone in hoping that all might not be lost, and that the King would 'not answer altogether the expectations of the Whigs'. For a few weeks Bolingbroke too fed himself on hopes, and even urged Swift to return. Swift refused, but only for the time being; 'I will not do it, till I am fully convinced that my coming may be of use.'[20] Bolingbroke's dismissal from office put an end to such hopes, and Swift wrote immediately to sympathize. Picturing Bolingbroke free at last to enjoy a pipe and a few glasses with the humdrum squires of Bucklebury, he struck the pose of one country gaffer to another. 'They tell me you have a very good crop of wheat, but the barley is bad.'[21] His main hope, he declared, was in a few months 'to grow as stupid as the present situation of affairs will require'. To that end, though denied the use of his dilapidated cabin at Laracor, he contrived to lead a 'country-life' in town, seeing no one but a few parsons, and going nowhere but to prayers. Yet he wrote to Arbuthnot in terms which led the doctor to assure Pope that his 'noble spirit' was undaunted, 'and tho' like a man knocked down, you may behold him still with a stern countenance, and aiming a blow at his adversaries'.[22] Bolingbroke's dismissal was the first indication of the new political realities, as the new Whig administration proceeded to settle some old scores. Writing to Ford, who still retained his post as Gazetteer, Swift confessed that he now expected 'the worst'.

> If anything witholds the whigs from the utmost violence, it will be only the fear of provoking the rabble, by remembering what passed in the business of Sacheverell.[23]

Secretive by nature, this instinct now became an obsession with Swift. He suspected, rightly, that his letters were being intercepted, and although he sometimes affected to laugh off such

[20] *Corr.* ii. 127.
[21] Ibid., p. 129.
[22] Sherburn, i. 251.
[23] *Corr.* ii, 131.

surveillance, he took care to suppress any dangerous political comments. Afraid to visit Vanessa on account of gossip, and afraid to write letters on account of interception, he grew increasingly morose, conversing only with God and the servants.

For want of better employment, he threw himself into his new role as Dean with a martyr's zeal. In the cathedral he redoubled his efforts to turn out his nine troublesome singing-men. At the Deanery he made a point of supervising all the stabling arrangements and became quite obsessive about the problems of finding adequate hay for his horses. He chivvied and chided his servants until they either did as he wished, or ran away. Always an exacting task-master, at this time he must have been almost unbearable. 'I am plagued to death with turning away and taking servants,' he groaned in October.[24] He did not trust his domestics to do anything right, and made himself more than ever a prisoner of his environment by supervising every detail of their activities. He spent weeks fussing over the bottling and storing of a great vessel of Alicante wine.

His only concession to the new reign was to light a bonfire on 19 October, the eve of the coronation, 'to save my windows' from the possible violence of a Whig mob.[25] Officially the event was celebrated throughout Dublin with fanfares, fireworks, and military honours. His sense of frustration turned naturally into attacks on those around him, and, after his servants, those most likely to suffer his barbs were his parishioners at Laracor. 'I can never imagine any man can be uneasy that has the opportunity of venting himself to a whole congregation once a week,' wrote Arbuthnot.[26] 'You may pretend what you will, I am sure you think so too.' Among his parishioners, Swift was aggrieved at Percival for encroaching on his land, but furious with Langford for continuing to support a Presbyterian conventicle at Summerhill. Swift took this as a personal affront and threatened to take all necessary means to stamp out the 'growing evil' of the conventicle.[27] But it proved to be an empty threat. Two years later, at Langford's death, the conventicle still flourished, and he left a substantial bequest for its continuance.

In November Swift travelled north to visit Chetwode at Wood-

[24] Ibid., p. 138.
[25] Ibid., p. 138.
[26] Ibid., p. 144.
[27] Ibid., p. 141.

brooke, where he enjoyed the country pleasures that the ruination of Laracor denied him. From the safety of Woodbrooke he wrote back to supervise the decorative work being done at the Deanery. 'Why the whole room painted? Is it not enough to have only the new panels & edges of the shelves painted? Do what you will, but pray let it be done before I come, that the smell may go off.'[28] He also continued to worry about his wine. Had the grooms taken care to prevent the frost getting at it? Had they avoided shaking the vessel? Gradually Chetwode and his wife managed to divert his mind from these obsessions. He went riding whenever he could, and approved of the plentiful woods and hedgerows in the area. In summer it would be 'a sort of England,' he thought, 'only for the bogs'.

He returned to Dublin to find more dismal news from London. Arbuthnot was to lose his position at Chelsea College. Ford had lost the *Gazette*, and there was every indication that Oxford would soon be 'struck at'. Behind these events were yet darker shadows. Ford was planning to go to France. That could mean only one thing. Arbuthnot's letter concluded with an ominous paragraph, referring to the Pretender's declaration of his 'disappointment' at the 'deplorable accident' of Anne's death.[29]

Jacobitism was the dreadful bogey of the period. It stimulated similar fears and antagonisms in the God-fearing Protestants of England to those that Communism stirred in those of McCarthyite America. At his great distance from the political centre, rumours were magnified and distorted by the time they reached Swift. In March 1715 he told Lord Harley that 'since I left England I have not seen a newspaper, nor have above three or four times heard anybody talk of what passes in the world'.[30] What he did hear dismayed and disturbed him. The Whigs were gradually strengthening their stranglehold on the administration. He had recently received a brief and urgent cautionary note from Lewis, 'that if you have not already hid your papers in some private place in the hands of a trusty friend, I fear they will fall into the hands of your enemies'.[31] Prior's papers had already been seized, and he himself had been impeached. From the highest to the lowest, Tories were being removed from office.

[28] Ibid., p. 145.
[29] Ibid., p. 144.
[30] *Corr.* ii. 159.
[31] Ibid., p. 156.

Chetwode was removed from the commission of the peace. 'I am sorry Tories are put out of the King's peace,' Swift complained. 'He may live to want them in it again.'[32] Nor was the Church free of royal and Whiggish intervention. In December the King issued directions for the 'preserving of unity in the Church . . . and also for preserving the peace and quiet of the State'. Swift regarded this as tantamount to a provocation. 'I saw in a print that the King has taken care to limit the clergy what they shall preach; and that has given me an inclination to preach what is forbid.'[33] Yet, for the moment, he felt too vulnerable to risk such a confrontation. Writing to an acquaintance in Italy, Swift feared that the Whigs 'sont tout à fait résolus de trancher une demy douzaine de têtes des meilleures d'Angleterre'.[34] In his alarmist mood, anything seemed possible. He asked Lord Harley to promise

> that whenever these gentlemen in power shall think fit to destroy the Church, and abolish bishoprics and deaneries as wicked and useless; you will settle on me £50 a year to live in Guernsey; for there I am determined; because wine and vittals are cheap in that island. . . . Perhaps I am not so much in jest as you may believe; for there is nothing too bad to be apprehended in my opinion, from the present face of things.[35]

He retreated further into his clerical chores. 'My notion', he told Chetwode, 'is that if a man cannot mend the public, he should mend old shoes if he can do no better; and therefore I endeavour in the little sphere I am placed to do all the good it is capable of.' What he did *not* want, was to get drawn into the brawling side-show of Irish politics. 'I hope I shall keep my resolution of never meddling with Irish politics,' he declared firmly in August 1714.[36] It took him six years to find a reason to change that resolution.

Suddenly, on 2 March 1715, Bolingbroke fled to France, and by his action split the Tory party down the middle. Letters sent between Swift and his Tory friends became even more guarded and discreet in their handling of political events. Chetwode, who kept in touch with several leading Tories in England, sent this cryptic summary of recent developments to Swift in April.

[32] Ibid., p. 154. [33] Ibid., p. 155. [34] Ibid., p. 157.
[35] Ibid., p. 160. [36] Ibid., p. 127.

Tommy, little Tommy, pretty Tommy is gone like Judas *ad locum proprium suum*; Galloway also stone dead; you and Gay Mortimer have brought a rot among the wicked.[37]

Decoded, the references are to the deaths of Wharton (Tommy) and Burnet (Galloway), 'Mortimer' is Oxford in whom neither Swift nor Chetwode now placed much reliance. Letters sent directly from England, however carefully dispatched, fared less well. At least two letters to Swift sent in May 1715 found their way into the hands of the Irish executive; more specifically into the hands of the two acting Lords Justices, Archbishop King and the Earl of Kildare, who sent them on to the Lord Lieutenant Sunderland, relaxing at Bath. One of the letters, from the printer John Barber, included some fairly incriminating remarks. 'Our friends here and all the kingdom over are in great spirit,' he wrote, 'we shan't always groan under the burden.'[38] Barber dismissed Oxford as 'the most contemptible creature in the kingdom' and was contemptuous of such other Tory weaklings as Prior and Arbuthnot for allowing their papers to be seized. Like a true fanatic he believed that the Whig enemy was in disarray, and that one last effort by Bolingbroke, Ormonde, and their friends would bring victory.

Ld Wharton's death hath extremely mortified the Whigs. Sunderland is very ill, has been mad for some time, and is going to the Bath. Walpole is in a very bad way, and Stanhope is the bully . . .

The other intercepted letter was scarcely less incriminating, since it came from Ormonde who fled to France the following month to join Bolingbroke. Swift affected to be amused by these interceptions. 'It is said there was another packet directed to me, seized by the government, but after opening several seals it proved only plumcake.'[39] The newspapers were full of rumours naming him as a wanted man, with a price of five hundred pounds on his head. King who, in his capacity as a Lord Justice, authorized the interceptions, summoned Swift before him to assure him of the efforts he had made to prevent him being sent for trial in England. In this he was not entirely candid. In May he had written to Secretary Stanhope, enclosing some letters and pamphlets which were due to have been delivered to Swift, and

[37] Ibid., p. 165. [38] Ibid., p. 169. [39] Ibid., p. 172.

which were, in King's opinion, treasonous. In Swift's favour, however, was the fact that nothing treasonable could be directly attributed to him. On the contrary, as King admitted, the letters 'seemed to acquit the Dean, by complaining of his not writing, which they interpreted as a forbidding them to write'.[40] Nevertheless, a pretty formidable circumstantial case could be mounted against Swift, and King was not slow to demonstrate the power he now had over him. However, in Swift's account of their interview, he refused to be cowed by King's veiled threats.

If I had been called before them, I would not have answered one syllable or named one person. He said that would have reflected on me. I answered I did not value that, and that I would sooner suffer more than let anybody else suffer by me, as some people did.[41]

As the Hanoverian Government attempted to root out all Jacobite sympathizers rumours were rife and informers numerous. 'It seems there is a trade going of carrying stories to the government and many honest folks turn the penny by it.'[42] In London it was rumoured that 'the famous libeller Swift' had gone into hiding, and so on 27 June he made a show of attending the court at Dublin Castle 'on purpose to show I was not run away'.

In the face of so many rumours, Swift sought to content himself with ignorance. 'I hate your account of one man, who saw another man, who saw a letter,' he told Chetwode.[43] If there were 'fewer rascals in the offices for letters' he might stand a chance of forming a judgement on present affairs. As it was, he did not greatly like what he heard, 'I am sometimes half-mad, he confessed. 'I should indeed think less, unless I could think better, or serve the world more.' In early July a letter did get through with grim news of Whig determination 'to carry matters to the highest extremes'. Prior, he heard, was kept in close confinement 'to force him to accuse Lord Oxford, though he declares he knows nothing, and that it is thought he will be hanged if he will not be an evidence'. On 16 July Oxford was committed to the Tower where he remained for two years, until proceedings against him were finally dropped. Swift wrote to him as soon as he heard of his imprisonment.

[40] *Corr.* v. 233.
[41] *Corr.* ii. 173.
[42] Ibid., pp. 174-5.
[43] Ibid., p. 179.

I . . . take the liberty of thinking and calling you, the ablest and faithfullest minister, and truest lover of your country that this age hath produced. And I have already taken care that you shall be so represented to posterity, in spite of all the rage and malice of your enemies.[44]

The letter has the formal defiance of a public statement addressed to the rascals in the Post Office, but also signifies to Oxford a reversal of roles; from now on his reputation will be in Swift's hands, just as Swift's hopes were formerly in his.

Even when Swift heard of Ormonde's flight, he could not bring himself to believe that he had gone to join the Pretender. The story was that 'upon his first arrival at Calais he talked of the King only as Elector etc.', but Swift insisted, 'this is laughed at, and is indeed wholly unlike him'.[45] However, when he added that some of Ormonde's friends still believed the Duke to be in England, he was guilty either of wishful thinking or of seeking to assure the Post Office of his own ignorance of Jacobite plans. By now fears of an invasion were being openly expressed. The Whigs referred to the Tories generally as Jacobites, a charge which every new defection seemed to confirm. The Habeas Corpus Act was suspended, as the Whigs put themselves upon a war footing. Arbuthnot alone, in this mad world, seemed cheerfully confident that 'the constitution is in no more danger than a strong man that has got a little surfeit by drunkenness'.[46]

Despite Swift's strong advice, Chetwode was determined to play his part on the political stage. In August, 'perplexed beyond measure' by a conversation with 'an old lady'[47]–code for a Jacobite contact–he finally resolved to follow her, but begged for Swift's opinion. Yet he feared that Swift would express himself in too 'politic and dark' a manner. Two months later Chetwode had taken the plunge, and was in England, from whence he continued to bombard Swift with letters which received no answers. Finally Chetwode expressed the 'ugly suspicion' that Swift had turned against him. Swift repudiated this charge indignantly. 'Generally I expect to be trusted, and scorn to defend myself.'[48] He protested that he knew nothing of their codes and signals and replied in deliberately loyal and ignorant terms.

[44] Ibid., p. 182. [45] Ibid., p. 183. [46] Ibid., p. 185.
[47] Ibid., p. 186. [48] Ibid., p. 191.

We are here in horrible fears, and make the rebels ten times more powerful and the discontents greater than I hope they really are; nay it is said the Pretender is landed or landing with Lord knows how many thousand.[49]

Swift had no intention of becoming implicated in an adventure for which he had only contempt. Already, in the Irish Parliament, all those who were not party-line Whigs were stigmatized as 'jacobites and dogs'. 'We are as loyal as our enemies, but they will not allow us to be so,' he protested angrily. Again, this may be a public protest, but it expresses genuine feelings. 'Pray keep yourself out of harm's way' was his advice to Chetwode, and was the policy which he adopted himself.

Bolingbroke's flight was an enormous tactical blunder, and could only be justified by his honest belief that his blood 'was to be the cement of a new alliance'. Oxford who had, prima facie, as much to fear, made an impressive contrast by staying to face his accusers. Having fled, Bolingbroke should at least have stuck to his original plan of avoiding the courts of Louis XIV and the Pretender, and retiring to his country estate at Bellevue, near Lyons. When in June he was impeached for treason his patience broke, however, and in a gamble which he was to regret ever afterwards, he accepted the post of Secretary of State to the Old Pretender.

From the start the Jacobite rising of 1715 was marked by misjudgements, mistrust and just plain bungling. Relations between Bolingbroke and James were never cordial, and money for their plans was scarce, particularly after the death of Louis XIV. In October and December Ormonde made two unsuccessful attempts to land an invasion force in Devon, while in November the Scottish Jacobites were defeated at Sheriffmuir.[50] By the end of December, when James himself landed at Peterhead, he found only 4,000 demoralized troops and his cause already in ruins. In March he dismissed Bolingbroke, and by April the whole invasion scare was over. Within the year Bolingbroke was seeking to ingratiate himself with the Hanoverians by offering to betray his former Jacobite associates.

Looking through Swift's deliberately restricted correspon-

[49] Ibid., p. 190.

[50] The English Jacobites were simultaneously defeated at Preston.

dence during the two years following Queen Anne's death, one comes to the uncomfortable discovery that, with the exception of Arbuthnot, Vanessa, and Archdeacon Walls, all those who wrote to him were either active or passive sysmpathizers with the Jacobite cause. It is no wonder that he was regarded with great suspicion, and it would be naïve to accept that he had no inkling of Jacobite plans. Yet there is nowhere, in any of his writings, the slightest evidence that he approved of Jacobitism. On the contrary, the defence of the Church of England, which Bolingbroke sought to impose as a policy upon a reluctant James, was a crucial element in Swift's philosophy. What for Bolingbroke was a mere political necessity, was for Swift an axiom of faith that transcended belief in God Himself.

The report of the Secret Committee appointed to investigate the peace negotiations was published in June, and Swift was pleased to see that not 'one word is offered to prove their old slander of bringing in the Pretender'. Writing to Pope he expressed his personal dismay at the disaster that former friends had brought upon themselves, and upon the whole Tory party.

> You know how well I loved both Lord Oxford and Bolingbroke, and how dear the Duke of Ormonde is to me: do you imagine I can be easy while their enemies are endeavouring to take off their heads.'[51]

In earlier days he would not let men run mad without warning them but now he was tired, and believed he must let them suffer the consequences of their own madness. He envied Pope his escape into Homeric lands, and sent this self-portrait to show that he too was a man far removed from state concerns.

> I live in the corner of a vast unfurnished house; my family consists of a steward, a groom, a helper in the stable, a footman, and an old maid, who are all at board wages, and when I do not dine abroad, or make an entertainment, (which last is very rare) I eat a mutton pie, and drink half a pint of wine; my amusements are defending my small dominions against the Archbishop, and endeavouring to reduce my rebellious choir.[52]

In Paris, Bolingbroke was offered 'all suitable hope and en-

[51] *Corr.* ii. 176. [52] Ibid., p. 177.

couragement'[53] of his eventual political rehabilitation, in return for betraying his Jacobite colleagues. His only stipulation was that this second treachery should be kept secret, 'to preserve his reputation with his friends'. So, with Government approval, he began earnestly dissuading his old Tory contacts from the Jacobite cause. In October 1716 he told Swift that the Tory Jacobites

> are got into a dark hole where they grope about after blind guides, stumble from mistake to mistake, jostle against one another, & dash their heads against the wall, & all this to no purpose for assure yourself that there is no returning to light, no going out but by going back.[54]

There were many who daily expected that Swift's name would be included among those whom Bolingbroke was denouncing. In November Archbishop King concluded a long letter with this insidious afterthought.

> We have a strong report that my Lord Bolingbroke will return here & be pardoned, certainly it must not be for nothing. I hope he can tell no ill story of you.[55]

Swift reacted fiercely to this innuendo.

> I should be sorry to see my Lord Bolingbroke following the trade of an informer, because he is a person for whom I always had and still continue a very great love and esteem. For I think as the rest of mankind do, that informers are a detestable race of people, though they may be sometimes necessary.[56]

His defence of his own conduct contains a not-so-veiled attack on the hypocrisy of the man who had been opening his letters all year.

> I am surprised to think your Grace could talk or act or correspond with me for some years past, while you must needs believe me a most false and vile man, declaring to you on all occasions my abhorrence of the pretender, and yet privately engaged with a ministry to bring him in, and therefore warning me to look to myself and prepare my defence against a false brother coming over to discover such secrets as would hang me.

[53] PRO State Papers Foreign (France), 78/160, f. 256. See H. T. Dickinson. *Bolingbroke*, 1970, p. 143.

[54] *Corr.* ii. 219.

[55] Ibid., pp. 227-8.

[56] Ibid., pp. 237-8.

Finally he makes this formal declaration of innocence.

Had there been ever the least overture or intent of bringing in the Pretender during my acquaintance with the ministry, I think I must have been very stupid not to have picked out some discoveries or suspicions.

One has the distinct feeling that Swift protests too much here. He found himself in the humiliating position of having to plead stupidity as a defence against the charge of treachery; only as a fool could he escape being labelled a knave. The realization was dawning upon him that there *had* been secret negotiations with James, from which he had been excluded. Both his honour and his status in the previous ministry were damaged by this recognition. Defending himself, he realized, could not simply be the same thing as defending the ministry. His concluding expression of loyalty is made in purely personal terms, with a weary disillusionment at finding that all his political mentors had, in one way or another, betrayed him.

If I were of any value, the public may safely rely on my loyalty, because I look upon the coming of the Pretender as a greater evil than any we are like to suffer under the worst Whig ministry that can be found.[57]

Respect for himself, not for King, draws this statement from him. He was a man accused, not openly, but by rumour and association. If defending himself meant accepting the fool's cap, then he was prepared to wear it as a badge of integrity.

III

SWIFT WAS determined at least to be master in his own house, and took unusual care to defend all the dignities and privileges that belonged to him as Dean. On several occasions this led to legal struggles with others, notably with King, who 'sometimes attempts encroachments on my dominions'. Swift's first trial of strength was with his Vicars Choral and he adopted an uncompromising tone.

I will immediately deprive any man of them who consents to any lease without the approbation aforesaid, and shall think the Church well rid of such men.[58]

[57] Ibid., p. 239. [58] *Corr.* i. 427.

'I am well instructed in my powers,' he warned them. A man trained up in the power struggles of St James's was not going to be out-manoeuvred in St Patrick's. 'I hear you will meet with great difficulty with your chapter,' observed Chetwode in December 1714,[59] but Swift affected to think otherwise. 'I hear they think me a smart Dean, and that I am for doing good.'[60] He must have made expert use of his deaf ear to hear such reports.

Both Swift and King were stubborn on points of dignity and precedent, and both had a tendency to prefer confrontation to compromise. As Metropolitan King had some influence in St Patrick's, and as a Lord Justice he had considerable powers of appointment, which he used to weaken Swift's authority within his own cathedral. King urged the appointment of Theophilus Bolton as Chancellor of St Patrick's, stressing Bolton's exemplary piety. More candidly, however, he told the Lord Lieutenant:

> I believe your Excellency knows Dr Swift the Dean of my Cathedral and what I am to expect from him, and except I have such a person as Dr Bolton in a station in the chapter, I am afraid my affairs there will not go very well.[61]

Bolton's appointment was approved, much to Swift's dismay, and he remained at St Patrick's until becoming Bishop of Clonfort in 1722. At that time Swift wrote that he 'was born to be my tormentor, he ever opposed me'.[62] King followed this success by returning to the old vexed issue of St Nicholas Without. Swift had earmarked the benefice for his friend Thomas Walls, but King favoured a man whose main virtue was 'not running in with the late managers' (i.e. the Tories). Swift described King's candidate as 'a man of very low parts . . . and party-mad into the bargain', and wrote defiantly to Walls in May 1715, 'they shall neither fool me nor you, at least in this point'.[63] However, as the Jacobite scare grew, Swift became more defensive and Walls was forced to accept a small sinecure in Co. Cork. Swift wrote in protest to King, complaining that Walls was being penalized merely for his friendship with Swift.

[59] Ibid., ii. 152.
[60] Ibid., p. 154.
[61] Landa, p. 78. King's correspondence, 29 Oct. 1714.
[62] *Corr.* ii, 434.
[63] Ibid., p. 170.

Those who are most in your confidence make it no manner of secret, that several clergymen have lost your Grace's favour by their civilities to me. I do not say anything of this by way of complaint, which I look upon to be an office too mean for any man of spirit and integrity, but merely to know whether it be possible for me to be upon any better terms with your Grace. . . . I cannot but think it hard, that I must upon all occasions be made uneasy in my station . . . by those who say they act under your Grace's direction.[64]

It is a partial admission of defeat for Swift to bring their covert antagonism into the open like this, yet he is still careful to leave a good deal unstated in this complaining non-complaint. His boast of having been, under the previous ministry, 'a continual advocate for all men of merit without regard of party' may have sounded rather disingenuous to the man who had lost the Primacy partly through Swift's opposition. King now sought to revive a prebend, dormant since the time of Elizabeth, as a means of insinuating another Whig into the Chapter of St Patrick's. 'He may have a thousand Whig curates in Dublin,' grumbled Swift. In a later letter he attempted to clamp on a mask of reconciliation, but his true feelings kept escaping in a series of bitter asides. Recalling that the Bishops of Ossory and Killaloe were rewarded for their part in obtaining the First Fruits, while actually hundreds of miles away, he observed, sardonically, 'it seemeth more reasonable to give bishops money for doing nothing, than a private clergyman thanks for succeeding where bishops have failed'.[65] Referring to Charles Whittingham, who had recently deprived Walls of another benefice, and who, in his ordination sermon, had called the clergy 'a thousand dumb dogs', Swift remarked: 'yet no notice at all shall be taken of this, unless to his advantage upon the next vacant bishopric, and wagers are laid already . . . but I forget myself'. He could not resist these barbs, intimating that although he had lost many battles, he was far from conceding the war. King replied with a long letter in which he argued that it was in their joint interest 'that there should be a good correspondence between us'. While commiserating with Swift over the First Fruits, he urged him to 'be not concerned . . . tis more honour for a man to

[64] Ibid., p. 206. [65] Ibid., p. 221.

have it asked why he had not a suitable reward to his merits, than he was overpaid'.[66]

Within a Chapter of twenty-three members, both sides could usually rely on ten votes. In such a finely balanced situation the revival of a long-dormant prebend could take on a decisive significance. Matters came to a head in 1716 with a direct challenge to Swift's power of veto. It was then that Swift appealed to Atterbury, complaining that a 'ringleader' among the Whigs in his Chapter 'has presumed to debate my power of proposing, or my negative, though it is what Deans of this cathedral have possessed for time immemorial'.[67] Arguing that the constitution of St Patrick's was modelled on that of Salisbury, Swift sought Atterbury's support, but the reply was hardly encouraging. Atterbury told him that the Dean of Salisbury had 'no extraordinary powers or privileges'[68] and advised against an argument based on precedent. 'A nice search into the peculiar rights of the Dean of Sarum will be needless, if not mischievous to you.' He cautioned Swift against insisting too forcibly on his powers, and particularly against taking the matter to the courts which 'it is ten to one (especially as things now stand) will go against you'. Swift was frankly disappointed by this cold douche of prudence. Replying to Atterbury he shifted tack, no longer relying on the precedent of Salisbury, but on 'several subsequent grants, from Popes, Kings and Archbishops' which had given the Dean of St Patrick's 'great prerogatives'.

The Dean can suspend and sequester any member, and punishes all crimes except heresy. . . . No lease can be let without him. He holds a court-leet in his district, and is exempt from the Lord Mayor, &c. No chapter can be called but by him, and he dissolves them at pleasure. He disposes absolutely of the petty canons and vicars-choral places.[69]

Yet even among these rights, so grandly asserted, he does not go so far as to claim an absolute veto, and after listing his many privileges, agrees to observe Atterbury's general advice to do his utmost to 'divert this controversy'. The struggles within the Chapter developed into a weary war of attrition. In July 1717 Swift again complained to Atterbury: 'to oppose me in everything relating to my station is made a merit in my chapter'.[70]

[66] Ibid., p. 224. [67] *Corr.* ii. 194. [68] Ibid., p. 195.
[69] Ibid., p. 198. [70] Ibid., p. 279.

Nevertheless Swift fought back, and as his personal prestige began to grow, gradually regained the initiative that King had stolen from him. By 1721, when the Dean of Ossory wrote to him for advice on handling a dispute within his own Chapter, Swift felt sufficiently confident to reply with a magnificent swagger. 'It is an infallible maxim, that not one thing here is done without the Dean's consent.'[71] He advised Ossory to argue that his constitution was modelled on that of St Patrick's, in which case, Swift would undertake to provide him 'with power and privileges enough'. He exaggerates, of course, but his exuberance is due to a sense of achievement at a victory finally won after a long and hard-fought campaign.

Swift's first two years back in Ireland were difficult years indeed. His emotional problems with Vanessa, the political embarrassment of the Jacobite rising, and the continual conflicts within his Chapter, left him feeling besieged and isolated. Yet gradually a new pattern of friendships developed to provide the social support that he needed. The families of the Walls in Dublin, the Raymonds at Trim, the Grattans at Belcamp, and the Chetwodes at their two country homes at Martry and Woodbrooke could be relied upon for hospitality and sympathy, at least until Chetwode left to join Ormonde. Swift continued to enjoy the exercise of walking and riding, and together with Chetwode, a lover of horseflesh, he discussed plans for building up his stable. In April 1715 Chetwode told him of a grey gelding for sale, 'strong, young, tolerably handsome, trots well, has good spirit'.[72] The horse belonged to a mad fellow, but not so mad that he didn't know how to drive a hard bargain. He wanted £16 for the horse. 'I believe the fellow rather thinks me mad than is mad himself; sixteen pounds–why it is an estate; I shall not be master of it in sixteen years.'[73] Madness had meanwhile struck nearer home. Poor Joe Beaumont, the draper at Trim, had lost his reason as a result of researches into the longitude, thereby confirming all Swift's suspicions of the new science. 'For God's sake do something to comfort Joe,' wrote Chetwode, 'keep him from melting his tallow . . . do anything to keep him alive.'[74] While Swift was in Dublin, Chetwode saw to it that 'the Dean's

[71] Ibid., p. 377.
[72] Ibid., p. 162.
[73] Ibid., p. 164.
[74] Ibid., p. 165.

field' at Athy flourished, clearing out a mile-long river walk, and supervising the mowing of the lawns.

In the spring of 1716 Swift was back in Laracor, where he found his gardens and grove to be 'sadly desolate'. He himself was suffering from a strained and enflamed thigh, caused by too much riding. From Trim, where the farmers were full of complaints about the weather as excuses for not paying their tithes, he went on to Martry, and from thence to visit Rochfort at Gaulstown, where he remained till the end of June. His marriage probably took place at Clogher sometime between 30 August and 4 October, when there is a gap in the correspondence. To mark the event Swift arranged to purchase from Percival the twenty acres that he needed to complete his grove at Laracor, using funds from the First Fruits for the purpose. He returned to Dublin to find a letter from Pope in which his young friend told him he looked on a friend in Ireland 'as upon a friend in the other world whom (popishly speaking) I believe constantly well-disposed towards me'.[75] It is a letter full of a healthy ecumenical raillery, and Swift replied in a similar vein, showing that he maintained a lively interest in literary matters, even while complaining that 'the scene and the times have depressed me wonderfully'. He turned Pope's conceit into a joke: 'You are an ill Catholic, or a worse geographer, for I can assure you, Ireland is not paradise', and in a throwaway conclusion he gave Gay the idea for his *Beggar's Opera*. 'What think you of a Newgate pastoral, among the whores and thieves there?'[76]

Swift spent much of the autumn at Trim, completing his transactions with Percival over the land at Laracor. Percival drove a hard bargain and Swift redoubled his efforts to gather in all the rents and tithes that were due to him. His letters to Walls at this time are full of urgent pleas to 'get in my arrears'. Both his agents, Gillespy and Parvisol, seem temporarily to have left him under the pressure of his continual complaints against their 'scurvy' inefficiency, and gradually Swift took more and more of his business affairs into his own hands, drawing up exact lists of his tenants and debtors. On the last day of the year he finally tracked down the Murphys of Summerstown, and relieved them

[75] *Corr.* ii. 211.
[76] Ibid., pp. 214-15.

of a long overdue bond for £11. All the while his constant refrain was 'I have not a farthing to bless myself'.

The ladies were away visiting Stearne at Dromore for Christmas. Their letters were full of reproaches for his absence, but as he explained, all the ridable horses seemed to have died. He was laying out on all sides for a decent horse, but in vain. So he remained at Trim, hoping at least to buy off their anger: 'Pray stop Mrs Johnson's mouth with £14.'[77]

With the New Year the atmosphere of political suspicion began to clear, and even Swift's relations with King thawed out a degree or two. In a conciliatory letter in November, Swift had attacked a group of high Tory Non-jurors as 'a parcel of zealots' whose activities were 'a complication of as much folly, madness, hypocrisy and mistake as ever was offered to the world'.[78] King seized enthusiastically on these remarks, adding his own condemnations of their heresies. In reply Swift described himself as having been 'always a Whig in politics'.[79] It was an unfortunate phrase. It soon became clear that King was displaying Swift's letter to friends in England as proof that Swift was about to desert the Tories and modify his opinions once more to suit those of the prevailing ministry. The first hint of these rumours reached Swift in a puzzled note from Lewis in January.

> I am told the Archbishop of Dublin shows a letter of yours reflecting on the high flying clergy. I fancy you have writ to him in an ironical style, & that he would have it other wise understood; this will bring to your mind what I have formerly said to you on that figure; pray condescend to explain this matter to me.[80]

This letter is illuminating in several ways. It was evidently a shared joke among Swift and his friends that he would dupe King with ironical remarks. Yet Lewis adopts the peremptory tone of a political commissar reproving a dangerous piece of artistic licence. Clearly this was not the first time that he had warned Swift against indulging in his favourite double-edged device. In July Swift was forced to defend himself again, this time to Atterbury, the loftiest of high-flyers. He wrote with self-righteous dismay that the rumours that he had 'wholly gone over to other principles more in fashion'[81] had been so readily

[77] Ibid., p. 249.
[78] Ibid., pp. 221-2.
[79] Ibid., p. 236.
[80] Ibid., p. 246.
[81] Ibid., p. 279.

believed by his friends. 'I should think the wishes not only of my friends, but of my party, might dispose them rather to think me innocent, than condemn me unheard.' Nevertheless he attempted to refute the charges in detail, making a valiant effort to recall the offending paragraphs exactly, but inevitably producing a version more acceptable to Atterbury than to King. As if to pre-empt any further criticism he declared, 'I confess my thoughts change every week, like those of a man in an incurable consumption.' Harried by pamphlets, pursued by innuendoes, and parodied by libellers, caution was still supremely necessary to him.

A ministerial crisis towards the end of 1716 brought a revival of Tory hopes. 'The silly Tories . . . are feeding themselves with foolish imaginations,' Swift told Walls.[82] Lewis was more optimistic. 'The division of the Whigs is so great, that, morally speaking, nothing but another rebellion can unite 'em.' According to Lewis, both Whig factions were now 'making their court to the Tories'.[83] The most that Swift hoped for from any political realignment was a greater tolerance that would permit him to visit, travel, and correspond without the constant fear of surveillance and arrest. The clearest indication that the period of Whig revenge was over came in July when the impeachment proceedings against Oxford were annulled by the unanimous vote of the House of Lords, and he was released from the Tower. Swift wrote immediately to congratulate him on his freedom, and to offer himself as 'a companion in your retirement', for some months in Herefordshire. 'I have many things to say to you, and to enquire of you, as you may easily imagine.'[84] Part of Swift's urgency stemmed from his desire to know the truth of the persistent allegations that there had been Jacobite sympathies within Oxford's administration. A week later, fearing that his first unsigned letter had miscarried, he wrote again. The idea of such a visit brought a brief relief from the slough of rent-rolls and Chapter squabbles. But Oxford's two years in the Tower had only refined his talent for procrastination. When he finally sent a brief, courteous acknowledgement of Swift's letters, his warmth was tempered by distance. 'My heart is often with you,' he declared, but 'our impatience to see you, should not draw you into

[82] *Corr.* ii. 243. [83] Ibid., p. 246. [84] Ibid., p. 276.

uneasiness; we long to embrace you–if you find it may be no inconvenience to yourself.'[85] The conditional clauses were meant to dampen Swift's enthusiasm. Probably Oxford was unsure how far to trust any of his former 'friends'; maybe he was simply tired, and unwilling to rake over old issues; maybe he was guiltily conscious of past duplicities and unwilling to suffer the scourging of Swift's righteous indignation. He would be visiting Wimpole, he told Swift, 'where you are too just not to believe you will be welcome before any one in the world'. But Swift was also too intelligent not to recognize dissuasion when he heard it. He stayed in Ireland that summer.

In January Swift's curate Thomas Warburton was appointed rector of Magherfelt, leaving Swift with the inconvenient chore of finding a replacement. 'I suppose I shall have offers enough, but I shall be hard to please.'[86] An acquaintance of Walls was unwise enough to apply for the post while Swift was suffering badly with a cold. Swift reacted angrily.

> He seems to have the least wit, manners, or discretion in his jesting, of any pretender to it I ever knew, I mean except he were drunk when he writ the enclosed, as in charity to his understanding I would willingly believe.[87]

Actually Swift soon found that there was no great rush of young men anxious to serve him and he was forced to preach in Laracor himself. At length a distant relative, Stafford Lightburne, sent a 'very foolish letter' applying for the post. Perhaps on account of the family connection, Swift turned him down, and it was not for a further five years that Lightburne finally received the curacy of Laracor in 1722.

Still busy with his improvement plans for Laracor, Swift took his own mad virtuoso, Joe Beaumont, with him to survey the ground and decide on the position and elevation of the house he designed to build. Progress was very slow, however, for which he mainly blamed the incompetence of his steward, the 'puppy' Parvisol. In April, Swift accompanied Stearne to his enthronement as Bishop of Clogher, and it amused him to observe Stearne's helplessness in domestic matters.

> Tis pleasant to see my Lord mustering up his goods upon leaving this place, and missing sheets, table-cloths, napkins, candle-sticks, by the

[85] Ibid., p. 282. [86] Ibid., p. 251. [87] Ibid., p. 245.

dozen, and bottles by the hundred, and all within half-a-year past. He is now persuaded to take a housekeeper if he can get a good one.[88]

No sheet escaped the meticulous inventories of Swift and his housekeeper Mrs Brent. But was this what he had come to? A tyrant of the linen-basket? When friends in England complained of his silence, he replied that there was simply nothing to write about. 'What have I to say, unless I should transcribe a sermon or a pamphlet? . . . We had been undone for talk during the north-west winds, if two ladies had not been carried off and ravished in the country.'[89] Far from cultivating the acquaintance of ministers and grandees, Swift presents himself 'casting about how to get acquaintance with one Boswell, a prentice boy who acts Punch to admiration'.[90] This pose of playing the teague was to be a favourite cover for the bitterness and frustration of exile.

He also began to see Vanessa once more. In an undated letter, probably of December 1716, he arranged a meeting, but when Vanessa told Provost Pratt that she would not be home that evening, Swift changed his mind; 'else he will think it was on purpose to meet me; and I hate anything that looks like a secret'.[91] Evidently the awkwardness and embarrassment continued. His main correspondent in England at this time was Prior, who had been released from prison to find himself penniless. A group of friends, led by Pope and Gay, proposed publishing a subscription edition of his poems to raise money, and Swift agreed to collect subscriptions in Ireland. Prior gratefully acknowledged his help, but his letters were guarded, as he took it for granted that 'whatever I write, as whatever is writ to me will be broke open'.[92] He assured Swift that Oxford was as dilatory and secretive as ever. 'Nothing can change him; I can get no positive answer from him, nor can any man else.' Thanking Swift for his assistance, Prior told him 'you are the happiest man in the world, and if you once got into la bagatelle, you may despise the world'.[93] If this judgement seems perverse to us now, it is worth remembering that it was as just such an exuberant, irreverent figure that Swift's intimate literary friends constantly saw him.

[88] Ibid., p. 269. [89] *Corr.* ii. 306. [90] Ibid., p. 307.
[91] Ibid., p. 239. [93] Ibid., p. 281. [93] Ibid., p. 291.

Swift contacted another old friend in England that summer–Joseph Addison. Following a number of ministerial reshuffles, Addison had now reached the dizzy eminence of Secretary of State for the Southern Department, and Swift wrote in July to congratulate him. Obviously his letter is capable of more than one interpretation. Taken in conjunction with his declaration to King of having been 'always a Whig in politics' it might have given his Tory friends serious cause for disquiet. Swift saw himself as a man isolated, exiled, betrayed, and suspected–and for what? The Jacobite episode had deeply disturbed his confidence in his former ministerial friends, and Oxford's obvious reluctance to come to any *éclaircissement* further dismayed him. It was hardly surprising that he should have sought to re-establish relations with a friend whose probity had never been questioned. 'I examine my heart, and can find no other reason why I write to you now, beside that great love and esteem I have always had for you,' he declared.[94] One has to say that he did not look very far into his heart if this was all that he found. The letter is a fishing expedition to see how he was now regarded by His Majesty's principal Secretary of State.

Whatever Swift's wavering inclinations may have been, Addison gave them no encouragement. It was not until the following March that he acknowledged Swift's 'obliging letter', four days after giving up the Secretaryship, on account of ill health. He was clearly determined that there should be no ambiguity in their relationship. 'Shall we never again talk together in laconic?' he asked,[95] in an elegiac tone, condoling with Swift for the loss of their mutual friend, St George Ashe. The curse of party meant that Addison did not feel he could trust Swift to maintain the valuable distinction between true friendship and mere patronage. His letter was almost that of a dying man, seeking a final reconciliation. The irony of it was noted by Swift in his endorsement; 'Mr Addison, just after resigning the Secretary of State office, 26 March 1718.'

[94] Ibid., p. 277.
[95] Ibid., p. 286.

IV

1718 was the year in which Swift came to terms with living in Ireland. He did so largely as a result of forming two new and lasting friendships, with Thomas Sheridan and Patrick Delany. Sheridan, twenty years younger than Swift, had been educated at Trinity College, taken holy orders, but then chosen to become a schoolmaster. He was, according to Swift, 'the best instructor of youth in these kingdoms, or perhaps in Europe'. Swift probably first encountered Sheridan at a performance of the Greek play in December 1717.[96] Thereafter the range of their common interests, particularly their shared delight in word-play, brought them into frequent contact. When Sheridan's *Art of Punning* appeared in 1719 under the name of 'Tom Pun-sibi' Swift was widely assumed to be the author, and for the next twenty years, authorship of their minor squibs and satires was to be a confused and contested matter. Initially Swift viewed the schoolmaster's efforts with some condescension, complaining that Sheridan bombarded him with verses that were 'out of all rules of raillery', but he was soon pleased to join him in a series of extensive 'pun-ic wars' as diversions from his decanal chores. Sheridan's main faults were irresolution and a nagging wife. In the two short 'Characters' of his friend that he drew, Swift dwelt at length on these failings. Sheridan was, he declared, 'the greatest deceiver of himself upon all occasions . . . the greatest dunce of a tradesman could impose upon him'.[97] In 1731, exasperated by his hapless friend's imprudence, Swift forced him to sign a confession of having been thirty times deceived by his servants and agents. But the clearest example of Sheridan's folly was his marriage. In 1710, while still an undergraduate, he had married, and within the week regretted it. His wife was 'a clog, bound to me by an iron chain, as heavy as a millstone'. To Swift he was the perfect example of a long-cherished theory. 'He acted like too many clergymen, who are in haste to be married when very young; and from hence proceeded all the miseries of his life.'[98] This judgement clearly echoes Swift's earlier attack

[96] See James Woolley's study 'Thomas Sheridan and Swift' in *Studies in Eighteenth Century Culture,* ix. 93–111.
[97] *PW* v. 223, 216.
[98] Ibid., p. 217.

upon those 'raw and ignorant scholars' who believe 'every silk petticoat includes an angel', and who 'entail misery on themselves and their posterity' by over hasty matches. It was unfortunate that Sheridan should have presented Swift with such a ready confirmation of what was already an *idée fixe,* since it encouraged Swift to become shrill and self-righteous. Here is his description of the Sheridans in 1729.

> He lets his wife (whom he pretends to hate as she deserves) govern, insult and ruin him, as she pleases. Her character is this: Her person is destestably disagreeable; a most filthy slut; lazy and slothful, and luxurious, ill-natured, envious, suspicious; a scold, expensive on herself, covetous to others.[99]

It is difficult to remember that these are real people that he is describing. Sheridan easily becomes another *exemplum,* like Strephon, or the Cambridge sophs Cassinus and Peter, of the disastrous path of uxoriousness. Such tirades reveal the sad unreality of the hope that Mrs Arbuthnot expressed that he might soon be 'well married'.[100] They also show something of the tyrannically fastidious standards by which Stella was constantly judged.

Delany was a Junior Fellow with chambers in Trinity College who joined happily in the verse games and riddles of Swift and Sheridan. Swift's first mention of him is in rhymes dated October 1718.

> To you whose virtues I must own
> With shame, I have too lately known.[101]

1718 was also the year in which the prodigals sought to return. The two Jacobite exiles, Ford and Chetwode, wrote with tentative enquiries from London about the political climate in Ireland. Swift was less than encouraging in his replies. 'I can give you no arguments to live in Ireland, but what the fox gave about his tail,' he told Ford in August, adding with a snarl of realism that 'the Tories have lived all this while on whipped cream, and now they have even lost that'.[102] Ford contented himself with a short visit to Dublin, and then returned to his congenial lodgings in Pall Mall. Chetwode had narrowly escaped prosecution,

[99] Ibid., p. 222.
[100] *Corr.* ii. 306.
[101] *Poems,* i. 214-19.
[102] *Corr.* ii. 292.

and his affairs were further complicated since his wife had been forced to let both their houses. Swift's advice to him was to 'linger out some time' living cheaply and 'picking up rent as you are able'. He was entirely opposed to Chetwode attempting to get 'into a figure all on a sudden'.[103] One suspects that Swift's advice may have been motivated less by prudence than by a sneaking feeling that Chetwode should suffer for his irresponsibility. There is more than a hint of self-righteousness in his description of the bleakness of Irish society.

> I doubt you will make a very uneasy change from Dukes to Irish squires and parsons, wherein you are less happy than I, who never loved great company, when it was most in my power.

For the first three years after his return to Ireland, the tone of Swift's complaints against the privations and miseries of his life are those of an outsider, an Englishman for whom the Irish are aliens. Here for the first time we find him identifying with the ordinariness of Irish life as a solid corrective to the volatile political fantasies of his friends.

> Whenever you talk to me of Regents or Grandees, I will repay you with passages of Jack Grattan and Dan Jackson. I am the only man in this kingdom who is not a politician. . . . Joe Beaumont is my oracle for public affairs in the country, and an old presbyterian woman in town.[104]

An obvious kind of inverted snobbery is at work here, leading Swift to the comic exaggeration of claiming a madman and a Presbyterian as his political oracles. Chetwode begged to disbelieve such hubristic assertions of humility, and Swift's reiteration that he strove to reduce his spirit below his fortune was both maudlin and peremptory. In fact he was never able to subdue his conjured spirit to accept his blighted fortune. His avowal of humility to Ford loudly proclaims its own opposite.

> I choose all the silliest things in the world to amuse myself, in an evil age, and a late time of life. . . . I do everything to make me forget myself and the world as much as I can.[105]

Swift is all too conscious of wasting his time in trivial concerns, because he still has a burning desire, even in this evil age, to

[103] Ibid., p. 293. [104] Ibid., pp. 293-4. [105] Ibid., p. 309.

involve himself in great affairs. He had not yet forgotten himself, or the world.

Such mellowness as there is, appears in Swift's refusal to be drawn into foolish schemes; in his building work at Laracor; in his simple friendships. It is an instinctive rather than a rational thing, a matter of acclimatization to the slower pace of Dublin. Moreover, one man's mellowness is another's defeat. Writing to his 'brother' Lord Harley, Swift's tone hovered between the two. 'I grow gentler every day, and am content only to call my footman a fool, for that which, when you knew me first, I would have broke his head.'[106] It was characteristic of Swift to see tolerance as a kind of moral apathy.

In October Arbuthnot sent a brief note, not knowing 'whether you care to hear from any of your friends on this side'.[107] His cheerful letter called forth an early reply in which Swift declared that good health and good horses made up the sum of his desires. 'There is 20 lords, I believe, would send you horses, if they knew how,' Arbuthnot wrote back.[108]

In January Swift wrote to Ford expressing a desire to visit England, but not London, though he recognized that the likelihood of such a visit was slight. 'My resolutions are like the schemes of a man in a consumption which every returning fit of weakness brings to nothing.'[109] He was distressed at the recent deaths of Parnell and the King of Sweden. He had once, amazingly enough, hoped to 'have begged my bread' at the court of Charles XII, and even dedicated his unfinished draft of a *History of England*[110] to the monarch whose meteoric rise and fall seemed to Johnson a classic example of the vanity of human wishes. In February he was in an ugly state of health, with disorders of the head which 'blister upon blister and pills upon pills will not remove'.[111] He was still denied his favourite remedy, since 'this whole kingdom will not afford me the medicine of an unfoundered trotting horse'. The equanimity of an Arbuthnot was still something for which he could only yearn.

[106] Ibid., p. 289. Lord Harley is Edward Harley who succeeded to his father's title of Earl of Oxford in May 1724.
[107] Ibid., p. 299.
[108] Ibid., p. 303.
[109] Ibid., p. 310.
[110] *PW* v. 11.
[111] *Corr.* ii. 311.

> It would be an admirable situation to be neither Whig nor Tory. For a man without passions might find very strong amusements. But I find the turn of blood at 50 disposes me strongly to fears, and therefore I think as little of public affairs as I can.[112]

His illness was so severe that it prevented him from finishing a letter to Bolingbroke for six weeks. Apparently it was worth the effort though, for the glad recipient declared that Swift's letter gave him a more sensible pleasure than any he had tasted for several years. He replied with a long self-consciously philosophical letter, full of Ciceronian and Horatian quotations, by which he sought to assure Swift that he was benefitting from exile by cultivating the stoic virtues of reflection and self-denial. He no longer repined at his misfortunes, for they had taught him to distinguish clearly between true merit, and the mere show of glamour. Yet amid all the high-minded philosophy, there remains a good deal of complacency. As Bolingbroke looks back on his past, he observes 'a multitude of errors, but no crimes'.[113] Nor can he resist an attack an Oxford as a man 'of whom Nature meant to make a spy . . . and whom fortune, in one of her whimsical moods, made a general'. Bolingbroke still saw himself as the injured party, the man whose superior talents had been consistently checked by mean and narrow rivals. It is a depressing confirmation of Swift's infatuation with Bolingbroke that he could be won over by the superficial charm of such a letter. Indeed, he seriously considered travelling to Aix-la-Chapelle to convalesce after his illness with his exiled friend. However, some 'paltry impediments' and the '*sang froid* of fifty' intervened to keep him at home. There may have been other reasons too, particularly a continuing nervousness about the political situation. 'You will not let us be quiet here one moment with your confounded invasions,' he complained to Ford.[114] Also he had finally managed to purchase a horse 'at a great price' and his head was already improving from the exercise, and the 'lazy remedy of Irish country air'. There is evidence too that his relationship with Vanessa was entering another active phase, though whether this should be counted as an argument for going or staying is debatable. It was at this time that he wrote to her his

[112] Ibid., p. 312. [113] Ibid., p. 315. [114] Ibid., p. 322.

most impassioned letter, entirely in French (which may suggest that Aix was in his head).[115]

Swift apparently deferred to Bolingbroke's view of himself, observing 'no men are used so ill, upon a change of times, as those who acted upon a public view, without regard to themselves'.[116] The suggestion that Bolingbroke's Jacobite adventure was a principled defence of public values, rather than an ill-judged and impatient piece of opportunism, is absurd. But one looks in vain through this inscrutable prose for the slightest trace of irony. When one reads his sketch of the relationship between Oxford and Bolingbroke, one sees again the admiration that Swift still felt for the charismatic younger man.

> When I think of you with relation to Sir Roger, I imagine a youth of sixteen marrying a woman of thirty for love; she decays every year, while he grows up to his prime; and, when it is too late, he wonders how he could think of so unequal a match, or what is become of the beauty he was so fond of.[117]

Instead of visiting France, he took a summer trip to Gaulstown, and to Lord Wexford's estate. He was particularly careful to leave Trim before Bishop Evans held his annual visitation. Evans, a violent Whig, had sought continual pretexts for quarrelling with Swift ever since his appointment to the see of Meath in 1716. At first Swift had tried to be friendly with Evans and was 'more than ordinary officious in my respects'. But the Bishop would not be mollified. He refused to license the curates Swift proposed, on the grounds that he would not assist Swift 'in keeping up the spirit of faction among the neighbouring clergy'.[118] In September 1717 Evans reported to Canterbury that 'Dr Swift with other dignitaries in several parts of this kingdom endeavour to keep up the spirits of their party by assuring them that the Whigs will soon be down & that their friends will sway etc.'.[119] As far as Evans was concerned, no 'honest man' in Ireland made any distinction between Tories and Jacobites. At the visitation in 1718 the antagonism between the two men came to a head when Swift defended three clergymen whom Evans declared 'were all of them very criminal'. Hence in 1719 Swift sent a

[115] Ibid., pp. 324-5. [116] Ibid., p. 319. [117] Ibid., p. 320.
[118] Landa, p. 182. [119] Ibid.

proxy in his place. Predictably, Evans refused to accept the proxy, precipitating a blistering reply from Swift.

You may remember if you please, that I promised last year never to appear again at your visitation; and I will most certainly keep my word, if the law will permit me: not from any contempt of your lordship's jurisdictions, but that I would not put you under the temptation of giving me injurious treatment, which no wise man, if he can avoid it, will receive above once from the same person. . . . I hope that . . . your lordship will please to remember, in the midst of your resentments, that you are to speak to a clergyman, and not to a footman.[120]

Before setting out on his country ramble, Swift wrote to Vanessa. No, he would not be away for three whole months, nor was he trying to avoid her, but travelling 'purement pour le retablissement de ma santé'. He recommended the same to her. 'J'espère que . . . vous vous promenerai à cheval autant que vous pouvez.'[121] He advised her to revise her verses, read stories to Moll, attend to her lawsuit, anything to keep her occupied while he was away. But nothing, he knew, would soothe her vacant hours so much as some flattering words of endearment from him. So here they are, in bad French, slid in guiltily between otherwise self-conscious references to the execrable style of his French prose.

Je vous jure qu'en faisant souvent la plus severe critique, je ne pouvais jamais trouver aucun defaut ni en vos actions ni en vos parolles. La coquetrie, l'affectation, la pruderie, sont des imperfections que vous n'avez jamais connu. Et avec tout cela, croyez vous qu'il est possible de ne vous estimer au dessus du reste du genre humain. Quelles bêtes en jupes sont les plus excellentes de celles que je vois semées dans le monde au prix de vous; en les voyant, en les entendant je dis cent fois le jour–ne parle, ne regarde, ne pense, ne fait rien comme ces miserables, sont ce du meme sexe–du meme espèce de creature?[122]

[120] *Corr.* ii. 327. Ehrenpreis argues persuasively (iii. 57) that Swift may have been responsible for a hoax obituary of Evans which appeared in a London newspaper in September 1721. See also *PW* xiv. 41.

[121] *Corr.* ii. 325.

[122] Ibid., p. 326. 'I swear to you that even when subjecting them to the most severe scrutiny, I have been unable to find the least fault in either your actions or your words. Coquetry, affectation, and prudishness are quite unknown to you. That being the case, how can I choose but consider you as superior to the rest of the human race? Compared with you, even the worthiest of the women one encounters in society appear like beasts in skirts. When I see and hear them I say to myself a hundred times a day–do not speak, nor look, nor think nor do anything like these miserable creatures. Do they belong to the same sex, or even to the same species as you?'

Evidently a gap of three months in their meetings was sufficiently unusual to call forth such an effusion. In a letter the following year he thought it 'inconvenient for a hundred reasons that I should make your house a sort of constant dwelling place', but the implication is that regular visits had been resumed.

He continued to suffer an irritating series of minor disorders, a broken shin, a giddy head, and warned his correspondents that he was 'forced to entertain you like an old woman with my ailments'.[123] Ill health and solitude made him increasingly bad-tempered; and bad temper in its turn made him less fitted for society. He confided to Ford a grievous truth, 'that when time brings a man to be hard to please, he finds the world less careful to please him'. In such a mood he believed it would be foolish even to contemplate a visit to England. 'The truth is, the fear of returning in ten times worse humour than I should go has been my strongest discouragement, as a prudent prisoner would not choose to be a day out of jail, if he must certainly go back at night.' Instead he continued to affect to avoid all public and political concerns, but the affectation was proving more difficult to sustain.

> . . . as the world is now turned, no cloister is retired enough to keep politics out, and I will own they raise my passions whenever they come in my way.[124]

By 1720 his attacks of deafness were becoming more frequent and prolonged. Hitherto they had been irritations, severe and debilitating when they occurred, but with long periods of intermission between attacks. Increasingly, from this time, deafness, combined with giddiness, became an endemic feature of his existence.

> I am hardly a month free from a deafness which continues another month on me, and dejects me so, that I cannot bear the thought of stirring out, or suffering anyone to see me.[125]

Such a condition was 'the most mortal impediment' to all thoughts of travelling. 'I should die with spleen to be in such a condition in a strange place.' To be helpless and dependent in his own house, with his own friends and servants to attend on him, was bad enough; to be so in a strange house, where he could not command, but must ask favours, filled him with

[123] Ibid., p. 329. [124] Ibid., p. 330. [125] Ibid., p. 342.

horror. Three days earlier, having invited several gentlemen to dinner, he suffered a severe fit of giddiness and was forced to take to his bed, while Grattan did the honours of the house in his stead. 'You healthy people cannot judge of the sickly.[126] He declined an invitation from Robert Cope to stay at Lough Gall in May, explaining that 'once in five or six weeks I am deaf for three or four days together; will you and Mrs Cope undertake to bawl at me, or let me mope in my chamber till I grow better?'[127] The ignominy of such dependence hurt him deeply.

In December Swift wrote again to Bolingbroke, renewing an earlier plea for an official memoir of the Tory ministry. He had previously suggested that Bolingbroke himself should write such a memoir, but now argued that it would be better for someone else to undertake the task. Obviously he did not want it to be a mere polemic against Oxford. He concluded his letter with another self-portrait, in which self-pity blends with self-mockery, and self-dramatization serves as a counterfeit for self-knowledge.

> I have gone the round of all my stories three or four times with the younger people, and begin them again. I give hints how significant a person I have been, and nobody believes me: I pretend to pity them, but am inwardly angry. I lay traps for people to desire I would show them some things I have written, but cannot succeed; and wreak my spite in condemning the taste of the people and company where I am. But it is with place as it is with time. If I boast of having been valued three hundred miles off, it is of no more use than if I told how handsome I was when I was young. The worst of it is, that lying is of no use; for the people here will not believe one half of what is true. If I can prevail on anyone to personate a hearer and admirer, you would wonder what a favourite he grows. He is sure to have the first glass out of the bottle, and the best bit I can carve.[128]

Comparing this with the 'Resolutions on When I come to be Old' which he had written twenty years earlier, one can see how conventional this apparently detailed portrait is. Swift was working at turning himself into the character that he had created for himself long before.

[126] Ibid., p. 342.
[127] Ibid., p. 348.

V

'I write nothing but verses of late,' Swift told Ford in December, 'and they are all panegyrics.'[129] This is hardly an accurate description, since even the first of his birthday verses to Stella which he wrote that year tempers flattery with condescension. He is typically imprecise as to age, making Stella, as he had done Vanessa, four years younger than she actually was.

> Stella this day is thirty-four
> (We won't dispute a year or more.)[130]
> (ll. 1-2)

Any thought that this is a piece of conventional gallantry is quickly dispelled as he remarks on her middle-aged spread.

> However, Stella, be not troubled
> Although thy size and years are doubled
> Since first I saw thee at sixteen
> The brightest virgin of the green.
> (ll. 3-6)

She was, as he recorded with unflattering scrupulousness, 'one of the most beautiful, graceful and agreeable young women . . . only a little too fat'.[131] No doubt her weight was something about which he teased her, a joke that he relished more than she. In his poem the following year he was equally unflattering.

> Now this is Stella's case in fact;
> An angel's face, a little cracked;
> (Could poets or could painters fix
> How angels look at thirty-six.)[132]

In one sense such unromantic comments are merely intended to debunk the high-flown idioms of conventional love poetry; but they also reveal Swift's instinctive tendency to reduce Stella's self-esteem by de-sexualizing her. She has passed from girlhood to tubby middle-age without being allowed a sexual prime. Swift's central conceit is highly revealing, for it imagines Stella split in two.

> Oh, would it please the Gods to split
> Thy beauty, size, and years, and wit,

[128] Ibid., pp. 333-4. [129] *Corr.* ii. 331. [130] *Poems,* ii. 721.
[131] *PW* v. 227. [132] *Poems,* ii. 734-5.

> No age could furnish out a pair
> Of nymphs so graceful, wise and fair.[133]
> (ll. 9-12)

It is as if Swift wished to break open the ageing, stout, cracked angel Stella, and find the little girl within who first delighted him. Actually, to be more accurate, Stella at thirty-six (or forty) would make four of the Stella that he first knew as 'the brightest virgin on the green'. But whereas dividing may pass for gallantry, quartering would seem a little like butchery. Besides, it seems clear that the idea of a *pair* of nymphs mirrored the ambiguous feelings that he experienced at the time.

> And then before it grew too late,
> How should I beg of gentle fate
> (That either nymph might have her swain,)
> To split my worship too in twain.
> (ll. 15-18)

It is probable that it was the reactivation of Swift's feelings for Vanessa which motivated him to begin his series of birthday verses for Stella in an attempt to restore the balance, and split his worship evenly between the two 'nymphs'. His poems to Stella emphasize the same masculine qualities of honour, fortitude, intelligence, and wit that he had previously noted in Vanessa. 'To Stella, visiting me in my Sickness' begins with lines that exactly recall the opening of 'Cadenus and Vanessa'.

> Pallas observing Stella's wit
> Was more than for her sex was fit . . .[134]

In both poems, Pallas argues that the combination of male wisdom and female beauty 'might breed confusion' rather than inspire admiration. Therefore, as a defence, she 'fixed honour in her infant mind'. Writing to Vanessa too, honour was the first quality that Swift mentioned.

> So honour animates the whole
> And is the spirit of the soul.
> (ll. 13-14)

Stella's honour is made up of integrity, loyalty, and courage. She will never deceive a friend, or give way to silly feminine fears.

[133] Ibid., p. 722. [134] Ibid., p. 723.

> For Stella never learned the art
> At proper times to scream and start . . .
> (ll. 71-2)

On the contrary, she was capable of shooting a burglar at point-blank range, without squeamishness. There is a quiet dignity in Swift's recital of her moral qualities, but finally it is as a patient nurse that he praises her most.

> When on my sickly couch I lay
> Impatient both of night and day,
> Lamenting in unmanly strains,
> Called every power to ease my pains,
> Then Stella ran to my relief
> With cheerful face, and inward grief.
> (ll. 97-102)

There is no romantic passion here, but a form of mutual dependence. For all the similarities between the two women, Vanessa's 'manliness' was not of this kind. Though she had nursed him through an attack of shingles, she was not such a self-effacing angel with the lamp, and her intellectual masculinity was combined with enough femininity of manner to make it impossible for him to describe her in these terms. 'Best pattern of true friends' was how he described Stella, and in the poem 'To Stella who collected and transcribed his poems', he gives an 'authorized version' of their long relationship which is a brief counterpart to 'Cadenus and Vanessa'. There is the same desire in both poems to set the record straight (that is, to falsify it).

> Thou Stella wert no longer young
> When first for thee my harp I strung:
> Without one word of Cupid's darts,
> Of killing eyes, or bleeding hearts:
> With friendship and esteem possessed,
> I ne'er admitted love a guest.[135]
> (ll. 9-14)

Once again he parodies the hackneyed phrases of love lyrics, but what exactly is the tone of this descent from sublimity to bathos?

[135] Ibid., p. 728.

A poet, starving in a garret,
Conning old topics like a parrot,
Invokes his mistress and his muse,
And stays at home for want of shoes:
Should but his muse descending drop
A slice of bread, and mutton-chop,
Or kindly when his credit's out,
Surprise him with a pint of stout . . .
Exalted in his mighty mind
He flies, and leaves the stars behind.
(ll. 25-32, 35-6)

The implication of these analogies is that Stella herself represents the same kind of unpretentious, homely comforts as a slice of bread or a mutton chop. This is not the tone that he uses with Vanessa, even though he forswears dainties in a letter to her.

A fig for partridges and quails
Ye dainties, I know nothing of ye,
But on the highest mount in Wales
Would choose in peace to drink my coffee.[136]

With a typical pose of impartiality, Swift feels that having praised Stella, if only as a well-darned stocking, he can now fairly 'expose her weaker side' and chastise her for her touchiness on certain subjects. When roused, her reason gives way to rage.

Perverseness is your whole defence:
Truth, judgement, wit, give place to spite,
Regardless both of wrong and right.[137]
(ll. 92-4)

He compares her anger to an eruption of Etna, in an extended simile whose elaboration does not conceal his stern disapproval.

Stella, for once you reason wrong;
For should this ferment last too long,
By time subsiding, you may find
Nothing but acid left behind.
(ll. 131-4)

It is an awkward, uncomfortable poem, for beneath the banter there is a magisterial assertion of control, and an uncompromising demand for obedience. The form of the poem itself is a test.

[136] Ibid., p. 733. [137] Ibid., p. 730.

Will she, like a penitent pupil, copy out these lines as a remedial exercise, just as she had formerly copied out his other poems?

> Dare you let these reproaches stand,
> And to your failings set your hand?
> (ll. 139-40)

The answer, apparently, was yes. In a birthday tribute the following year, this time from Stella to Swift, she acknowledges her indebtedness to him in dutiful lines that reveal no spark of combativeness. There is unconscious pathos in the way that she thanks Swift for having stamped out her youthful frivolity.

> When men began to call me fair,
> You interposed your timely care;
> You early taught me to despise
> The ogling of a coxcomb's eyes;
> Shewed where my judgement was misplaced;
> Refined my fancy and my taste.[138]
> (ll. 9-14)

Several of Swift's poems develop the contrast between the idealistic notion of women as angels in petticoats and the cynical rejection of them as beasts in skirts. Celia, in 'The Progress of Beauty' (1719), is the first in a series of female characters whose painted faces cannot conceal the rottenness within.

> To see her from her pillow rise
> All reeking in a cloudy steam,
> Cracked lips, foul teeth, and gummy eyes,
> Poor Strephon, how would he blaspheme![139]
> (ll. 13-16)

Satirists of the Restoration and Augustan period were fascinated by cosmetics which turned women's faces into testing-grounds for the conflicts between art and nature, truth and falsehood. Belinda's toilette in 'The Rape of the Lock' is a fine example of art achieving a divine grace beyond the reach of nature.

> First, robed in white, the nymph intent adores
> With head uncovered, the cosmetic powers.

[138] Ibid., p. 737. [139] Ibid., i. 226.

A heavenly image in the glass appears,
To that she bends, to that her eyes she rears . . .[140]

Celia is considerably less adept at the cosmetic arts, and there is a comic element in the gargoyle effect created by her botched painting.

For instance; when the lily slips
Into the precincts of the rose,
And takes possession of the lips,
Leaving the purple to the nose.
(ll. 25-8)

More than any other satirist, with the possible exception of Juvenal, Swift took a delight in scrubbing and flaying the human epidermis until it smarted with goodness. In his verses to Vanessa he did not omit this familiar reminder.

Nymph, would you learn the only art
To keep a worthy lover's heart?
First, to adorn your person well
In utmost cleanliness excel.[141]

By contrast, Celia's first recourse is to the paintbrush, not the scrubbing-brush. There is an obvious suggestion of a whited sepulchre as she applies white paint over her cracks.

Love with white lead cements his wings
White lead was sent us to repair
Two brightest, brittlest earthly things,
A lady's face, and china ware.[142]
(ll. 61-4)

The waning moon provides a further predictable analogy for the progressive syphilitic decay of Celia's looks.

Each night a bit drops off her face . . .

When Mercury her tresses mows
To think of oil and soot is vain,
No painting can restore a nose,
Nor will her teeth return again.
(ll. 109-12)

[140] Pope, *The Rape of the Lock,* ed. G. Tillotson, Twickenham Edition, 1940 (re-set 1962), pp. 155-6.
[141] *Poems,* ii. 732.
[142] Ibid., i. 227.

Swift's birthday poem to Stella (1721) expresses similar themes, though in appropriately modified tones. Her 'angel's face a little cracked' is compared to a familiar inn sign.

> And though the painting grows decayed,
> The house will never lose its trade.[143]

This is contrasted with a hypothetical rival 'Chloe' whose doll-like face is an advertisement 'as fine as dauber's hand can make it'. Chloe is represented as crowing over Stella's increasing years.

> Then Chloe, still go on to prate
> Of thirty-six and thirty-eight
> Pursue thy trade of scandal picking,
> Thy hints that Stella is no chicken.
> (ll. 39-42)

In reply to these taunts, Swift berates 'Chloe' in moralistic tones.

> No bloom of youth can ever blind
> The cracks and wrinkles of your mind,
> All men of sense will pass your door
> And crowd to Stella's at fourscore.
> (ll. 55-8)

Once again Swift's lines read like an evasive fictionalization of a real situation. He is clearly attempting to reassure Stella that her fears of a younger rival are groundless, but one finds it hard to equate the artificial doll 'Chloe' with the real-life Vanessa, and the reassurance carries little force. If Stella were only half as intelligent as he claimed, she must have realized that Vanessa's charms were not simply those of fresh paint and teenage bloom. Yet in her penitential birthday tribute to Swift that year, Stella apparently accepts all his precepts.

> Your lectures could my fancy fix
> And I can please at thirty-six.
> The sight of Chloe at fifteen
> Coquetting, gives not me the spleen.[144]
> (ll. 41-4)

As one reads through these platitudes, delivered, as learnt, by rote, and notes the grateful self-abnegation of her tone, it is clear that Stella's rebellion is over.

[143] Ibid., ii. 735. [144] Ibid., p. 738.

St Patrick's dean, your country's pride,
My early and my only guide,
Let me among the rest attend
Your pupil and your humble friend.
(ll. 1-4)

The rumblings of Etna have been silenced. Stella has been chastised for her tantrum, has apologized, and been forgiven.

Vanessa's situation continued to be both awkward and melancholy, and although some of her difficulties were of her own making, Swift could not be insensible of her plight. She had few friends in Dublin, and was cut off from the circle of his. The complications of her estate seemed insoluble, and her young sister Moll was clearly dying. It is true that she did little to encourage new friends, turning down the attentions of both the deans Winter and Price. 'Your age of loving parsons is not yet arrived,' commented Swift with wry humour.[145] At some point in 1720 she moved from Dublin to her country house at Celbridge, but the loneliness of that place, and her sister's illness, made her more desperate than ever. 'I have asked you all the questions I used ten thousand times, and don't find them answered at all to my satisfaction,' she wrote in a short sulky letter.[146] She affected surprise that he should write at all, chiding him for breaking his promise to write in a week and indulging her ever-ready refuge of self-pity.

I own I never expected to have another letter from you for two reasons. First, because I was so very ill that I thought I should have died . . .

Swift wrote back immediately in mock-exasperation, which served as a cover for his real impatience.[147]

If you knew how many difficulties there are in sending letters to you, it would remove five parts in six of your quarrel. But since you lay hold of my promises, and are so exact to the day, I shall promise you no more.

'You were born chiding into the world,' he complained, and recommended her to 'leave off huffing'. Since she was being 'so exact to the day' he also ticked her off for a slip of the pen. 'One would imagine you were in love by dating your letter August 29th, by which means I received it just a month before it was

[145] *Corr.* ii. 351. [146] Ibid., p. 352. [147] Ibid., p. 352-3.

written.' As a further caution, he encouraged her to use a new code of dashes for endearments.

I wish your letters were as difficult as mine; for then they would be of no consequence if they were dropped by careless messengers, a stroke thus -- signifies everything that may be said to Cad.

The inclusiveness of that secret stroke suggests more than simple prudence; it also signifies Swift's desire to restrain her emotional outpourings, and to save himself the embarrassment of reading her expressions of passion. As to those questions, he continued to squirm out of answering by mocking her moodiness.

You do not find I answer your questions to your satisfaction, prove to me first that it was possible to answer anything to your satisfaction so as that you would not grumble in half an hour.

Swift was a master of the syllogistic side step, and here with his taunting 'prove to me first . . .' he defies his pupil to beat him at his own game. Vanessa seized on his idea of the stroke with great enthusiasm, and strewed her letters with these cryptic symbols of affection.

-Cad-you are good beyond expression, and I will never quarrel again if I can help it.[148]

She still wanted to argue, but was ready to surrender at his first frown. 'I am not so unreasonable as to expect you should keep your word to a day, but six or seven days . . .' The unpunctuated gush of her prose shows the torrent of her emotions, like a series of blushes.

I am sorry my jealousy should hinder you from writing more love letters for I must chide sometimes and I wish I could gain by it at this instant as I have done and hope to do is my dating my letters wrong the only sign of my being in love pray tell me did not you wish to come where that road to the left would have led you . . .

Replying, Swift spoke of accompanying Charles Ford (alias Glass Heel) on a visit to her, while continuing to recommend a course of country walking which would make her 'a most prodigious scholar, a most admirable housekeeper, a perfect housewife, and

[148] Ibid., p. 354.

a great drinker of coffee'.[149] He spoke of carving names on trees, and in a passage calculated to give her the greatest pleasure, proposed writing a History of Cad and–'through all its steps from the beginning to this time'.

It ought to be an exact chronicle of 12 years; from the time of spilling the coffee to drinking of coffee, from Dunstable to Dublin with every passage since. There would be the chapter of the blister, the chapter of madam going to Kensington, the chapter of the colonel's going to France . . .[150]

There is an obvious pleasure in recounting these happy moments from a shared and secret past, and this catalogue of memories shows a lover-like fascination with the minutiae of a romance. Vanessa's reply was tremulous with gratitude and excitement. She could hardly contain the pleasure that this treasury of moments had given her, and instinctively wondered whether it was merely a ploy to pacify her. 'Tell me sincerely, did those circumstances crowd in upon you, or did you recollect them only to make me happy?' She was delighted too that he was coming to see her at Celbridge, a prospect which called forth a whole battery of strokes.

– –, – –, – –, Cad, is it possible that you will come and see me, I beg for God's sake you will, I would give the world to see you here.[151]

From Swift's point of view, though, he would be chaperoned by Charles Ford, and he probably believed that a visit to Celbridge was safer than permitting Vanessa to make a special journey to see him in Dublin as she had planned.

Both of the main emotional relationships in Swift's life are enveloped in mysteries of his own creation, yet this does not mean that we are seriously ignorant of their nature. His relationship with Vanessa was sexual, in that it was characterized by an awareness, sometimes anxious, sometimes coy, of sexuality by both parties. The notorious 'coffee' references that occur so often in their correspondence probably do not refer to intercourse, but rather to a special sexually-charged sense of intimacy that originated at their first meeting at Dunstable when Vanessa spilt some coffee. Had she been, physically, his mistress and Stella his wife, the relationship could not have had a more

[149] Ibid., p. 355. [150] Ibid., p. 356. [151] Ibid., p. 357.

furtive, guilty, adulterous flavour; it would merely have been a more conventional form of infidelity. For that reason it seems more likely that Swift tantalized both himself and the world by rejecting the fruit which gossips everywhere believed him to be enjoying. On the other hand, his relationship with Stella was not sexual, whether or not they were married. She occupied the posts respectively of pupil, friend, confidant, housekeeper, and nurse and was indeed the 'best pattern of true friendship'. Quite possibly it was because he reposed such deep trust in her that he was unable to feel any passion for her. She became a symbol of goodness; her virtue was a compound of chastity and reason. To think of such a woman sexually would have seemed impossibly dirty, the behaviour of a yahoo.

In October Swift wrote to Vanessa a hurried letter, in intervals snatched from 'impertinent business'. He had heard that Moll was weaker yet, and with his almost fanatical belief in the efficacy of violent exercise, he urged it as 'infallible that riding would do Molkin more good than any other thing'. He himself complained of an ill head from want of exercise, and coffee, in Dublin. All round him people talked of nothing but the collapse of the South Sea Bubble. 'Everybody grows silly and disagreeable, or I grow monkish and splenetic, which is the same thing.'[152] It is a genial, but hardly intimate note that concludes with a hail of strokes too profuse to be intense.

Three letters written by Vanessa to Swift in 1720 exist, although no exact date can be assigned to them. All are similarly plaintive, and so utterly unguarded in expression that it is little wonder that Swift urged her to 'make more difficulties' about what she wrote. Her tone of morbid self-pity was by now only too familiar.

O -- -- -- how have you forgot me you endeavour by severities to force me from you . . . for heaven's sake tell me what has caused this prodigious change in you which I have found of late. If you have the least remains of pity for me left, tell me tenderly. No, don't tell it so that it may cause my present death, and don't suffer me to live a life like a languishing death which is the only life I can lead if you have lost any of your tenderness for me.

. . . was I an enthusiast still you'd be the deity I should worship what marks are there of a deity but what you are to be known by you are

[152] *Corr.* ii. 361.

present every where your dear image is before my eyes some times you strike me with that prodigious awe I tremble with fear at other times a charming passion shines through your countenance which revives my soul . . .[153]

No further correspondence exists until Swift's short grief-stricken note of commiseration to Vanessa on Molkin's death at the end of February 1721. 'For God's sake get your friends about you to advise and to order everything in the forms,' he counselled,[154] implicity warning her not to come to him for this advice. 'I want comfort myself in this case, and I can give little.' Not for the first or last time we see him shrinking away from an emotional demand of which he recognized the justice.

After Moll's funeral, Swift visited Vanessa once or twice in Dublin in an effort to cheer her up. She was now more isolated than ever, although being no longer tied to an invalid sister, she also had more freedom to visit or entertain a wider society. He was still very circumspect about visiting her, and did not write again until July, when he was staying at Gaulstown. 'It was not convenient, hardly possible to write to you before now, though I had a more than ordinary desire to do it."[155] Gradually Swift was trying to extricate himself from a relationship that now caused him a great deal more pain than pleasure. Once more he was forced to answer those old familiar questions of hers.

I answer all your questions that you were used to ask Cad–and he protests he answers them all in the affirmative [he] . . . assures me he continues to esteem and love and value you above all things, and so will do to the end of his life.

The value of this statement is considerably reduced by Swift's characteristic shift into the third person disguise of 'Cad'. What she needed to remedy attacks of the spleen was, predictably enough, good healthy exercise, fresh air, and company. Above all she needed to stop brooding, 'Cad' entreated her 'that you would not make yourself or him unhappy by imaginations'. The more she whined and pleaded, the more she destroyed what moments of happiness they might still have.

The wisest men of all ages have thought it the best course to seize the minutes as they fly, and to make every innocent action an amusement.

[153] Ibid., pp. 363-4. [154] Ibid., p. 378. [155] Ibid., p. 392.

This *carpe diem* theme sounds unusual from Swift's pen, and tells us more about the status of his relationship with Vanessa than about his real philosophy. If her questions still included the question of marriage, then his evasions are more explicable. 'Cad' may love 'Van . . . to the end of his life', but Swift and Hester Vanhomrigh could only enjoy some fleeting moments of drinking coffee together. He encouraged her to see more people, and finally he advised her to leave Ireland altogether.

> Settle your affairs, and quit this scoundrel island, and things will be more as you desire. I can say no more being called away, mais soyez assuré que jamais personne du monde a été aimée, honorée estimée, adorée par votre amie, que vous.[156]

What was Vanessa supposed to make of that? He tells her that if she returns to England, things will be as she desires. Then he breaks off enigmatically, with that transparent device of being 'called away'. The implication, to a woman anxiously seeking an answer to the question of his feelings for her, is surely that he might join her in England. Yet there is no evidence that Swift hoped or intended to visit England at this time. He was, quite simply, deceiving her, and thus, for his final promise of love, honour, esteem, and admiration he breaks into French. It may be that French is the language of gallantry. It may be a forlorn attempt at secrecy; but it seems most likely that a promise in French is rather like a promise with fingers crossed. Somehow it doesn't quite count. Otherwise what he says here is frankly dishonest. His esteem and honour for Stella were always greater than for Vanessa. It is hard to read this letter as anything other than a flattering dismissal, an attempt to free himself from a flirtation he had long outgrown. Her reply was confused and pathetic. Her law-case still went badly. The Master of Chancery behaved 'abominably' to her. She was tired and ill, but most of all she missed Swift. It was very easy for Cad to advise her 'to seize the moments as they fly' but 'those happy moments always fly out of the reach of the unfortunate'.[157]

They continued to meet at the home of a bookbinder in Dublin, until in the early summer of 1722 Swift went for a lengthy ramble through the north of Ireland. It was several

[156] Ibid., p. 393. [157] Ibid., p. 394.

weeks before he wrote to her, and then his message was the same as ever. She should take more exercise 'grow less romantic, and talk and act like a man of the world'.[158] He recalled some of the headings of his proposed 'History' of Cad and Van, but there was something perfunctory about the familiar catalogue, which had become part of his patter for cheering her up.

. . . remember the china in the old house, and Rider Street and the Colonel's journey to France and the London wedding . . .

He mentioned coffee no less than four times, swearing that he would rather spend three or four hours drinking coffee tête à tête, than travelling round Ulster in miserable weather.

Remember that riches are nine parts in ten of all that is good in life, and health is the tenth; drinking coffee comes long after, and yet it is the eleventh, but without the two former you cannot drink it right.

Though Swift's arithmetic here is a little baffling, his message to Vanessa to readjust her 'romantic' and 'splenetic' priorities was clear enough. She did try. She rode and read and forced herself to go more into society 'because you commanded me'. But the damage was already done. The female paragon so praised in 'Cadenus and Vanessa' for despising the frivolities of the beau monde could not change now, even upon command. At an assembly she found the company like baboons and monkeys, 'they all grinned and chattered at the same time'. It seems that she had already read an early draft of Gulliver's voyage to Brobdingnag.

One of these animals snatched my fan and was so pleased with me that it seized me with such a panic that I apprehended nothing less than being carried up to the top of the house and served as a friend of yours was.[159]

Poor Vanessa seems to have been the first proselyte of Gulliver's view of the world. 'I have so little joy in life that I don't care how soon mine ends,' she declared. 'For God's sake write to me soon.' He wrote when he could, but constant travelling made posting a letter difficult, and it was not until he reached Lough Gall on 13 July that he sent one. There was near despair in his criticism that she had become impossible to please, since he sensed,

[158] *Corr.* ii. 427. [159] Ibid., p. 428.

rightly, that he had contributed to make her such a social misfit. But he resented the implied blame that she charged him with for her indulgence in self-pity. 'We differ prodigiously in one point,' he told her, 'I fly from the spleen to the world's end. You run out of your way to meet it.'[160] Three weeks later he wrote again. It is the last letter that survives between them; a bluff traveller's letter full of cabins, bogs, tick-bites, and saddle-sores. When not riding, he was reading books of history and travels as background for *Gulliver's Travels.* He assured her that he often thought of the old familiar places, Chelsea, the sluttery, etc., 'especially on horseback'.[161] It was a self-consciously active and energetic enactment of his prescriptions for her, and takes its cue from Hotspur's words to Kate: 'And when I am o'horseback I will swear I love thee infinitely.'[162]

Vanessa died on 2 June 1723. A month earlier she made her will, which contains no mention of Swift. He is not even included in the list of friends to whom she bequeathed 'twenty-five pounds to buy a ring'. Instead, much to his own amazement, the philosopher George Berkeley was a chief beneficiary. He expressed his surprise in a letter to a friend.

> Mrs Hester Van Omry, a lady to whom I was a perfect stranger, having never in the whole course of my life, to my knowledge, exchanged a single word with her, died on Sunday night: yesterday her will was opened, by which it appears that I am constituted executor, the advantage whereof is computed by those who understand her affairs to be worth £3,000.[163]

The perversity of this bequest offers strong circumstantial confirmation of the theory of a sudden catastrophic rift between Vanessa and Swift in the months before her death. She knew of Swift's sceptical disregard for Berkeley's Utopian scheme for founding a college in the Bermudas, and there is a strong whiff of spite about this bequest. Once again, Sheridan is the main source for the legends of a last terrible meeting between Vanessa and Swift at Celbridge. According to this story, Vanessa was tortured by the rumours that Swift was secretly married to Stella.

[160] Ibid., p. 429.
[161] Ibid., p. 433.
[162] *Henry IV,* Part I, Act I, sc. iii.
[163] 4 June 1723. Quoted by Murry, p. 309.

Impatient of the torments which this idea gave her, she determined to put an end to all farther suspense, by writing to Mrs Johnson herself upon this head. Accordingly she sent a short note to her, only requesting to know from her whether she was married to the Dean or not. Mrs Johnson answered her in the affirmative, and then inclosed the note she had received from Miss Vanhomrigh to Swift. After which she immediately went out of town without seeing him, or coming to any explanation, and retired in great resentment to Mr Ford's country-seat at Wood Park. Nothing could possibly have excited Swift's indignation more than this imprudent step taken by Miss Vanhomrigh. He knew it must occasion great disturbance to Mrs Johnson, and give rise to conjectures fatal to her peace. Her abrupt departure, without so much as seeing him, already shewed what passed in her mind. Exasperated to the highest degree, he gave way to the first transports of his passion, and immediately rid to Celbridge. He entered the apartment where the unhappy lady was, mute, but with a countenance that spoke the highest resentment. She trembling asked him, would he not sit down? No– He then flung a paper on the table, and immediately returned to his horse. When, on the abatement of her consternation, she had strength to open the paper, she found it contained nothing but her own note to Mrs Johnson. Despair at once seized her, as if she had seen her death-warrant: and such indeed it proved to be.[164]

How Sheridan gained such a detailed knowledge of this highly charged scene we can only guess. In the final stages of her own illness, Stella may have confided things to his father which Swift himself would hever have done. Circumstantial evidence, however, offers some partial confirmation of elements in the story. As a final act of retaliation Vanessa 'laid a strong injunction upon her executors that immediately after her decease, they should publish all the letters which passed between Swift and her, together with the poem *Cadenus & Vanessa*'. Manuscript copies of the poem were already in circulation in the months before Vanessa's death, and even reached Stella in her retreat at Wood Park. A guest there, unaware of Stella's situation, remarked that Vanessa must have been an extraordinary woman to inspire the Dean to write so well. To which Stella retorted, 'she thought that point not quite so clear, for it was well known that the Dean could write finely on a broomstick'.[165]

As soon as he heard of Vanessa's death, Swift made a hurried

[164] Sheridan, pp. 330-1. [165] Delany, p. 58.

departure from town. He sent a note to Chetwode sometime after midnight on 2 June, after receiving the news, 'I am forced to leave the town sooner than I expected.'[166] He took a long journey to the South of Ireland, eventually reaching Clonfert in August. From there he wrote to Sheridan, enquiring whether the ladies were in town or still in the country. 'If I knew, I would write to them, and how are they in health?'[167] Evidently there had been no communication between Stella and himself for some considerable time. Inevitably the circumstances of Vanessa's death left him with a terrible burden of guilt. He could no doubt rationalize her suffering as the result of an over-active spleen, and insufficient exercise, but could not repress the recognition that his own rejection of her had been a crushing final blow. For her part, although she reciprocated that rejection in her amended will, there is no evidence that she was able to erase from her heart the marks of the deity that she worshipped. 'You are present everywhere your dear image is before my eyes. . . . Is it not more reasonable to adore a radiant form one has seen than one only described?' The Bishop of Meath, Swift's old enemy, gave this account of her heretical end.

> Tis generally believed that she lived without God in the world. When Dean Price offered her his services in her last minutes she sent him word, no Price, no prayers, with a scrap out of the Tale of a Tub . . . and so she died.[168]

With that last defiant gesture she remained faithful to the God who had deserted her.

VI

In 1720 Swift finally decided to break his political silence. He had signalled his change of mood to Ford the previous December; 'as the world is now turned, no cloister is retired enough to keep politics out'. However, there was an important difference about this new phase of political writing. Hitherto, all Swift's political concerns, since his return to Ireland, had been bound

[166] *Corr.* ii. 457.
[167] Ibid., p. 464.
[168] Quoted in Temple Scott's edition of the *Prose Works of Jonathan Swift* (1897-1908), 12 vols., xii. 95.

up with the last Tory ministry. Now for the first time he turned his attention to current political issues, and did so from a specifically Irish viewpoint. After five years of suffering the deprivations of a quasi-colonial existence, Swift ceased to attack or apologize for Ireland, and began to ask why? Why should life in Dublin be so far inferior to life in London? Why should the Irish Parliament be a mere seal for decrees of the English Parliament? Why should Irish trade be suppressed in the interests of English trade?

The occasion for his renewed involvement in politics was a Bill making its way through Parliament in London that asserted the complete dependence of the Irish legislature on England. Swift's reaction was a brief, passionate piece of polemic as powerful as anything he had written. *A Proposal for the Universal Use of Irish Manufacture* is vibrant with the sense of injustice. It is a powerfully sustained cry of pain and accusation, which marshals a devastating array of charges against English indifference and Irish inaction. The proposal, that the Irish should ban imported goods, was not new. Three times at least in the first decade of the century, the Irish Commons had resolved that 'it would greatly conduce to the relief of the poor and the good of the kingdom, if the inhabitants thereof would use none other but the manufactures of this kingdom in their apparel and in the furnishing of their houses'. Yet Swift's purpose goes further than to reiterate this plea for import controls. He indicts the whole tendency of English exploitation of Ireland. As Murry remarks, a hasty reading of the title page makes the point effectively, since the last words in bold type are, *Utterly Rejecting and Renouncing . . . ENGLAND.*[169]

Unlike his *Examiner* pieces, Swift's Irish writings present no clear party-line. On the contrary, they signal a rejection of party allegiances, following his disillusionment with the prevarications and deceits of his former colleagues. Yet any hint of criticism by him of the Tories was gleefully seized upon by his enemies. In 1719 he was maliciously identified as the author of a panegyric on King George by the author of *A Letter to Dean Swift*, who declared, 'as it is well known you were never a slave to constancy and principle, we can easily account for this your

[169] Murry, p. 319.

behaviour'. Though he had published nothing for five years, he was still the target for such smears.

He sought instead for a personal constituency for a campaign that was as much moral as political. Yet in a country split not only between Whigs and Tories, but further segregated into Catholics, Anglicans, and Dissenters, each with very different constitutional rights, it was no easy matter to cross the lines. Swift, after all, was a great believer in most of those lines which made the country so easy for an absentee landowning class to control. Yet he had few illusions about the personal qualities of the members of the ruling Anglican establishment, and often despaired of the trivial, narrow-minded, brawling disputes which took place in Parliament and Convocation. However, the question, as he wrote to Ford, was 'whether people ought to be slaves or no'. It was little wonder that Irish debates were such squabbling side-shows when they were frustrated by laws which reduced them to near irrelevance.

> You fetter a man seven years, then let him loose to show his skill in dancing, and because he does it awkwardly, you say he ought to be fettered for life.[170]

The voice which Swift cultivated in his Irish pamphlets was that of a lone crusader.

> Whoever travels this country, and observes the face of nature, or the faces, and habits, and dwellings of the natives, will hardly think himself in a land where either law, religion, or common humanity is professed.[171]

His *Proposal* was timed to appear on the eve of George I's sixtieth birthday on 28 May. Six years had elapsed since the appearance of Swift's last major pamphlet, the *Public Spirit of the Whigs,* but silence had brought no immunity, and this pamphlet suffered the same fate as its predecessor. Waters, the printer, was prosecuted for publishing a 'scandalous, seditious and factious' work, and Swift's old enemy Brodrick, now Lord Midleton, was determined to obtain a conviction. When the Grand Jury, which Swift claimed had been packed with Whigs, brought in a verdict of Not Guilty, the Lord Chief Justice Whitshed sent them back no less than nine times to reconsider

[170] *Corr.* ii. 342. [171] *PW* ix. 21.

their decision. Finally, in a combination of weariness and despair, the jurors brought in a 'special verdict' which meant, in effect, a re-trial. During the trial, according to Swift, Whitshed went so far as to declare that the author's design in the *Proposal,* was to bring in the Pretender, although 'there was not a single syllable of party in the whole treatise'.[172] Before another trial could take place Swift was successful in influencing the Lord Lieutenant Grafton to quash any further proceedings. Yet even before Grafton's decision was known, Swift produced an *Excellent New Song*[173] in which he taunted Whitshed, and deliberately flaunted the open secret of his authorship of the *Proposal.* Having found a cause capable of sustaining his fierce moral indignation, he seemed almost to invite prosecution as a means of obtaining a platform from which to make his denunciations.

In the final paragraph of his *Proposal* Swift delivered a warning shot against a new scheme for the establishment of a national bank in Dublin. Though probably a sound enough plan, it was associated in Swift's mind with the irresponsible power of 'moneyed men' and stock-jobbers, whose influence was most spectacularly to be seen in the sudden rise in the value of South Sea stock from 150 per cent in January to 1,000 per cent in August. He composed several pieces attacking the national bank, making the identification of money and faeces with an explicitness to delight the hearts of Freudians. The King himself approved of the idea of a national bank for Ireland. 'Bankrupts are always for setting up banks,' observed Swift sardonically.[174] However, when in September the South Sea Bubble finally burst, so did public confidence in paper credit of all kinds. South Sea shares quickly toppled back to their original value of 150 per cent. 'Conversation is full of nothing but South Sea, and the ruin of the kingdom,' Swift observed to Vanessa in October.[175] The following year the Irish Parliament rejected the idea of a bank.

Meanwhile economic conditions throughout Ireland steadily deteriorated. In the first few months of 1721, unemployment, poverty, and starvation became acute, made worse by the spate of bankruptcies in the wake of the South Sea Bubble. In April, Arch-

[172] *Corr.* ii. 368.
[173] *Poems,* i. 236-8.
[174] *Corr.* ii. 405.
[175] Ibid., p. 361.

bishop King estimated the number of distressed families belonging to the weaving trade as 1,700, and Swift calculated that in Dublin alone the number of those who 'are starving for want of work amounts to above 1,600'.[176] The Government's reaction was a special grant for relief purposes – of £100. Collections were made in churches, and a special performance of *Hamlet* was given in April, on behalf of the weavers, for which Swift and Sheridan wrote a prologue and epilogue urging the ladies to forego their silks in favour of honest Irish wool.[177] But the play brought in only £73. The sums collected by such charitable initiatives were all sadly inadequate.

However, in addition to using his pen in the weavers' defence, Swift also demonstrated his sympathy and support in a more practical manner. From about this time he began his practice of lending money to necessitous artisans and tradespeople at negligible rates of interest. In the words of Sheridan, 'the first five hundred pounds which he could call his own, he leant out to poor industrious tradesmen in small sums of five and ten pounds, to be repaid weekly at two or four shillings, without interest'.[178] Mrs Brent kept the records of these transactions, and the only charge was to provide her with a small gratuity. These schemes were a powerful example of Swift's belief in self-help. He was always scathing in his denunciations of simple charity, which he regarded as demeaning to both donor and recipient. His intention was to provide just enough money to prime the pump of an individual's own survival instinct. Johnson however, asserted that the schemes were a failure, for reasons associated with the ineradicable ambiguities in Swift's attitudes to his fellow men.

> He took no interest . . . but required that the day of promised payment should be exactly kept. A severe and punctilious temper is ill qualified for transactions with the poor; the day was often broken, and the loan was not repaid. This might have been easily foreseen; but for this Swift had made no provision of patience or pity. He ordered his debtors to be sued. . . . The clamour against him was loud, and the resentment of the populace outrageous; he was therefore forced to drop his scheme, and own the folly of expecting punctuality from the poor.[179]

Meanwhile Swift worked on a defence in case of prosecution,

[176] Ibid., p. 380.
[177] *Poems*, i. 273-6.
[178] Sheridan, pp. 270-1.
[179] Johnson, iii. 45.

and the resulting arguments appeared later in the form of a *Letter to Pope.* In this *Letter* his description of his actions during the Tory ministry has a familiar ring: how he sought endlessly to reconcile the feuding ministers; how he endeavoured to defend Whig writer from Tory rage; how there was never any plot to bring in the Pretender. When he asserts 'in my conscience I think I am a partaker in every ill design they had',[180] the rhetoric sounds a little shrill, however. If he still believed that, then he had a remarkable capacity for self-deception. Similarly, when he states that 'Mr Addison's friendship to me continued inviolable' he is guilty of bending, if not breaking the truth. It is important to note that he insists he is not recalling his support for Whig writers 'with any view towards making my court'. On the contrary, 'the new principles fixed to those of that denomination, [Whigs] I did then, and do now from my heart abhor, detest and abjure'. If there had been a moment in 1717 when he had flirted with the idea of renewing a connection with the Whigs, it was all over now. The miseries of Ireland were too firmly associated in his mind with the present ministry for him to feel anything but detestation for them. The *Letter to Pope* indicates that Swift's concern for the plight of Ireland was leading him to articulate and defend a new set of libertarian principles. For example, the two justifications that he offers for his loyalty to the Protestant succession are the law, and the opinion of the people. 'For necessity may abolish any law, but cannot alter the sentiments of the vulgar.'[181] It is rather surprising to find Swift presenting the opinion of 'the vulgar' as the best guarantee of stability. Parliaments may be bought, corrupted, or packed, but the people cannot be bribed. One should not make too much of this apparent appeal to democracy. It merely brings Swift into line with an axiom of Lockean contractualism more often articulated than accepted. But it is a further indication of his growing identification with the Irish people, and a step towards the appeal of M. B. Drapier to 'The Whole People of Ireland'. His other political principles are familiar, but consolidate a critical posture. He is against a standing army; he is in favour of such 'gothic institutions' as annual Parliaments; he abominates the influence of 'moneyed men'; he makes an impassioned plea

[180] *Corr.* ii. 370. [181] Ibid., p. 372.

against the practice of suspending *Habeas Corpus* in times of crisis, a law 'upon which the liberty of the most innocent persons depended'. Altogether, Swift's letter is a package of whose liberal principles John Locke could have been proud, and illustrates the general truth that, during the long period of the Whig supremacy, it was the Tories who took up the libertarian cause of limiting the powers and prerogatives of the executive.

The *Letter to Pope* is an eloquent statement of Swift's beliefs at the beginning of a new phase of his career. The contemporary world, as it seemed to him, was dominated by 'politics and South-sea, and party, and operas and masquerades'. Even Dublin abounded in political periodicals, the natural consequence of wretched times, he drily observed; 'When people are reduced to rags they should turn them to the only use that rags are proper for'.

Meanwhile he continued to assure Chetwode that he had forsworn politics forever. This was a diplomatic lie, for Chetwode was still under threat of prosecution, and Post-Master Manley had lost none of his enthusiasm for opening Swift's mail. In October 1722 Swift wrote a passage of ironically glowing praise for King George 'whose clemency, mercy and forgiving temper have been so signal, so extraordinary, so more than humane, during the whole course of his reign'.[182] This is clearly aimed at the Post-Master, who had recently intercepted a letter to Swift, and re-sealed it 'in a very slovenly manner'. Such surveillance had increased during 1722, as there were renewed fears of a Jacobite invasion. In April a plot was revealed to seize the Tower, the Bank of England, and the Royal Exchange. Soon afterwards, Atterbury, who had hailed the birth of an heir to the Old Pretender as 'the most acceptable news which can reach the ears of a good Englishman', was arrested.[183] 'Every plot costs Ireland more than any plot can be worth', complained Swift, as Walpole suspended *Habeas Corpus,* and sent more tax gatherers to Ireland to collect 'more money by three times than is now in the hands of the treasury'.[184] Swift kept safely out of the way all that summer, visiting the Copes at Lough Gall, and the Sheridans at Quilca.

[182] Ibid., p. 436.
[183] Letter to James, 6 May 1720. See *Corr.* ii. 434.
[184] *Corr.* ii. 435.

Swift suffered several bouts of severe deafness that year, which he sought to remedy by placing a clove of garlic, steeped in honey, in his ears. In March he was very melancholy at the death of 'the best servant in the world',[185] Alexander McGee, known as 'Saunders'. He was, Swift insisted, 'One of my best friends as well as the best servant in the kingdom'. He erected a monument in St Patrick's to commemorate McGee's 'discretion, fidelity and diligence'. The tablet had been intended to read as from a 'grateful friend and master' until a more status-conscious friend prevailed on him to change it to 'his grateful master'.[186] It is an interesting sidelight on Swift's awkwardness in matters of status that he should give way to such a snobbish gesture. He has often been seen as something of a domestic tyrant, yet the corollary of the exacting demands that he made on his servants was a correspondingly intense care and concern for those whose loyalty and integrity met his high standards.

At Christmas a short, affectionate letter from John Gay plunged Swift into a mood of depression and self-criticism. He had not heard from Gay for eight years, and yet the younger man dropped easily into the style of an old friend. 'You find I talk to you of myself,' he wrote. 'I wish you would reply in the same manner.'[187] This clear and unexpected voice from the past forced Swift to ask himself what he had done with all those intervening years. He did not come up with any very cheerful answers.

> I am towards nine years older since I left you, that is the least of my alterations: my business, my diversions, my conversations are all entirely changed for the worse, and so are my studies and my amusements in writing.[188]

He wasted his time, as he explained to Pope, with companions 'of least consequence and most compliance', reading 'the most trifling books I can find' and writing upon 'the most trifling subjects': subjects, as he told Ford, such as 'a country of horses' and 'a flying island'.[189]

185 Ibid., p. 422. 186 Delany, pp. 194-6. 187 *Corr.* ii. 439.
188 ibid., p. 442. 189 Ibid., iii. 5.

VII

THE TEMPERAMENTAL disqualifications which, according to Johnson, vitiated Swift's schemes for assisting needy tradesmen, can be clearly discerned in his sermons. Since it is difficult to assign most of these to any particular date, it may be convenient to deal with them here as a group. The sermons demonstrate many of Swift's positive beliefs, and they are hardly of a kind to warm the hearts of humanitarians. It is true that the sermons do not give us the whole man, being specifically tailored for the needs of a small country parish. When Swift discovered that his successor at Kilroot had found some of his old sermons there, and was proposing to use them, he tried to dissuade him. 'They are what I was firmly resolved to burn. . . . They will be a perfect lampoon upon me, whenever you look upon them, and remember they are mine'.[190] Although not quite lampoons on Swift, they certainly reveal his sober side. As Landa writes:

> Swift had a modest and conservative conception of what preaching should accomplish – 'to tell the people what is their duty, and to convince them that it is so' and to do this without oratory or displays of learning, by 'sober sense and plain reason . . .'[191]

To these principles was added a more prudent strain of pragmatism counselling orthodoxy, since he had already recognized that nothing was so fatal to a clergyman's hopes of preferment 'as the character of wit, politeness in reading or manners . . . these qualifications being reckoned by the vulgar of all ranks to be marks of levity, which is the last crime the world will pardon in a clergyman'.[192]

Swift's sermons are homilies on social, rather than spiritual topics. They seek to encourage dutiful behaviour and orthodox opinions by eschewing theological problems and recommending instead a simple, deferential code of conduct to his parishioners. The sermon on the 'Causes of the Wretched Condition of Ireland'[193] lists four main causes of distress: vanity, luxury, idle-

[190] *Corr.* i. 31.
[191] Landa. 'Introduction to the Sermons', *PW* ix. 101. See my own article 'Swift and the Beggars', *Essays in Criticism,* xxvi (1976), pp. 218-35.
[192] *The Intelligencer,* No. V, *PW* xii. 40.
[193] *PW* ix. 199-210.

ness, and the avarice of landlords. The dominant strain that runs through the sermon is a puritan insistence upon the virtue of self-reliance. He attacks women for their extravagance, and beggars for often choosing to beg and steal 'rather than support themselves with their own labour'. It is an axiom of Swift's philosophy that social ills have their origins in moral failings, and that social stability depends on hard work and self sufficiency – moral virtues in their own right. As a partial solution to present miseries he proposes the establishment of charity schools where children may be taught 'to think and act according to the rules of reason, by which a spirit of industry and thrift and honesty would be introduced among them'. He goes on:

> In all industrious nations, children are looked on as a help to their parents, with us, for want of being early trained to work, they are an intolerable burthen at home, and a grievous charge upon the public, as appeareth from the vast number of ragged and naked children . . .

'People are the riches of a nation' – so ran the pious commonplace; but in real life this was not always so – at least not all people. Gregory King's census figures for England in 1688 drew a deliberate and invidious line between those families which increased the wealth of the nation, according to contemporary notions, and those which impoverished it. The number below the line was disconcertingly high – 849,000 families decreasing the wealth of the nation, and only 511,586 families above the line, striving to increase that wealth. In fact the greater per caput income of those above the line left the nation with a healthy surplus of £1,825,100 – but the implication of that sharp line drawn between getters and spenders was a sombre one. And this was in England. Very few of the substantial families living in Ireland could make up much of the deficit created by the thousands of beggars throughout the country. Both Swift in his sermon, and the 'author' of *A Modest Proposal* address themselves to the problem of raising the children of the Irish poor over that line, so that they might cease to be a national burden, and become a national asset.

> I think it is agreed by all parties, that this prodigious number of children in arms, or on the backs, or at the heels of their mothers, and frequently of their fathers, is in the present deplorable state of the kingdom, a very great additional grievance; and therefore whoever could find out a fair, cheap, and easy method of making these children sound and useful

members of the commonwealth, would deserve so well of the public, as to have his statue set up for a preserver of the nation.[194]

Swift hopes that a sound moral education will do the trick. His Modest Proposer's ideas are considerably more radical; yet it is worth noting that their economic premisses are essentially the same. Neither invokes compassion or conscience as relevant considerations in analysing the situation of the poor. There are strict limits to the application of Christian charity in social matters, and Swift goes to some length to justify the division of the world into rich and poor.

But, before I proceed farther, let me humbly presume to vindicate the justice and mercy of God and his dealings with mankind. Upon this particular He hath not dealt so hardly with his creatures as some would imagine, when they see so many miserable objects ready to perish for want: For it would infallibly be found, upon strict enquiry, that there is hardly one in twenty of those miserable objects who do not owe their present poverty to their own faults; to their present sloth and negligence; to their indiscreet marriage without the least prospect of supporting a family, to their foolish expensiveness, to their drunkenness, and other vices, by which they have squandered their gettings, and contracted diseases in their old age.[195]

In his sermon 'On the Poor Man's Contentment', which is devoted specifically to this topic, Swift strains the quality of his charity even further.

Perhaps there is not a word more abused than that of the poor, or wherein the world is more generally mistaken. Among the number of those who beg in our streets, or are half starved at home, or languish in prison for debt, there is hardly one in a hundred who doth not owe his misfortunes to his own laziness or drunkenness or worse vices.

To these he owes those very diseases which often disable him from getting his bread. Such wretches are deservedly unhappy; they can only blame themselves; and when we are commanded to have pity on the poor, these are not understood to be of their number.[196]

There is something unpleasantly pharisaical about the smug formula 'deservedly unhappy', which Swift applies to ninety-nine in every hundred of the beggars he meets. It is a distasteful inversion of the parable of the lost sheep when Swift's concern is not

[194] *PW* xii. 109. [195] *PW* ix. 206. [196] Ibid., p. 191.

for the one in a hundred who has fallen, but exclusively for the one in a hundred who is pure.

For by the poor I only intend the honest, industrious artificer, the meaner sort of tradesmen, and the labouring man, who getteth his bread by the sweat of his brow . . .

It is by these clichés – the sweating brow rather than the idle drunkard – that Swift regulates his conscience and his charity. He is determined to identify financial distress with moral culpability, and to see poverty as the outward and visible sign of sinfulness. Hence compassion is ruled out, as an accessory to the original vice that has brought an individual to penury. Even the old and infirm have no real claim upon his charity.

For the artificer or other tradesman, who pleadeth he is grown too old to work or look after business, and therefore expecteth assistance as a decayed housekeeper; may we not ask him, why he did not . . . make some provision against old age, when he saw so many examples before him of people undone by their idleness and vicious extravagance.[197]

Only when he has clearly established this moral position does Swift concede that there may be weaker brethren whose distress is not wholly of their own making, and who do not entirely deserve their misery. It is these 'whom it is chiefly incumbent on us to support'. There is a complete lack of spontaneity about that response. Relief is admitted as a duty since 'we are commanded to have pity on the poor', but Swift's reactions are coloured by his belief that, at root, fecklessness and vanity are the causes of hunger. His charity therefore is suspicious and grudging. Economic ills, like all other hardships, are part of the lot which fallen man must endure. To err on the side of generosity is to err on the side of vice, and Swift does all in his power to ensure that charity is only made available to that one in a hundred of the truly deserving poor. The scheme which he proposed in this sermon, and subsequently introduced within the liberty of St Patrick's, was thoroughly penal in style, since his main objectives were to enforce humility, while freeing the city authorities of responsibility for the strolling beggars who were properly the concern of outlying parishes.

[197] Ibid., p. 206.

If every parish could take a list of those begging poor which properly belong to it, and compel each of them to wear a badge, marked and numbered, so as to be seen and known by all they meet, and confine them to beg within the limits of their own parish . . . driving out all interlopers from other parishes, we could then make a computation of their numbers, and the strollers from the country being driven away, the remainder would not be too many . . .[198]

The parable of the 'good Samaritan' has no place in Swift's philosophy. The neighbour whom he feels enjoined to love must reside within the pale of Dublin city.

From within these limits, Swift found it impossible to believe he was actually required to love his neighbour. This was a commandment which caused him real heart-searching, conflicting as it did so directly with the self-interest of his 'favourite', La Rochefoucauld. In his sermon 'On Doing Good', Swift faces up to this question with some bold opening assertions:

We are . . . commanded to love our neighbours as ourselves, but not as well as ourselves. The love we have for ourselves is to be the pattern of that love we ought to have towards our neighbour: But as the copy doth not equal the original, so my neighbour cannot think it hard, if I prefer myself, who am the original, before him, who is only the copy.[199]

This is a highly revealing piece of scriptural exegesis, and as fine an example of apologetic reasoning as one could hope to find in the works of any of the neo-Aristotelian church fathers. In another context – in *A Tale of a Tub* for instance – one would applaud such a passage as a neat ironic parody of clerical casuistry. For what we have here is the conversion of a commandment to altruistic behaviour, into a confirmation of egocentricity. The centrifugal force of the commandment has been made centripetal. It is a valuable glimpse of Swift's conscience at work, trying to reconcile his dogmatic faith with his psychology which did not, could not cease to put self first. He would not admit that his own deepest instincts, instincts which he believed were common to all men, were contrary to a central article of the Christian faith. So he offers this ingenious reinterpretation of the commandment to de-sublime its demands. The word 'as' in 'love thy neighbour as thy self' is made to do quite as much work as the word 'end' in the virtuoso punning performance on the injunction to 'regard thy end' in the

[198] Ibid., p. 207. [199] Ibid., p. 232.

'Digression in praise of Digressions'. But here the full weight of a defensive obsession is felt in a way we only dimly suspect at the back of the earlier satire.

This interpretation agrees with the tendency of the sermons as a whole, which is to dismiss or deride theological disputes while enjoining practical conformity within the Anglican Church, and stressing the heavenly rewards for such conformity. Swift wished to make it easier for both himself and his parishioners to believe that which it was necessary for them to believe for their own salvation. 'Human nature is so constituted, that we can never pursue anything heartily but upon hopes of a reward,'[200] he writes in his sermon 'On the Excellency of Christianity', where consequently his endeavour is to promote the Anglican Church as a 'best buy'. He is particularly scathing about the more sophisticated philosophers who seek to identify the moral recompense of virtue itself.

> ... some of the philosophers ... pretended to refine so far, as to call virtue its own reward, and worthy to be followed only for itself: Whereas if there be anything in this more than the sound of the words, it is at least too abstracted to become an universal influencing principle in the world, and therefore could not be of general use.

So crude does this argument of 'general use' appear to us in this spiritual context, that we look for a hint of irony in those quasi-chemical terms, 'abstracted' and 'refined'–terms which Swift put to such good effect in several of his satires. But there is none of that playfulness here. In the interests of general use and orthodoxy Swift is prepared to deride Thales, Solon, Aristotle, Plato, Zeno, and Epicurus as benighted heretics caught up with the casuistry of virtue being its own reward when only membership of the Anglican Church can ensure eternal happiness. The sermons present us with the voice of Swift the churchman; it is the voice of a deliberately narrow and shallow orthodoxy. In his literary and political satires we see Swift defending positions of orthodoxy with a bewildering range of unorthodox devices. Few of his religious writings show anything like the same versatility, and the inference is inescapable that it was in his relationship with God that Swift felt most uneasy.

[200] *PW* ix. 244.

Swift's *Letter to a Young Gentleman lately entered into Holy Orders,* dated January 1720, is a handbook for the aspiring young clergyman. Unremittingly practical in its prescriptions and concerns, it gives further insight into Swift's conduct and priorities as a churchman. Having marvelled that anyone should willingly enter a career that holds out the prospect of an annual stipend of £30 or £40 for life unless 'some bishop, who happens to be not over-stocked with relations'[201] should prove favourable, he moves on to the subject of preaching. The young clergyman should first practise out of town, at some desolate chuch, and get some 'intimate and judicious friend' to hear him and point out his failings. His main advice on sermons is to keep them simple. 'Proper words in proper places makes the true definition of a style.'[202] The young clergyman should avoid all 'hard words'; if in doubt whether a word will be intelligible to his congregation, he should consult a chambermaid, 'not the waiting-woman, because it was possible she might be conversant in Romances'. Theological complexities of all kinds should be avoided. 'I do not find that you are anywhere directed in the canons, or articles to attempt explaining the mysteries of the Christian religion.' Instead he should stick to the official line, and 'deliver the doctrine as the Church holds it'. Above all, the young clergyman should avoid the sin of wit.

> I cannot forbear warning you, in the most earnest manner, against endeavouring at wit in your sermons: Because by the strictest computation, it is very near a million to one, that you have none; and because too many of your calling have consequently made themselves ridiculous by attempting it.[203]

Well intentioned, and almost totally unironic (that 'very near a million to one' is a momentary lapse) this *Letter* shows how utterly restricted, intellectually and imaginatively, Swift believed that the outlook of the exemplary, honest, useful clergyman should be. It is a description so completely at odds with his own predilections for irony, erudition, and *la bagatelle,* that one senses not only Swift's deliberate self-mortification in returning to his pastoral chores, but his deeper sense of a temperamental–even a spiritual–unsuitability for the post that he held.

[201] Ibid., p. 64. [202] Ibid., p. 65. [203] Ibid., p. 72.

VIII

On 12 July 1722, While Swift was enjoying the rural calm of Lough Gall, William Wood, an English manufacturer, obtained a patent from the Crown to provide Ireland with a new copper coinage. More accurately, the Duchess of Kendal obtained the patent, which she sold to Wood for £10,000. Even at that price Wood reckoned to make a handsome profit of some £3,000 per annum throughout the fourteen years' duration of the grant. He was empowered to supply Ireland with £100,800 of copper coins at a time when the entire Irish currency only amounted to £400,000, and there were many who feared that this massive increase in copper coins would cause a sudden inflation and drive out all the gold and silver from the kingdom. Already it took thirteen Irish pennies to make one English shilling, and there seemed little guarantee that Wood would not debase his coins to increase his profits. Moreover the patent had been granted without the consent of any Irish representative, either elected or appointed. Official protests against Wood's patent began immediately, and by the following year a concerted opposition, similar to that which had defeated the plan for a national bank, had been formed. Swift was away from Dublin at the time, on his long southern journey following the death of Vanessa. He returned to find the House of Commons demanding an investigation of Wood's patent, and begging to inform his Majesty that Wood had practised 'a most notorious fraud' in having imported great quantities of coins 'of much less weight than was required by the patent'. An inquiry into these complaints was initiated as opposition to Wood grew steadily.

1724 began quietly enough for Swift. Together with Stella and Dingley he spent Christmas at Quilca with the Sheridans, where Tom Sheridan composed a poem 'To the Dean of St Patrick's' lamenting Swift's descent from high life.

> O what a mighty fall is here!
> From settling governments and thrones
> To splitting rocks and piling stones.[204]
>
> (ll. 22-4)

[204] *Poems,* iii. 1040.

He left Quilca in time for the annual visitation to his cathedral in January, and promptly fell ill with a town cold. He felt frustrated, harassed, without time to write or think, and he wrote to Ford, full of complaints.

Caesar was perhaps in the right, when he said he would rather be the first man in some scurvy village than the second in Rome, but it is an infamous case indeed to be neglected in Dublin, when a man may converse with the best company in London.[205]

It was such feelings of personal neglect, coupled with a sense of public indignation, which motivated Swift to lead the resistance to Wood in the persona of M. B., draper of St Francis Street. The first of M. B.'s letters, *To the Shopkeepers Tradesmen, Farmers and Common People of Ireland,* appeared in February. Writing to Ford a few weeks later, Swift made no secret of his authorship.

I do not know whether I told you that I sent out a small pamphlet under the name of a draper, laying the whole villainy open and advising people what to do; about 2,000 of them have been dispersed by gentlemen in several parts of the country, but one can promise nothing from such wretches as the Irish people.[206]

Swift entered the controversy in the firm conviction of being supported by all right-thinking Irishmen, including Archbishop King and a majority of both Houses of the Irish Parliament. The *Letter* was published in a deliberately cheap format by John Harding, who had printed material opposing the bank, and who now ran a weekly *Newsletter.* It opens with stern admonitions, more in the tone of the Dean than the draper. These halfpence are no trivial matter.

What I intend now to say to you, is, next to your duty to God, and the care of your salvation, of the greatest concern to yourselves and your children; your bread and clothing, and every common necessary of life entirely depend upon it.[207]

Usually in his *Drapier's Letters* Swift attempts to match his idiom to M. B.'s character and status, but sometimes, as here, he takes a magisterial line, berating the Irish for their apathy and ignorance.

[205] *Corr.* iii. 3. [206] Ibid., pp. 9-10. [207] *PW* x. 3.

It is your folly, that you have no common or general interest in your view, not even the wisest among you; neither do you know or enquire, or care who are your friends, and who are your enemies.

The Draper plays expertly on his readers' provincial pride and prejudices. How is it that 'a mean ordinary man, a hardware dealer' such as Wood, could gain the King's patent? He offers to make such mysteries very plain.

We are at a great distance from the King's Court, and have nobody there to solicit for us. . . . But this same Mr Wood was able to attend constantly for his own interest; he is an Englishman and had Great Friends, and it seems, knew very well where to give money . . .[208]

With a neat statistical sleight of hand, he develops a surrealistic image of the nightmare consequences of inflation: farmers requiring three horse-loads of copper to bring their half-year's rent to town; squires requiring anything up to two hundred and fifty horses to collect in such rents. The remedy for such madness is to boycott Wood's 'trash', but to be effective, such a boycott must be total. The Privy Council and Parliament had refused to accept the currency, but M. B. addresses himself to those most vulnerable to pressure, the 'poorer sort of tradesmen' who 'seldom see any silver'.

You may take my word, whenever this money gains footing among you, you will be utterly undone. . . . Do you think I will sell you a yard of tenpenny stuff for twenty of Mr Wood's halfpence? No, not under two hundred at least.[209]

Colloquial and emphatic, spiced with proverbs and scriptural tags, this first letter is as fine a piece of political rhetoric as Swift ever wrote. It is uncompromising but never monotonous; didactic but humane. It spells out Irish rights in a language that tradesmen could understand, and exhorts them to actions which they were well able to carry out.

The *Letter* had the intended result of strengthening resistance to Wood's coinage. Walpole soon realized that a more astute politician than Grafton would be needed to deal with this situation, and he appointed Carteret, whom he had just succeeded in removing from the Secretaryship, as Lord Lieutenant. This was a

[208] Ibid., p. 5. [209] Ibid., p. 11.

typical instance of Walpole's political expertise, since if his rival succeeded in this difficult post he would merely strengthen the English administration, and if he failed, he would destroy himself. Either way, Walpole would be satisfied.[210]

Swift was delighted to hear of Carteret's appointment, since he regarded him as an 'old friend' from his days in London. He wrote immediately to welcome him to his post and to put him straight on a few key points. 'There is not one person of any rank or party in this whole kingdom, who does not look upon this patent as the most ruinous project that ever was contrived against any nation.'[211] Enclosing a copy of the *Letter*, he remarked that although 'suited to the vulgar' it was generally 'thought to be the work of a better hand'. After six weeks of waiting, not very patiently, for a reply, Swift wrote again, hinting that perhaps Carteret was now too high and mighty to remember old friends. Carteret remained in London all summer, but in June Thomas Tickell, one of Addison's protégés, arrived in Dublin as Carteret's secretary. He and Swift dined together, but restricted their conversation to literary matters. Soon afterwards, Swift received a gracious apology from Carteret, who was pleased to recall their former friendship, and hoped that it might continue. On the subject of Wood he maintained a diplomatic silence, but acknowledged a 'general aversion which appears to it in the whole nation'[212] and hinted at conciliation.

In July Primate Lindsay died, and it is a good indication of the new patriotic alliance between Swift and King, that Swift wrote immediately to him with hopes that he might succeed. 'I would never be at an end if I were to number up the reasons why I should have your grace in the highest station the crown can give you.'[213] However King was right to suspect that he had missed his chance. Hugh Boulter, Bishop of Bristol, was appointed to head the Irish Church, a trusted English Whig who could be relied upon to support English interests and halfpence.

[210] Walpole told Newcastle, 'I should not be for sending [Carteret] over now if I did not think it would end in totally recalling him. We shall at least get rid of him here.' (See Ballantyne. *Lord Carteret: A political Biography, 1690-1763*, 1887, p. 70.)

[211] *Corr.* iii. 12.

[212] Ibid., p. 17.

[213] Ibid., p. 20.

On 1 August the report of the Privy Council committee examining Wood's coinage was published. Newton's assay of the coins had found no debasement or irregularities, and the committee cleared Wood of all charges against him. However, in deference to the continued opposition, they offered a compromise proposal: only £40,000 of the coins would be minted, instead of £100,800; and no one would be obliged to accept more than $5\frac{1}{2}$*d.* in these coins in any one payment. It was to reject this compromise that Swift wrote his second *Drapier's Letter* 'To Mr Harding'. Again he sounded a dictatorial note.

If I tell you there is a precipice under you, and that if you go forwards you will certainly break your necks: if I point to it before your eyes, must I be at the trouble of repeating it every morning?[214]

He picks up points in the committee's report for travesty and denunciation, rather than for reasoned reply. Newton had found that Wood 'had in all respects performed his contract'. His contract! With whom? 'Was it with the parliament or people of Ireland?' Swift shifts his ground, rather than challenge the findings of the assay head-on. The neatest deflection is when he seizes on the Government's concessions and turns them into marks of tyranny. Wood's promise that no one would be 'obliged' to receive more than $5\frac{1}{2}$*d.* in the coins at a time, is described as 'perfect high TREASON'.

Mr Wood will oblige me to take five-pence halfpenny of his brass in every payment. And I will shoot Mr Wood and his deputies through the head . . . if they dare to force one farthing of their coin on me in the payment of an hundred pounds. It is no loss of honour to submit to the lion! But who, with the figure of a man, can think with patience of being devoured alive by a rat?[215]

His most persuasive argument is a more subtle piece of insinuation, where he reminds his readers how much the English government takes annually from Ireland.

England gets a million sterling by this nation; which, if this project goes on, will be almost reduced to nothing: And do you think those who live in England upon Irish estates, will be content to take an eighth or tenth part, by being paid in Wood's dross?[216]

[214] *PW* x. 22. [215] Ibid., pp. 19-20. [216] Ibid., p. 22.

The cynical implication, that it is only by appealing to English self-interest that one can hope to relieve Irish miseries, was to become a familiar argument in Swift's Irish pamphlets. He feigns the voice of the quisling as a way of indicating how far English rule had already corrupted the spirit and identity of Ireland. Writing of Swift's unpublished *Verses on the Union* (1707), Ehrenpreis argues that Swift himself had caught the same debilitating disease. 'Just as the victims of long persecution tend to adopt the standards of their persecutors, so Swift, whose failure to win the posts he desired was partly due to his Irish birth and education, clung to the picture of himself as entirely English.'[217] Towards the end of his life, it is true that Swift's wistful fantasies of having been born in Leicester and educated at Oxford betray the same insecure longing that one finds in those immigrants who seek to be more English than the English. But throughout the period of his major Irish pamphlets Swift's infection with this strain seems to have acted as an immunizing dose. During the 1720s he ceased to identify himself with England–or at least, with English politics. Yet he had too much pride ever to acknowledge himself as Irish. He was an island of self-interest, whose political concerns were born of public outrage and personal disappointment.

In his third *Drapier's Letter,* addressed to the nobility and gentry of Ireland, a deferential M. B. argued that Wood had a long record of fraud and crime behind him. The whole conflict is presented as being between one rapacious individual and the whole Irish nation.

> The high priest said, *it was expedient that one man should die for the people*; and this was a most wicked proposition. But that a whole nation should die for one man, was never heard of before.[218]

Such an unequal conflict of interests would never be tolerated for a moment in England, and it is this discrepancy between the treatment of issues in England and Ireland that stimulates his most stirring rhetoric.

> Were not the people of Ireland born as free as those of England? . . . Is not their parliament as fair a representative of the people, as that of England? . . . Does not the same sun shine over them? And have they not the same

[217] Ehrenpreis, ii. 175. [218] *PW* x. 41.

God for their protector? Am I a free-man in England, and do I become a slave in six hours by crossing the channel?[219]

If you prick me, do I not bleed? The accents of protest against prejudice have a familiar ring. The main obstacle for those like Molyneux and Swift who wished to assert that Ireland was a sister kingdom with the same independent rights under the Crown as England, was Poyning's Law, which asserted that no Act could pass in the Irish Parliament, which had not been previously approved by the monarch and Privy Council in England. The Declaratory Act of 1719 had further asserted the right of the British Parliament 'to make laws and statutes . . . to bind the kingdom and people of Ireland'. In writing that 'I do not understand that Poyning's Law deprived us of our liberty'[220] and by neglecting to mention the Declaratory Act at all, Swift is rather writing about the situation that he wished to exist, than the one that did. In reality, Ireland in 1719 had become a colony in all but name; and it was a colonial system which the *Drapier's Letters* were written to challenge. But like all conservatives, even the most radical, Swift could not avow so revolutionary an intention, and preferred to argue instead in terms of precedent and tradition. Even the royal prerogative, he argued, was 'bounded and limited by the good and welfare of his people'. Although a monarch 'can by law declare *anything* to be current money by his letters patents',[221] the Drapier cannot bring himself to believe that a 'just and merciful' monarch would behave in the same tyrannical manner as King James who 'obliged his subjects here to take a piece of copper under the value of half a farthing, for half a crown'. The honest Drapier is convinced that King George has the interests of his loyal subjects in Ireland at heart, and has been mistakenly advised by greedy people to approve Wood's patent. He concludes with a memorable image of Wood as another Goliath, defying the armies of the living God.

For Goliath had a helmet of brass . . . and the weight of the coat was five thousand sheckles of brass, and he had greaves of brass upon his legs. . . . In short, he was like Mr Wood, all over brass.[222]

Shortly afterwards an effigy of Wood was paraded through the streets of Dublin and ceremonially burned. With a campaign of

219 Ibid., p. 31.
220 Ibid., p. 39.
221 Ibid., p. 37.
222 Ibid., p. 48.

processions, pamphlets, and poems the anti-Wood forces kept up the momentum of their attack until the arrival of the new Lord Lieutenant on 22 October. But the centrepiece of the campaign was Swift's fourth, and most devastating *Drapier's Letter*, which he addressed 'To the Whole People of Ireland' on the day that Carteret set foot in Dublin. This time Swift sought to go beyond the normal political establishment, to make a direct, demagogic appeal to the whole population. In the opening paragraph of this fourth *Letter* he sounds the note that raised this particular grievance to the level of a general principle–the note of liberty.

> A people long used to hardships, lose by degrees the very notions of liberty; they look upon themselves as creatures at mercy.[223]

Leaving aside considerations of the coins themselves, of Newton's assay, of Wood's attacks on his opponents as disaffected Jacobites, Swift establishes the central issue of the struggle as being the political status of Ireland itself. In recent years it had grown the custom to speak of Ireland as a 'depending kingdom'. M. B. utterly rejects this description, affirming his loyal independence with a neat ironic twist.

> I declare, next under God, I depend only on the king my sovereign, and on the laws of my country, and I am so far from depending upon the people of England, that, if they should ever rebel against my sovereign, (which God forbid) I would be ready at the first command from his majesty to take arms against them; as some of my countrymen did against theirs at Preston.[224]

In this triumphant reversal of Wood's Jacobite smear, Swift seeks to make a virtue of a disability, and, as Molyneux had done before him, attempts to present Poyning's Law as a privilege, rather than a punishment. According to this view, the sovereign had a closer relationship with Parliament in Ireland than in England, since in Ireland he had a positive voice in the framing of legislation, whereas in England he could only exercise the power of veto. It is a clever argument that seeks to weld together radicalism and tradition, loyalty and independence. The Drapier presumes so far as to commend the new Lord Lieutenant to the people of Ireland as 'a man of great accomplishments, excellent learning' from whom great things might be hoped. But it is a double-edged

223 Ibid., p. 53. 224 Ibid., p. 62.

compliment, for he goes on to remind his readers of the numbers of office-holders and court pensioners who are maintained at their expense. Lord Palmerston, First remembrancer, £2,000; Mr Dodington, Clerk of the Pells, £2,500; four under-treasuryships, worth £9,000 per ann., etc., etc. As he views this list, the Drapier is almost tempted to hope that Wood's project might succeed, since these office-holders would then be reduced to a 'jolly crew . . . of beggars' and would 'prance about in coats of mail' for want of clothes, and 'eat brass as ostriches do iron, to compensate for want of meat'. By now, all sorts of rumours about Wood's coinage were rife. Some reported that the patent had been withdrawn; others, that large quantities of the coins had already been shipped to Ireland. Most colourfully, the *Dublin Flying Post* of 12 October had reported that 'if the people of Ireland persisted in their refusal of the brass coin, Walpole would make them swallow it in fire-balls'. The Drapier treats this last threat with a farcical mock-seriousness. Demonstrating once again his remarkable flair for exploiting figures and statistics, Swift treats us to this incongruous arithmetical proof.

> Now, the metal he hath prepared, and already coined, will amount to at least fifty millions of half-pence to be swallowed by a million and a half of people; so that allowing two halfpence to each ball, there will be about seventeen balls of wildfire a-piece, to be swallowed by every person in the kingdom: and to administer this dose, there cannot be conveniently fewer than fifty thousand operators, allowing one operator to every thirty; which, considering the squeamishness of some stomachs, and the peevishness of young children, is but reasonable.[225]

Rhetorically, Swift's argument has all the exuberance of a celebration. He is playing with Wood and Walpole, confident of popular support, and of eventual victory.

Such an overt attack could not be overlooked. At a meeting of the Privy Council on 27 October, Carteret denounced the *Letter*, and ordered that the printer, Harding, should be prosecuted. A reward of three hundred pounds was offered for the discovery of the author. However, the members of the Council, who had recently signed a declaration against Wood's coinage, were unwilling to denounce the *Letter* as a whole, and the proclamation was only issued against some 'seditious and scandalous

[225] Ibid., p. 68.

paragraphs' in it. Even then, some members of the Council, including Archbishop King, refused to sign. The situation was not without elements of farce, since all of them knew the author of the *Letter* to be Swift. Writing to a friend in November, Tickell described it as 'Swift's pamphlet' and mentioned that 'great endeavours are using to get Dr Swift the freedom of the city in a gold box.' He went on: 'The enclosed quotation out of scripture has been got by rote, by men, women and children, and, I do assure you, takes wonderfully.' The enclosed quotation was this.

And the people said unto Saul, shall Jonathan die, who hath wrought this great salvation in Israel? God forbid: as the Lord liveth, there shall not one hair of his head fall to the ground, for he hath wrought with God this day. So the people rescued Jonathan that he died not.[226]

Actually Swift felt himself in more danger from the pains and giddiness in his head than from the punishment of Saul. He complained of having 'the noise of seven watermills in my ears',[227] and apologized quite coolly to Tickell for failing to greet the new Lord Lieutenant, on the grounds that his illness made him 'fit for nothing but to mope in my chambers'.

On 7 November Harding was arrested, but refused to divulge the Drapier's name. He was content to stand trial, confident that the Grand Jury would acquit him. On the 11th Swift published his *Seasonable Advice to the Grand Jury*, in which he urged the jurors to consider 'very maturely . . . what influence their finding the Bill may have upon the kingdom'. If they were to condemn the Drapier for some 'inadvertent expressions' then they would undermine the force of his arguments in all four *Letters*. In other words, to find against the Drapier meant finding *for* Wood. There was no middle way. Carteret promptly asked the Attorney General to indict this *Seasonable Advice* as an 'insolent libel'. He was, Swift told Oxford, 'acting the most unpopular part I ever knew, though I warned him against it . . . before he came over'.[228] When on 21 November the Grand Jury refused to condemn the *Letter*, Whitshed discharged them in a rage and ordered a new jury. The piquancy of the situation may well have been increased by Swift's own presence in the court, since he claimed that it was on that

[226] Ibid., pp. xix–xx. (The quotation is 1 Samuel 14:45.)
[227] *Corr.* iii. 36.
[228] Ibid., p. 41.

very day that he noticed the motto on Whitshed's coach, *libertas et natale solum,* liberty and my native land, 'at the very point of time when he was sitting in his court, and perjuring himself to betray both'. He registered this irony in some sharp verses.

> *Libertas et natale solum;*
> Fine words, I wonder where you stole 'em.
> Could nothing but thy chief reproach
> Serve for a motto on thy coach.[229]
>
> (ll. 1-4)

Swift also circulated copies of an extract from a book of Commons' debates, which contained a resolution declaring that the discharging of a Grand Jury was 'arbitrary, illegal . . . and a means to subvert the fundamental laws of this kingdom'.

Whitshed's arbitrary actions provided Swift with a splendid opportunity to broaden the appeal of his polemics. His fifth *Drapier's Letter,* in December, was addressed to Lord Molesworth, a retired Whig whose writings had earned him a reputation as a great libertarian. There is a holiday atmosphere in this fifth *Letter* as Swift taunts the legal authorities to do their worst. He affects a mock-penitence, saying 'I WENT TOO FAR' in having imbibed the dangerous notions of Locke, Molyneux, and Molesworth himself, 'that freedom consists in a people being governed by laws made with their own consent, and slavery in the contrary'.[230] In a typical piece of oblique self-congratulation, the Drapier admits that he has only himself to blame for running into difficulties, since he was warned by 'a certain Dean' not to trust to 'the general good will of the people'. Nevertheless, Swift ended the year buoyed up by that good will. 'I hardly thought such a spirit could ever rise over this whole kingdom.'[231]

Unbeknown to Swift, Carteret had already recommended that the patent should be cancelled. Boulter, the new Primate, agreed; 'while the fear of these halfpence hangs over the nation', he argued, 'It is impossible to have things easy'.[232] It was in January that the long-awaited confrontation of the two central figures in this drama, Swift and Carteret, took place. Public interest in the event was enormous, as Swift later told the Lord Lieutenant, 'the town has a thousand foolish stories of what passed between us,

[229] *Poems,* i. 347-9.
[230] *PW* x. 85-7.
[231] *Corr.* iii. 44.
[232] *PW* x. p. xxv.

which indeed was nothing but old friendship without a word of politics'.[233] According to Sheridan's dramatically heightened account, the meeting took place on the day following the proclamation against the Drapier. Swift pushed his way through the crowds at Carteret's levee, and harangued him 'with the voice of a stentor that re-echoed through the room, "So, my Lord Lieutenant, this is a glorious exploit that you performed yesterday, in issuing a proclamation against a poor-shopkeeper, whose only crime is an honest endeavour to save his country from ruin . . ." ' Sheridan went on:

> The whole assembly were struck mute. . . . The titled slaves, and vassals of power, felt and shrank into their own littleness, in the presence of this man of virtue. He stood super-eminent among them, like his own Gulliver amid a circle of Lilliputians.[234]

Carteret himself, however, was sufficiently composed to reply with a verse from Virgil: 'Res durae, & regni novitas me talia cogunt moliri.' (Hard fortune, and the newness of my reign, compel me to such measures.) Both men had a good deal of respect for each other, but Carteret was not prepared to be bullied by Swift. Sheridan records how he would parry Swift's denunciations with maddening coolness and restraint, until Swift cried out in exasperation: 'What the vengeance brought you among us? Get you back, get you back, pray God Almighty, send us our boobies again.'

In March Swift told Ford that he was thinking of coming to London to settle the matter of Wood there. He dismissed the warnings and apprehensions of his friends, with a hubristic boast that this would not be the first time he had risked martyrdom.

> There was a time when in England some great friends looked on me as in danger, and used to warn me against night-walking &c; but I thought it was a shame to be afraid of such accidents and looked as if a man affected to be thought of importance. Neither do I find that assassinations are things in fashion at present.[235]

However, he did not travel to England. Stella's health had been deteriorating for several months, and her appetite had declined until she was now eating barely two ounces of food a week. Together they decided to spend the summer in the Sheridans'

[233] *Corr.* iii. 49. [234] Sheridan, p. 247. [235] *Corr.* iii. 53.

house at Quilca, in the hope that a change of air might prove beneficial. But Quilca proved to be a less than ideal spot for convalescence. It was a wild and dismal place, eight miles from Kells. Swift strongly discouraged Chetwode from attempting a visit. 'We have hardly room to turn ourselves, and . . . I never thought I could battle with so many inconveniences."[236] Still, there was a sense in which Swift relished the privations of Quilca, and he compiled a three-page list of the 'Blunders, Deficiencies, Distresses and Misfortunes of Quilca' in the same ironic spirit that later led him to compose his *Holyhead Journal.*

> But one chair in the house fit for sitting on, and that in a very ill state of health. . . . Not a bit of mutton to be had in the country. . . . Want of beds, and a mutiny thereupon among the servants. . . . An egregious want of all the most common necessary utensils. Not a bit of turf this cold weather . . .[237]

And so on; this meticulous inventory of discomforts was drawn up with an eye to details that would do credit to the shopkeeper of St Francis Street. But Swift's tone is self-mocking, rather than genuinely melancholic, and he admitted to Chetwode that, for all the deficiencies of Quilca, 'there are some agreeablenesses in it'. It was at least a refuge from the intolerable social awkwardness caused by his continuing deafness, as he told the young Lord Oxford.

> I fled to avoid company in frequent returns of deafness . . . for while I am thus incommoded, I must be content to live among those whom I can govern, and can make them comply with my infirmities.[238]

Typically, Swift had no sooner arrived at Quilca that he began making improvements, 'levelling mountains and raising stones, and fencing against inconveniencies of a scanty lodging, want of victual, and a thieving race of people'. Stella assisted Swift in these landscaping schemes, making use of the little axe that she wore in her belt, and after a few weeks of this bracing exercise, Swift was able to report that both he and Stella were 'generally much better'.[239] At the beginning of April, Swift had been made a Freeman of the City of Dublin, but it suited him better to battle it out with the elements, rather than enjoy the tawdry honour of being a Free-

[236] Ibid., p. 61.
[237] *PW* v. 219–21.
[238] *Corr.* iii, 84.
[239] Ibid., p. 89.

man of an enslaved city. He could only be happy, he told Chetwode 'by never coming near Dublin, nor hearing from it'.[240] His only concern in life now, he declared, was the preservation of his fortune, for 'the ruin of a man's fortune makes him a slave, which is infinitely worse than loss of life'.

Swift's protestations of unconcern at public affairs was no idle paradox. It was Gulliver's contention that 'the yahoos appeared to be the most unteachable of all animals' and although M. B. Drapier had for once inspired the yahoos of Ireland to recognize and defend their own interests, Swift knew that this was a temporary, and mainly symbolic victory. His concern for his financial situation was equally genuine. He had recently spent all his money to build a 'cursed wall' around Naboth's vineyard, a private garden sanctuary just south of the Deanery in Dublin. His fears of imminent ruin were dispelled by John Worrall, his vicar at St Patrick's, in July, who provided him with a very welcome statement of his resources, which showed that his prudent management had actually saved him a little over £1,000.[241]

In July Swift received a letter from Bolingbroke suggesting that they should emulate Berkeley and emigrate to some new Atlantis. But perhaps, wondered Bolingbroke, Swift was already booked for voyages to 'those countries of giants and pigmies from whence he imports a cargo I value at an higher rate than that of the richest galleon'.[242] As the summer ended, it was time to forget such costly cargoes, and return instead to the matter of brass money. With the prospect of a return to Dublin, both Swift and Stella experienced a renewal of their infirmities. Deaf and giddy, he prepared a *Humble Address to both Houses of Parliament*, to be published on the day that Parliament reassembled. During the summer, his original draft of this pamphlet had undergone several revisions and had been 'half spoiled' by the 'cowardly caution' of Grattan.[243] The six months' period of the proclamation against the Drapier had now expired, but as Swift had no wish to have a further price

[240] Ibid., p. 60.

[241] Ibid., p. 74. As Ehrenpreis demonstrates (iii, 326-7), the financial disaster which Swift anticipated was due to the imprisonment of John Pratt, the Deputy Vice-Treasurer of Ireland, to whom Swift had entrusted large sums of money for investment.

[242] *Corr.* iii. 82.

[243] Ibid., p. 91.

placed on his head, he agreed, reluctantly, to accept his friend's modifications. However, just as he was completing a final version of this *Humble Address* he received an anonymous letter which informed him that 'he had got a sop to hold his tongue'.[244] On 19 August the Lords Justices of England told Carteret that Wood's patent had been cancelled, and they directed him to announce the cancellation in his speech from the throne at the opening of Parliament. They also required him to emphasize 'this extraordinary instance of His Majesty's condescension to the desires and petitions of his people in Ireland'.[245] Swift wrote immediately to stop publication of the *Humble Address.* 'I would have it taken back, and the press broke, and let her be satisfied. The work is done and there is no more need of the Drapier.'[246] 'Her' refers to Mrs Harding, whose husband, the printer, had died in prison only a week before, a martyr to the Drapier's cause.

Parliament finally assembled on 21 September, and Carteret made his announcement, requesting that suitable acknowledgements should be made for His Majesty's gracious concession. The Commons complied immediately, but in the Lords, Archbishop King added a little comic coda to the story, by proposing an acknowledgement of the King's wisdom, as well as his condescension in cancelling the patent. This amendment was passed, whereupon the Archbishop could not resist pointing out, mischievously, that if it was wisdom to surrender the patent, it must have been foolishness to have granted it in the first place. Once the implication of this sunk in 'it was urged that this word was removed as an affront either to his Majesty or his ministers, and therefore would be improper'.[247] At once King triumphantly retorted that nothing could be more of an affront than to deny his Majesty's wisdom in this matter. All through the campaign against Wood, the Archbishop proved an invaluable ally to Swift, a fact which Swift acknowledged by celebrating this latest sally of King's polemical wit in a short poem 'On Wisdom's Defeat in a Learned Debate'.[248]

Naturally there was great rejoicing in Dublin at this climb-down by the English Government. According to Bishop Nicolson, Swift became 'the darling of the populace: his image and superscription

[244] *PW* x. p. xxvi. [245] Ibid., p. xxviii. [246] *Corr.* iii. 93.
[247] *PW* x, pp. xxix-xxx. [248] *Poems,* i. 340-3.

on a great many signposts in this city and other great towns'.[249] But Swift did not stay in Dublin to enjoy his moment of triumph, preferring to retire to Quilca where he dug ditches, drained bogs, and wrote to his friends in England insisting that 'principally I hate and detest that animal called man'.[250]

The *Drapier's Letters* brought Swift his greatest personal triumph in the political arena. In the *Examiner* and the *Conduct* he had articulated an agreed ministerial line. As the Drapier, he formulated, led, and directed a national campaign, largely on his own initiative. Inevitably, for those whose main interest is in Swift rather than in Ireland, the *Drapier's Letters* bulk large. If they cannot justify his own vainglorious boast,

> Fair Liberty was all his cry
> For her he stood prepared to die . . .[251]

they still show him developing a sustained polemical campaign that is inventive, ingenious, and successful. But we should resist the temptation to see the campaign against Wood's halfpence purely in terms of Swift's career. The victory was of a strictly limited kind, and made little difference to the general endemic exploitation of Ireland by absentee English authorities. The English administration was more surprised than anything else, at this well-organized resistance to a measure which was merely an addition to the overall system of colonial rule. Nor should we enquire too closely into the details of Swift's case. We have no way of knowing whether Wood's coins were in fact debased. The report of the examining Committee, and of Newton's assay, seem restrained and reasonable documents; whereas Swift's figures are unreliable, and his tone is frequently bullying. The degree of corruption involved in granting Wood's patent was hardly greater than usual for the period; nor was his own rapacity any more remarkable than that of other holders of government offices and concessions. In the end, Walpole was happy to sacrifice Wood, to consolidate his administration's traditional control over the neighbour kingdom.

Swift's appeal to 'the whole people of Ireland' to reject the notion that their native land was a 'depending kingdom' was a potentially radical call. Yet although the rhetoric was directed to

[249] *PW* x, p. xxxi. [250] *Corr.* iii. 103. [251] *Poems,* ii. 566.

the 'whole people', his reliance was not upon his army of teague supporters, but upon the reasonableness of his 'old friend' Carteret, and the authority of his new friend the Archbishop. As a political demagogue, Swift was seriously limited by his complete lack of faith in the Irish people, to whom, ostensibly, he appealed. The real target audience for his *Letters* was the small political élite, both in England and Ireland. He wished to show Walpole that he could mount, and maintain such a campaign, though he was indeed surprised by the response to his call. He admitted to Chetwode that he 'hardly thought such a spirit could ever rise over this whole kingdom'. Not since the Partridge episode had he demonstrated so efficiently the power of the press. Yet he did not seek to exploit the new political situation that he had created, but rather hid away from it at Quilca. He did not wish to liberate the slaves, merely to make the terms of their slavery more acceptable. As his agreement to accept Grattan's modifications to his *Humble Address* indicated, he did not in fact 'boldly stand alone', but took prudent measures to avoid prosecution. It was the forgotten printer, John Harding, not the celebrated Dean, Jonathan Swift, who was the real martyr of the campaign against Wood.

IX

SWIFT CONTINUED to seek to influence Carteret's policies, often urging upon him the importance of preferring Irish clergymen to Irish posts, and, in particular, of finding a living for Tom Sheridan. The Lord Lieutenant complied, and appointed Sheridan as one of his chaplains, and as the incumbent of Rincurran, Co. Cork. Swift immediately drew up a careful list of rules for his friend, insisting that he should 'observe all grave forms for the want of which both you and I have suffered'. Above all, he should be thrifty. Swift did not want to hear 'of one rag of better clothes for your wife or brats, but rather plainer than ever'.[252] But alas, all his good advice was in vain. Sheridan was no sooner installed at Rincurran when, on 1 August, the anniversary of the accession of George I, he abandoned all 'grave forms' and preached a sermon on the text 'Sufficient unto the day is the evil thereof'. When Carteret came to hear of this, he had no choice but to strike Sheridan's name from

[252] *Corr.* iii. 68.

his list of chaplains. Swift was exasperated at his friend's imprudence. 'It is safer for a man's interest to blaspheme God, than to be of a party out of power,'[253] he commented bitterly, and fell back on berating the lamentable influence on Sheridan of his shrewish wife. However, the incident in fact turned out to Sheridan's benefit. Archdeacon Russell, in whose church Sheridan had preached the offending sermon, felt partially responsible for the ruin of a man 'loaded with a numerous offspring' and made him a gift of the manor of Drumlane in Cavan, worth £250 a year. It was a perfectly Irish resolution to the situation. It filled Swift with despair that in his native land, trusting to luck should seem as rational as hard work, prudence, and self-denial.

As soon as *Gulliver's Travels* was completed Swift began to think seriously about visiting England to see old friends, and arrange for publication. His letters for this period are teasing, provocative, presenting a foxy guise of misanthropy. 'You should think and deal with every man as a villain,' he counselled Sheridan; 'the devil take that animal who will not offend his enemy when he is provoked.'[254] Swift's aphorisms so often invert New Testament tenets, that one cannot help sensing that he and his Saviour would have had very little in common. At the end of September he sent to Pope his much anthologized letter, describing the motives that lay behind *Gulliver's Travels.*[255] His 'chief end', he declared, was 'to vex the world, rather than divert it . . . when you think of the world, give it one lash the more at my request'. He grandly announced his hatred of 'that animal called man', but qualified this by confessing a hearty love for individuals, 'John, Peter, Thomas and so forth'. And for one individual in particular: hearing of Arbuthnot's recent illness, he declared, 'O, if the world had but a dozen Arbuthnots in it, I would burn my *Travels.*' Then, as though embarrassed at being caught in a charitable pose, he added, 'However, even he is not without fault.'

Meanwhile, Swift's deafness grew worse and the noise in his ears, which he had formerly described as like the sound of seven water-mills, increased to the level of 'a hundred oceans rolling in my ears'. With a characteristically puritan instinct for self-mortification, he still attributed his condition to a surfeit of golden pippins and tantalized himself every summer by refraining from

[253] Ibid., p. 93. [254] Ibid., pp. 94, 101. [255] Ibid., pp. 102-5.

the soft fruits that he loved so much. More recently it has been suggested that he was actually suffering from Meniere's syndrome, or labyrinthine vertigo. It is possible that he might have played a more prominent part in the public triumph over the cancellation of Wood's patent, had not his instinctive desire to hide his affliction driven him to Quilca. 'Anything', he told Ford, was better than 'the complaint of being deaf in Dublin'.[256]

It was a consciousness of the social awkwardness that his deafness produced which restrained him, even now, from attempting a visit to England. As the old Scriblerian friends began to take the first tentative steps towards a reunion after a gap of a dozen years, a distinct apprehension mingled with their pleasure at the prospect. 'Life is good for nothing,' Swift wrote to Pope, 'otherwise than for the love we have to our friends.' Pope looked forward not only to meetings in England, but also in heaven. 'I say, that we should meet like the righteous in the millenium, quite in peace, divested of all our former passions . . . and content to enjoy the kingdom of the just in tranquillity.-[257] Bolingbroke spoke of their meeting like Stoic philosophers in some new Atlantis, and Arbuthnot pictured 'some of our old club met together like mariners after a storm'.[258] But beneath the sentimental imagery were real fears. The old Scriblerians were now unwell, and each feared he might prove dull or burdensome to the others. A meeting might shatter the golden haze that hung over their memories. Swift was not entirely guilty of false modesty when he claimed that twelve years in Ireland had disqualified him for civilized society. Worst of all was the fear of death that haunted them. Pope's often expressed hope of a meeting in heaven was an attempt to preempt this fear, but Swift had no such cosy optimism about the hereafter. He was obsessed with the loneliness of old age, and the agony of losing his friends. Almost selfishly, his fond desire was to die first, rather than be left friendless, without a barrier between himself and the yahoo world. Pope's reply to Swift indicates the differences between the two men. Although he claimed to 'enter as fully as you can desire into your principles of love of individuals' he could not refrain from attempting to convince Swift of the indissoluble links between self-love and social. 'I think the way to

[256] Ibid., p. 89.
[257] *Corr.* iii. 107-8.
[258] Ibid., p. 110.
[259] Ibid., p. 108.

have a public spirit, is first to have a private one.'[260] Whether consciously or not, he confronted Swift's philosophy head-on, when he proposed writing 'a set of maxims in opposition to all of Rochefoucauld's principles'. One can imagine Swift's wry grin as he read that. Pope often betrays a certain anxiety, resulting from his uncertainty about the seriousness of Swift's assertions of misanthropy. He was moved to pour balm over his friend's indignation, reminding him, with all due deference, of the mercy, charity, and forgiveness enjoined by the gospels. In reply Swift merely redoubled his pronouncements of gloom, adding some melodramatic flourishes. 'Drown the world!' he declared, 'I am not content with despising it, but I would anger it if I could with safety.'[260] The last hint of prudence, 'with safety', seems a way of poking fun at his own assertion, and of indicating to Pope that he is only mad north-north-west. Teasing him further he insisted that La Rochefoucauld was his favourite, 'because I find my whole character in him', and he delivered a telling blow against the providential 'optimism' of Pope and Bolingbroke. 'I tell you after all that I do not hate mankind; it is *vous autres* who hate them because you would have them reasonable animals, and are angry for being disappointed.'[261] Swift declared that he was no more angry with Walpole, than with a kite which had flown off with one of his chickens, 'and yet I was pleased when one of my servants shot him'. There is a chilling ruthlessness about the facts of nature as Swift presents them. Most chilling of all is his assertion that these were not recent conclusions, brought on by age and disillusionment, but were his settled opinions throughout life. 'I have credible witnesses ready to depose that it [his attitude to the world] hath never varied from the twenty-first to the f——ty eighth year of my life.' From the moment in adolescence when that great fish had got away from his line, he was convinced that the world was in a conspiracy against him.

As he grew older, Swift's expressions of gloom could often develop into exhibitions of bullying. At the begining of 1726 he deliberately picked a quarrel with Viscount Palmerston, Sir William Temple's nephew, over a disputed tenancy of some chambers in Trinity College.[262] The rights and wrongs of the case are unclear, but Swift evidently suspected a typical piece of Temple

[260] Ibid., p. 117. [261] Ibid., p. 118. [262] *Corr.* iii. 122-3.

family arrogance, and wrote to Palmerston with the same tone of bristling righteousness that he had used to Lady Giffard in 1709. Palmerston replied with pained condescension; 'I should not give myself the trouble to answer your polite letter, were I as unconcerned about character and reputation as some are.'[263] However, when he complained of Swift's ingratitude 'to a family you owe so much', he touched on a raw spot. Swift's reply was truly vitriolic. 'If my letter was polite,' he began, 'it was against my intention.' As far as he was concerned, his debts to the Temple family had been paid many times over in the neglect and humiliations he had suffered at Moor Park, and since.

I hope you will not charge my living in his family as an obligation, for I was educated to little purpose, if I retired to his house on any other motive than the benefit of his conversation and advice, and the opportunity of pursuing my studies. For, being born to no fortune, I was at his death as far to seek as ever.[264]

This letter was curtly endorsed by Palmerston, 'not answered'.

In December a letter from Pope and Bolingbroke invited Swift to join them in erecting a hospital for 'despisers of the world', though Bolingbroke admitted that the pose of *contemptus mundi* might often be merely a cover for envy; 'the language of a slighted lover, who desired nothing so much as a reconciliation'.[265] Having finally been allowed to return to England, Bolingbroke had quickly cast off his own stoic guise, and was pursuing his political ambitions once more. Since resignation had never been anything more than an affectation in him, it was natural that he should suspect the sincerity of others who adopted the pose. All the Scriblerians were united in one thing, however. They all eagerly awaited Swift's return to England.

X

ONE DAY in early March, full of anticipation, Swift sailed back to England. Travelling via Oxford he reached London about the middle of the month, and took up familiar lodgings in Bury Street. He was soon joined by Pope, who found him in perfect health and

[263] Ibid., p. 123. [264] Ibid., p. 125. [265] Ibid., p. 121.

5 Esther Johnson ('Stella') by an unknown artist of the 'Irish School'

6 Wood's Halfpence 1724

spirits, 'the joy of all here who knew him as he was'.[266] The next fortnight was taken up with a happy series of reunions as Arbuthnot took him on 'a course through the town'.[267] He spent time at Bolingbroke's newly acquired residence, Dawley, and visited Pope at Twickenham. He observed the 'wild boy', a natural yahoo found walking on all fours in a German forest, who had been placed under Arbuthnot's care. Most of his meetings were taken up with old friends and reminiscences, but the *Drapier's Letters* had made Swift a major political figure, and Bolingbroke was keen to enlist his aid in a new political offensive. The Prince of Wales, having quarrelled with his father, had established a rival court at Leicester House, where he flattered the hopes of Tories and disaffected Whigs. Princess Caroline, and Mrs Howard, the Prince's mistress, were both professed admirers of the Dean, and plied him with invitations to visit them. As usual, he played hard-to-get, and later boasted that the Princess had sent for him eleven times before he consented to obey her summons. His reluctance was not entirely feigned, for although his friends did their best to make him feel at home, it was some weeks before he was re-accustomed to this glittering world. Habituated to disappointment, he tested all these new protestations of friendship carefully for any tell-tale taint of hypocrisy. 'Your people are very civil to me', he wrote to Tickell in April, 'and I meet a thousand times better usage from them than . . . in Ireland.'[268] His intention was to exploit that good usage for the benefit of Ireland. Early in April he dined with Walpole 'to represent the affairs of Ireland to him in a true light'. Their meeting was immediately the subject of rumours by cynics who believed that Swift had come to make his peace with the ministry at the price of an English bishopric. Walpole listened, but was very much on his guard. He may well have intercepted a letter from Swift to Arbuthnot in which the Dean had mentioned an intention of deceiving him with flattery, and both men approached each other with wary distrust. Afterwards Swift told Peterborough, 'I failed very much in my design', for Walpole had 'conceived opinions from the examples and practices of the present and some former governors, which I could not reconcile to the notions I had of liberty'.[269]

[266] Sherburn, ii. 372.
[267] Portland MSS VII, p. 431.
[268] *Corr.* iii. 128.
[269] Ibid., p. 132.

Nevertheless Swift composed a careful statement of Irish grievances for Peterborough to pass on to Walpole.

There is not one farmer in a hundred through the kingdom who can afford shoes or stockings to his children, or to eat flesh, or drink anything better than sour milk or water, twice in a year; so that the whole country ... is a scene of misery and desolation, hardly to be matched on this side Lapland.

Swift had timed his absence from Dublin well. No sooner had he left than copies of 'Cadenus and Vanessa' began to circulate, and with them various ugly rumours. Chetwode wrote to warn him of the gossip in early April. Replying, Swift took such pains to minimize the matter that we can tell a nerve had been touched.

It was a task performed on a frolic among some ladies, and she it was addressed to died some time ago. ... I am very indifferent what is done with it, for printing cannot make it more common than it is; and for my own part, I forget what is in it, but believe it to be only a cavalier business.[270]

His language is characteristically evasive; the plural form 'some ladies' seeks to diminish the embarrassing particularity of the poem, while the description of it as a 'task', a 'frolic', and 'a cavalier business' attempts, unsuccessfully, to pass it off as a conventional piece of flattery. His declaration 'I forget what is in it' is clearly false; he could forget nothing of that painful relationship. More candidly he wrote to Tickell that the poem's appearance was proof 'how indiscreet it is to leave any one master of what cannot without the least consequence be shown to the world' and 'ought to caution men to keep the key of their own cabinets'. However, the most damaging effect of the poem was not on Swift's reputation, but on Stella's health, which suffered a serious relapse.

By the end of April the first optimistic phase of Swift's visit was over. His hopes of relieving Ireland's ills had met with a rebuff, and he was increasingly concerned by the rumours that reached him of Stella's ill health. He was soon tired of all the bustle and stir that his friends made to keep him entertained. From May till July he was forever on his rambles, staying first with Pope and Gay, and then with Bathurst. Pleasant as it was to renew so many old acquaintanceships, he was soon complaining of leading 'so rest-

[270] *Corr.* iii. 130.

less and visiting, and travelling and vexatious a life'.[271] He had heard but twice from Stella since reaching England, and worried about her greatly. Tickell had visited her one morning, but been denied admittance. 'I wonder how you could expect to see her in a morning,' Swift rebuked him, 'which I, her oldest acquaintance, have not done these dozen years, except once or twice in a journey.'[272] In July he confessed to Sheridan, 'I have had two months of great uneasiness at the ill accounts of Mrs Johnson's health.'[273] Nevertheless, his sense of guilt inhibited him from writing out his fears in full to Sheridan, who had become a close friend of Stella himself. It was in letters to his vicar John Worrall that he unburdened himself of his feelings. 'Since I left you my heart hath been so sunk, that I have not been the same man, nor ever shall be again, but drag on a wretched life till it shall please God to call me away.'[274] He begged Worrall to send him regular weekly bulletins on Stella's condition, in place of the vague and optimistic platitudes that he had been receiving from Mrs Dingley. 'I would not for the universe be present at such a trial [as] seeing her depart.' Rather than return to Dublin in time to suffer the agony of witnessing her death, he would stay in England in some place of retirement, far from London. His words to Worrall were repetitive and confused, indicating the agitation of his feelings. Again he asserted. 'I am determined not to go to Ireland to find her just dead or dying. Nothing but extremity could make me so familiar with those terrible words applied to such a dear friend.' Lamely he attempted to defend this decision: 'She will be among friends that upon her own account and great worth will tend her with all possible care, where I should be a trouble to her and the greatest torment to myself.' In reality, he could not justify his behaviour. His fear of a deathbed scene was a terror of some final emotional demand. The nearer her death approached, the more barriers he needed between them. 'I must be where I can govern,' Swift declared, when illness made him unfit for normal social life. Illness and the approach of death threatened to bring a final *éclaircissement* to those discrete patterns of dependence that held them together. At the end of a long and tortured letter he made his most guilty demand, that the ladies should no longer be

[271] Ibid., p. 137.
[272] Ibid., p. 138.
[273] Ibid., p. 138.
[274] Ibid., p. 141.

lodged at the Deanery, 'which . . . cannot but be a very improper thing for that house to breath her last in'. His syntax completely broke down, under the pressure of intense guilt and misery. He commanded Worrall 'to burn this letter immediately, without telling the contents of it to any person alive'. All that Worrall was to tell Stella was that Swift had bought her 'a repeating gold watch for her ease in winter nights'. It seems a cruel paradox that he should give a watch to her when her time was all but run out, but perhaps he intended her to count over her last minutes, like a miser counting halfpence. There was a pathetic earnestness in his insistence that Worrall should tell her of this gift, 'that she may see my thoughts were always to make her easy'.

Although he dreaded her death, he wished, for both their sakes, that her dying agony would not be long protracted. His letters to friends in Ireland all included phrases with the elegiac formality of obituary statements. 'We have been perfect friends these 35 years,' he wrote. To Worrall he confided his melancholy opinion 'that there is not a greater folly than to contract too great and intimate a friendship, which must always leave the survivor miserable'.[275] He repeated these sentiments to Stopford a week later, and the repetition gives the remark the formality of a public pronouncement on the human condition. He concluded an anguished letter to Worrall with another desperate demand for secrecy. 'When you have read this letter twice and retain what I desire, pray burn it, and let all I have said lie only in your own breast.' Worrall disobeyed this instruction, and the letter stands as eloquent testimony to Swift's obsessive desire to distance himself from all emotional claims. Swift's uneasiness made him very weary of London.

> This is the first time I ever was weary of England, and longed to be in Ireland, but it is because go I must; for I do not love Ireland better, nor England, as England, worse; in short you all live in a wretched dirty dog-hole and prison, but it is a place good enough to die in.[276]

In fact he did not have to go. He had received 'the fairest offer . . . of a settlement here that one can imagine . . . within twelve miles of London, and in the midst of my friends'. It is not known what this settlement was, but the strongest likelihood is that it was as a

[275] *Corr.* iii. 142. [276] Ibid., p. 140.

chaplain to one of his aristocratic friends. Such a post would inevitably revive his feelings of dependency, and although he said that 'if I were ten years younger I would gladly accept', he was now 'too old for new schemes, and especially such as would bridle me in my freedom and liberalities'. He would undoubtedly have accepted a bishopric in England, if offered, but a private chaplaincy, however lucrative, and however benign the patron, would not answer his needs for independence and authority. Rather be king in a dog-hole, than a spaniel in a palace.

His letters to Sheridan showed continuing gloom. 'I have been long weary of the world.' Without Stella, life would be almost unbearable. 'I fear while you are reading this, you will be shedding tears at her funeral.'[277] His remarks to Stopford stressed a proprietary, paternalistic care: 'This was a person of my own rearing and instructing, from childhood, who excelled in every good quality.'[278]

Swift's Scriblerian friends did all in their power to divert his thoughts, inviting him to dinners, to the opera, and dubbing him 'professor of divine science, la bagatelle'. A supremely inappropriate letter from Bolingbroke concluded 'mirth be with you'.[279] Swift was relieved to retire from all this to Twickenham, but unfortunately Pope too was ill, and caused Swift further anxiety in his frail condition. Pope was so weak that his least extravagance, 'if it be only two bits and one sup more than your stint',[280] was liable to have severe repercussions. Soon Swift could no longer bear to witness the daily struggles of his well-intentioned host, and in August he returned to London where he lodged with Gay in Whitehall, and made final preparations for the publication of *Gulliver's Travels.* He still expected to hear 'the worst' from Ireland by every post, but had steeled himself to return on the 15th, though he anticipated 'a very melancholy reception'. He sent a last message to Worrall insisting that if Stella should die before his return, he should send word immediately to Miss Kenah's inn at Chester, so that Swift could spend a month in Derbyshire or Wales, recovering from the shock.

On 8 August 'Richard Sympson', cousin to Lemuel Gulliver, sent one book of his cousin's *Travels* to the publisher Benjamin Motte. In return he demanded that 'a bank bill of two hundred pounds'[281]

[277] Ibid., p. 147. [278] Ibid., p. 145. [279] Ibid., p. 147.
[280] Ibid., p. 149. [281] *Corr.* iii. 153.

should be delivered 'to the hand from whence you receive this, who will come in the same manner exactly at 9 o clock at night on Thursday'. Motte was pleased with this first book, but protested that he was unable to find so much money at short notice. He promised to publish the whole work within the month, and to pay up within six months. 'Sympson' agreed, and a few nights later, Motte received the rest of the *Travels* 'he knew not from whence, nor from whom, dropped at his house in the dark, from a hackney coach'. It is not clear how far Swift felt these clandestine manoeuvres were necessary, but 'Sympson' admitted that 'the following volume may be thought in one or two places to be a little satirical', and Swift was evidently not in a mood to take chances. As soon as this final piece of business was transacted, he packed his bags and stole away from London as quietly as possible. In a final note to Worrall he wrote, 'I expect to be very miserable when I come, but I shall be prepared for it.'[282]

His English friends were certainly very miserable that he had gone. To Pope it was like 'a limb lopped off', but he consoled himself by imagining Swift's presence in all their familiar haunts. He could not see 'one seat in my own garden, or one room in my own house, without a phantom of you, sitting or walking before me'.[283] Swift returned this visionary compliment. 'I can every night distinctly see Twickenham, and the grotto and Dawley.' He invited Pope to visit him in Dublin, boasting that he was 'almost twice as rich as you, and pay no rent, and drink French wine as cheap as you do port'.[284]

The situation in Dublin was less gloomy than Swift had feared, and Stella's health gradually revived during the autumn, though she remained 'very lean and low'. Although this was a great relief to Swift, Dublin still felt like a prison.

> Breed a man a dozen years in a coal-pit, he shall pass his time well enough among his fellows, but send him to light for a few months, then down with him again; and try what a correspondent he will be.[285]

In November he had the agreeable task of degrading a parson 'who couples all our beggars'.[286] Taking a stern Malthusian line, Swift threatened the man with the gallows if he continued to officiate at

[282] Ibid., p. 156. [283] Ibid., pp. 156-7. [284] Ibid., pp. 158-9.
[285] *Corr.* iii. 171. [286] Ibid., p. 189.

such matches which could only increase the numbers of the poor. He had not been long back in Dublin when he received a rather surprising letter from Pope, who had been discussing his situation with Walpole.

He said he observed a willingness in you to live among us; which I did not deny; but at the same time told him, you had no such design in your coming this time, which was merely to see a few of those you loved; but that indeed all those wished it, and . . . wished you loved Ireland less, had you any reason to love England more.[287]

This was a pretty broad hint to the Great Man, who, naturally enough, promised nothing. But Pope continued to hope that they might live together as philosophers, contemptuous of the world, if not in England, then 'in Wales, Dublin or Bermudas'. In fact Swift's hopes of a return to England depended less on Walpole than upon the new leaders of the Opposition, Pulteney and Bolingbroke. Swift had met Pulteney briefly during the summer, and had written to him just before he left. Pulteney replied in fulsome terms. 'I would rather not have you provided for yet, than be provided for by those that I don't like,' he declared in typical courtier's style.[288] 'When we meet again,' he went on, 'I flatter myself we shall not part so soon, & I am in hopes you will allow me a larger share of your company.' He promised to provide hospitality to suit Swift's exacting tastes.

. . . you shall not have one dish of meat at my table so disguised, but you shall easily know what it is, you shall have a cup of your own for small beer and wine mixed together, you shall have no women at table if you don't like them, & no men but such as you like . . .

Evidently Swift's distress at Stella's condition had made him increasingly uneasy in any female company that summer. Pulteney was convinced that any letter from him to Swift was bound to be intercepted, so he would only hint at 'something which I would willingly have communicated'. Almost certainly what he had in mind was the new Opposition paper *The Craftsman*, the first issue of which appeared in December. Both he and Bolingbroke were keen to have Swift's help with this new enterprise. Meanwhile arrangements went ahead for the publication of a set of *Miscellanies* by Swift and Pope, over which he granted Pope

[287] Ibid., p. 160. [288] Ibid., p. 162.

'despotic power to tear as many as you please'. Pope was also busy with the poem which he still referred to as his 'Dulness'; Gay's *Fables* were completed, and awaiting publication; *Gulliver's Travels* was due to appear at the end of October. At long last it seemed that the Scriblerian satirists were about to take over the literary world.

It was at this time that Swift's correspondence with Mrs Howard began, after their meetings at Leicester House. She was a quick-witted woman who had made the transition from bedchamber woman to mistress of the Prince of Wales without alienating the affections of his wife, the Princess Caroline. In 1724, at the Prince's expense, she had a villa built at Marble Hill, near Pope's home at Twickenham. Swift always regarded her as 'a most unconscionable dealer',[289] a charming but deceitful politician. In the portrait of her which he composed the following year, he observed: 'The credit she hath is managed with the utmost thrift; and whenever she employs it, which is very rarely, it is only upon such occasions where she is sure to get much more than she spends.' It was as a fellow-sufferer from deafness that she first approached Swift, advising him, in her best governessy manner, to buck up and stop grumbling.

> I hear you are melancholy because you have a bad head and deaf ears; these are two misfortunes I have laboured under several years; and yet was never peevish with myself or the world. Have I more philosophy and resolution than you?[290]

Though he did not entirely relish this patronizing tone, Swift replied in a facetious vein, but insisted that his pains were 'ten times greater' than any she might suffer. In return for a ring which Mrs Howard gave him, Swift sent her a length of Irish plaid, 'wherein our workmen here are grown so expert, that in this kind of stuff they are said to excel that which comes from the Indies'. However, he joked that she must not 'tell any parliament men from whence you have this plaid, otherwise, out of malice they will make a law to cut off all our weavers' fingers'.[291] He hoped to make the wearing of Irish cloth the fashion for that season, and told the Princess that if she would only purchase some of the material 'at the terrible expense of eight shillings and three pence a yard', he

[289] *PW* v. 214. [290] *Corr.* iii. 232. [291] Ibid., p. 176.

would have her health drunk by five hundred weavers as 'an encourager of the Irish manufactury'. At first it seemed that his plan might succeed, and the Princess soon appeared at court 'clothed in Irish silks'. 'Are you determined to be national in everything?' asked Pope, 'even in your civilities? You are the greatest politician in Europe at this rate.'[292] But the fashion did not catch on, despite the Princess's example. Few of the court ladies could be persuaded to exchange their fine Indian silks for the coarser Irish variety.

With the beginning of November, Swift waited anxiously for news of *Gulliver's* reception. Arbuthnot was the first to write. The book, he said, was 'in everybody's hands', and prophesied it would have 'as great a run as John Bunyan'.[293] He described to Swift a London taken by storm by Gulliver, just as it had been by Bickerstaff. One sea-captain claimed to be 'very well acquainted with Gulliver, but that the printer had mistaken, that he lived in Wapping, and not in Rotherhithe'. Pope too declared that the book would be 'the admiration of all men'. The worst that he had heard said of it was that it contained 'rather too bold, and too general a satire', but 'no persons of consequence or good judgement' had any faults to find. Gay wrote that 'the whole impression sold in a week', and told Swift that 'from the highest to the lowest it is universally read, from the cabinet-council to the nursery'.[294] For the next few weeks Swift's correspondence was full of Gulliverian allusions and Lilliputian word-play. One letter from Mrs Howard so abounded in these, that Swift pretended he was unable 'to comprehend three words of it together'. It was not until a bookseller told him of Captain Gulliver's *Travels* that he could decipher her meaning. 'I thought it hard to be forced to read a book of seven hundred pages in order to understand a letter of fifty lines.'[295] Peterborough wrote to him of the new fashion of pronunciation on the Houyhnhnm model, and informed him that 'there is a neighing duetto appointed for the next opera'.[296] Meanwhile in Godalming a woman claimed to have given birth to rabbits, inspiring Mrs Howard to wonder whether, in due course, 'our female yahoos will produce a race of Houyhnhnms?'[297] Not everyone was so appreciative. One anonymous clergyman anticipated the Victo-

[292] Ibid., p. 181. [293] Ibid., p. 179. [294] Ibid., p. 182.
[295] Ibid., p. 187. [296] ibid., p. 191. [297] Ibid., p. 186.

rian horror at the fourth book of the *Travels,* when he declared that 'a man grows sick at the shocking things inserted here; his gorge rises; he is not able to conceal his resentment; and closes the book with detestation and disappointment'.[298]

There was a general expectation among Swift's friends that he would return to England the following year to bask in the glory of his *Travels.* Gay hoped he would come *'cum hirundine prima,* which we modern naturalists pronounce ought to be reckoned . . . the end of February'.[299] 'Going to England is a very good thing,' Swift told Pope, 'if it were not attended with an ugly circumstance of returning to Ireland.'[300] He added, 'it is a shame you do not persuade your ministers to keep me on that side, if it were but by a court expedient of keeping me in prison for a plotter.' In fact Swift was in two minds about visiting England. He had not greatly enjoyed his visit the previous year, and can have entertained few realistic hopes that the ministry would grant this year what it had denied last. He asked Mrs Howard to tell Walpole 'that if he does not use me better next summer than he did last, I will study revenge, and it shall be *vengeance ecclésiastique*',[301] but this was mere bluff. Nevertheless, England still excited him. In Ireland he constantly had the feeling that he was missing out on things. His visit in 1726, despite all its disappointments, had rekindled his desire for excitement again. At this late stage of his career, he relished the chance to prove that he was as good as the best of them, politicians, satirists, or peers. He was the author of the year's best-seller, the patriotic hero of a nation, and was eager to show off. Meanwhile he and Pope continued to congratulate each other on seeking no special favours from the great. 'We care not three pence whether a prince or minister will see us or no,' declared Swift, with ringing disingenuousness.[302]

In February he busied himself encouraging weavers to produce more Irish gowns that he could carry across to England. Just before embarking he received an enthusiastic note from Bolingbroke, enclosing the first issues of the *Craftsman.* 'Your deafness

[298] 'A letter from a Clergyman to his Friend', 1726. See *Swift: The Critical Heritage,* ed. K. Williams, 1970, pp. 66-70.
[299] *Corr.* iii. 184.
[300] Ibid., p. 189.
[301] Ibid., p. 196.
[302] Ibid., p. 193.

must not be a hackney excuse to you,' he insisted. 'What matter if you are deaf? . . . You are not dumb, and we shall hear you, and that is enough.'[303] The hail-fellow tones of Bolingbroke's letter were not likely to please Swift, especially when he added that his wife was sending some Gulliverian fans 'which you will dispose of to the present Stella, whoever she be'. Happily promiscuous, and having recently rid himself of his wife in order to marry his mistress, Bolingbroke had little inkling of the anxieties and guilt which Swift experienced in his own personal relationships. With Stella still in a delicate state of health, such cavalier remarks were hardly a welcome foretaste of London life. Letters from Pope and Gay were more congenial. Pope informed him that their *Miscellany* was printed, 'in which methinks we look like friends, side by side, serious and merry by turns, conversing interchangeably, and walking down hand in hand to posterity'.[304] Gay had been unwell during the winter, but had high hopes of a recovery in the spring, aided, of course, by Swift's company. The political climate grew constantly warmer. 'If you were with us, you'd be deep in politics,' said Pope, sounding a warning note. 'People are very warm, and very angry, very little to the purpose.' For his own part, Pope remained quietly at Twickenham, 'without so much as reading newspapers'. It is noticeable how insistent all the Scriblerians were about their lack of interest in politics, just as the political situation seemed set to suck them in.

On 7 April, Swift requested a six months' leave of absence, planning, if possible, to travel on to Aix-la-Chapelle after his stay in England. A proxy was arranged for the next visitation; a rogue of a Deanery proctor, who had cost Swift some £600, was dismissed; the ladies were happily installed in the Deanery for the summer, and on the 9th Swift set sail for England.

XI

As soon as he arrived in London he found himself plunged into politics. The *Craftsman* group had taken 'a firm settled resolution to assault the present administration, and break it, if possible'. Bolingbroke had been admitted to an audience with the King, and Swift found himself treated with 'high displeasure' by Walpole,

[303] Ibid., p. 200. [304] Ibid., p. 201.

who suspected him of a far greater involvement in Opposition journalism than was actually the case. As usual, Swift found that his letters were being intercepted; but now there were even more drastic threats to his safety.

I am advised by all my friends not to go to France (as I intended for two months) for fear of their vengeance in a manner which they cannot execute here.[305]

The implication is very sinister indeed. Yet, though a ruthless opportunist, there is no evidence that Walpole went in for this kind of Jonathan Wild tactic. No doubt Swift's friends magnified the dangers of travel, as a means of keeping him in England, where he might assist them with their campaign. 'You will wonder to find me say so much of politics,' he wrote to Sheridan, 'but I keep very bad company, who are full of nothing else.'[306] Actually he took very little part in the agitation of those months. He did write a *Letter* for publication in the *Craftsman*, which concentrated on attacking Walpole's hired pressmen for prostituting themselves 'to the service of injustice, corruption, party rage, and false representations of things and persons'.[307] Bolingbroke liked the *Letter*, but suggested a number of corrections. 'I would have you insinuate there, that the only reason Walpole can have . . . is the authority of one of his spies.'[308] At this late stage in his career Swift was not prepared to be supervised as closely as this, even by an old friend, and a rift soon developed between the two men. His revised version of the *Letter* was submitted in early June, but still did not suit. It was, said Bolingbroke, in a note dashed off quickly, 'extremely well, but . . . '.[309] Thereafter, Swift supplied a number of hints and 'heads' of argument for the *Craftsman* writers to work up, but his heart was never wholly in the enterprise.

The news from Ireland soon brought fresh anxieties. Stella had caught a cold, which further weakened her condition. He wrote to her frequently, advising her to spend more time in the country, 'and only now and then visit the town'. Archbishop King also caused difficulties by refusing to accept Swift's proxy at the visitation. The temporary alliance between Swift and King had not succeeded in making them friends, and Swift could never forgive

[305] *Corr.* iii. 207.
[306] Ibid., p. 208.
[307] *PW* v, pp. xii-xiv. 93-8.
[308] *Corr.* iii, 211.
[309] Ibid., p. 212.

those years of snubs and slights when King had obstructed him in every way he could. Suddenly his pent-up anger exploded in a letter to King, strongly reminiscent of his letter to Palmerston the year before. The letter seems to suggest that Swift may indeed have entertained hopes of an English post, since there is a smell of burning bridges about it, stoked on a fire of home truths.

From the very moment of the Queen's death, your Grace hath thought fit to take every opportunity of giving me all sorts of uneasiness, without ever giving me, in my whole life, one single mark of your favour, beyond common civilities. . . . This hath something in it the more extraordinary, because during some years when I was thought to have credit with those in power, I employed it to the utmost for your service, with great success, where it could be most useful against many violent enemies you then had.[310]

King, evidently astounded by this sudden aggressive display, sent a terse reply, agreeing that 'if a proxy come any time this month, it will do'.[311]

Realizing that Swift would not be of any great assistance to him, Bolingbroke ceased to feed him dismal stories of the dangers of travelling in France, and offered him letters of introduction instead to people who might receive him with honour. Voltaire, who was in London at the time, also furnished him with further introductions. *Gulliver's Travels* had just been translated into French, and Swift had high hopes of repeating his London triumph in Paris. However, just as he was about to leave, news came that altered his plans. King George, on a journey to Hanover, had died at Osnabrück. Bolingbroke wrote immediately to insist that 'there would not be common sense in your going into France at this juncture'.[312] His letter trembled with barely controlled excitement. Swift must on no account contemplate 'such an unmeaning journey . . . when the opportunity of quitting Ireland for England is, I believe, fairly before you'. Although Bolingbroke made a statutory nod towards stoicism, he could not conceal his hunger for power.

. . . to hanker after a court is fit for men with blue ribbands, pompous titles, gorgeous estates. It is below either you or me, one of whom never made his fortune, and the other's turned rotten at the very moment it

[310] Ibid., p. 210. [311] Ibid., p. 212. [312] Ibid., p. 215.

grew ripe. But without hankering, without assuming a suppliant dependent air, you may spend in England all the time you can be absent from Ireland, *& faire la guerre à l'orgueil.* There has not been so much inactivity as you imagine.[313]

For a moment Swift was tempted, but Bolingbroke's allusion to the trauma of 1714, when he had held power for just one week, was a prophetic warning. Bolingbroke was always the first proselyte to his own enthusiasms and Swift already detected a familiar strain of make-believe in the Tory euphoria. However, he allowed himself to be dissuaded by the 'great vehemence' of the entreaties of Bolingbroke and Mrs Howard, from leaving London. He wrote to Sheridan of his hopes for 'a moderating scheme, wherein nobody shall be used the worse or better for being called Whig or Tory',[314] but continued to insist, 'I design soon to return into the country.' Meanwhile Bolingbroke was troubled by an untimely 'defluxion of rhume' in both eyes, which roused him to Hamlet-like musings.

Good God, what is man? Polished, civilized, learned man? A liberal education fits him for slavery, and the pains he has taken give him the noble pretensions of dangling away life in an ante-chamber . . . or of making his reason & his knowledge serve all the purposes of other men's follies and vices.[315]

'We have made our offers,' wrote Swift, 'and if things do not mend it is not our fault.' Even if the political situation had not detained him in London, he would have remained there for a time, as he was determined to take his grievances against King to court. 'I will spend a hundred or two pounds,' he declared, 'rather than be enslaved, or betray a right which I do not value three-pence, but my successors may.' A week later he was more pessimistic about the political situation. 'Here are a thousand schemes wherein they would have me engaged, which I embrace but coldly, because I like none of them.' The tense atmosphere upset his health. 'I have been these ten days inclining to my old disease of giddiness, a little tottering.' As a diversion he wrote home to Sheridan about such homely matters as Mr Synge's crops of celery and fennel. He denied any remaining ambitions in a fierce outburst to Mrs Howard, assuring her that he had 'not the least interest with the

[313] Ibid., pp. 215-16. [314] Ibid., p. 219. [315] Ibid., p. 216.

friend's friend of anybody in power, on the contrary I [have] been used like a dog for a dozen years by every soul who was able to do it, and were but sweepers about a court'.[316] The *Craftsman* now called upon the new monarch to replace Walpole, and the Tories put forward the Speaker of the Commons, Spencer Compton, as their candidate to head a new administration. However, this genial nonentity proved unable to form a ministry, or even to draft a King's speech, without the assistance of Walpole. Throughout the summer, Walpole worked hard at re-establishing his political position; he promised the King an increased Civil List, he secured the new Queen Caroline as an ally, and gradually clawed his way back to power. Tory contempt for the incompetent Speaker can be glimpsed in Swift's letter to Mrs Howard which pokes fun at his own deafness and ill handwriting, as a means of making some satiric points. 'I make nothing of . . . knight of a share for knight of a shire, monster for minister; in writing speaker I put an "n" for a "p" and a hundred such blunders.'[317] Once again Bolingbroke found that his fortune turned rotten as soon as it ripened. This political humiliation coincided with the worst bout of deafness that Swift had ever experienced. He became unhappy and restless at Twickenham, where it pained him to watch Pope's awkward attempts to entertain him. In desperation he resolved to move to Greenwich, where he could be nursed by his relations the Lancelots. He was particularly disturbed when he heard that the new King and Queen were planning to visit Marble Hill, where he was expected to meet them.

In mid-August he received a further devastating blow in a letter from Sheridan. Stella's health had taken a serious turn for the worse. Feeling wretched, Swift wrote that he would be 'perfectly content if God [should] please to call me away at this time'. He begged Sheridan, in despair, to 'tell me no particulars, but the event in general: My weakness, my age, my friendship will bear no more.'[318] Just as the year before he determined to avoid Dublin while her life was in danger. 'I will either pass the winter near Salisbury plain, or in France.' This was a panic reaction. Only a few days later, when staying with his cousin Lancelot, he declared that if his health were tolerable he would 'go this moment to Ireland'.

[316] *Corr.* iii. 223. [317] Ibid., p. 233. [318] Ibid., p. 234.

But his deafness was worse than ever and he staggered as he walked 'like a drunken man'.

On 2 September he received another letter from Sheridan, which he kept unopened in his pocket for an hour, terrified that it contained 'the worst news that fortune could give'. He tried to strike a pose of elegiac dignity, to heighten and generalize his sense of loss. 'The last act of life is always a tragedy at best,' he wrote,[319] but it remained a bitterly personal blow. Just as the year before, he felt anxious and guilty; just as before he wrote to Worrall to insist that Stella should not lodge 'in domi deanus';[320] just as before he sought to justify his need to be absent from her deathbed. He made something of a virtue of his own illness, as a kind of sympathetic mortification, or punishment.

> I know not whether it be an addition to my grief or no, that I am now extremely ill; for it would have been a reproach to me to be in perfect health, when such a friend is desperate.[321]

All his own hopes for life were now blasted away. 'What have I to do in the world? I was never in such agonies as when I received your letter, and had it in my pocket.' He repeated again that Sheridan should not 'enlarge upon that event; but tell me the bare fact'. There was no one to whom he could confide his fears and miseries in full. After his sudden departure, Pope wrote that he feared 'some disagreeable news from Ireland'[322] had added to the torments of Swift's illness. Clearly Swift had said nothing explicit about Stella's condition, and Pope was too tactful to enquire. In September, when renewing his licence of absence, he was still undecided about his movements, and requested permission to travel in Great Britain or elsewhere. Apparently he still had thoughts of visiting France, if Stella should die, and if his own health improved. Yet just two days later he had changed his mind again, and told Mrs Howard that since he had a home in Dublin, he was determined to return there 'before my health and the weather grow worse'.[323] He spent his last week in London moving hither and thither, from Hammersmith to Whitehall, from Greenwich to Bond Street, in a despairing panic that drove him to avoid the consolation of well-meaning friends. Finally, on 18 September,

[319] Ibid., p. 236. [320] Ibid., p. 237. [321] Ibid., p. 236.
[322] Ibid., p. 241. [323] Ibid., p. 238.

deaf, giddy, exhausted, and wretched, he left London for the last time in his life.

A week later he was in Holyhead, awaiting a wind. For six days he was marooned there, mid-way between England and Ireland, and almost, one feels, between life and death. To exorcize the guilty fears that tormented him, he wrote his *Holyhead Journal,* and composed his 'Holyhead Verses'.

> . . . rather on this bleaky shore
> Where loudest winds incessant roar
> Where neither herb nor tree will thrive,
> Where nature hardly seems alive,
> I'd go in freedom to my grave
> Than rule yon isle and be a slave.[324]
> (ll. 29-34)

XII

TEN DAYS after the publication of *Gulliver's Travels* Gay reported, 'from the highest to the lowest it is universally read, from the cabinet-council to the nursery'. It continues to enjoy this wide-ranging appeal. The recent leader of the Labour party recommended 'that everyone standing for political office in Dublin, the United States or London, should have a compulsory examination in *Gulliver's Travels*'.[325] And, in its abridged and illustrated form, it maintains its place as a nursery classic. *Gulliver's Travels* encapsulates many of the paradoxes of Swift's career: the competing impulses of authoritarianism and libertarianism; the assertions of self-interest that motivated a career of social satire; the fastidiousness that delights in excremental fantasies; the rage for order, expressed in images of anarchy. Where *A Tale of a Tub* drew attention to such paradoxes with its rhetorical display, *Gulliver's Travels* presents a deceptive simplicity. The *Tale* carries the provocatively ironic epigraph 'Written for the Universal Improvement of Mankind', but Gulliver's introductory letter to his cousin Sympson forswears all reformist intentions.

[324] *Poems,* ii. 420-1.

[325] Reported in the *Guardian,* 20 October 1980. Michael Foot earlier wrote a study of Swift and Marlborough, *The Pen and The Sword,* 1957.

I should never have attempted so absurd a project as that of reforming the yahoo race in this kingdom; but I have now done with all such visionary schemes for ever.[326]

This apparent simplicity is an open invitation to summary judgements. 'When once you have thought of big men and little men, it is very easy to do all the rest,'[327] declared Johnson, confident that formulaic representations of mankind through the microscope, the telescope, and the looking-glass formed the basis of Swift's satire. In the nineteenth century an equally forthright, but far less charitable judgement was given by Thackeray. The work, he declared, is written in 'yahoo language'.

A monster gibbering shrieks, and gnashing imprecations against mankind – tearing down all shreds of modesty, past all sense of manliness and shame; filthy in word, filthy in thought, furious, raging, obscene.[328]

Gulliver's Travels tempts its readers to such blurted explanations and execrations with its teasingly simple formulas, yet elusive force. Gulliver, the most fully developed of Swift's personae, embodies many of the work's ambiguities in his own character. Though no modern critic would commit the blunder of confusing Gulliver with Swift, there are still various opinions concerning his reliability as a narrator. Biographical critics, noting that he was educated at Temple's old college, Emmanuel, see this as a source for Swift's ambivalence towards him.[329] Freudian critics note with glee that he is apprenticed to 'good master Bates', which may offer an explanation for some of his sexual and excremental obsessions.[330] Others believe that Gulliver's profession, as a surgeon, indicates that his travels should be seen as exploratory operations on the posteriors, if not the consciences, of the world.

Initially, in Lilliput, we are encouraged to trust Gulliver's reactions. We share his amusement at the exaggerated vanities of these tiny creatures, regarding them as naturalists might a species of insects whose social organizations minutely parody our own.

[326] *PW* xi. 8.

[327] Boswell. *The Life of Johnson,* '24 March, 1775'. Oxford Standard Authors edition, 1969, p. 595.

[328] Thackeray. p. 40.

[329] See Elias, p. 203. Also Ehrenpreis (iii, 456) suggests that 'at points we might even think of him [Gulliver] as a humorous reincarnation of Sir William'.

[330] See Phyllis Greenacre. *Swift and Carroll: A Psychoanalytical Study of Two Lives,* New York, 1955.

Noting the reductive analogies with English court and social life, we find no difficulty in enjoying and endorsing Gulliver's description of the scramble for office as a form of circus spectacle.

> Whoever performs his part with most agility, and holds out the longest in leaping and creeping, is rewarded with the blue-coloured silk; the red is given to the next, and the green to the third, which they all wear girt twice round about the middle.[331]

A certain amount of recent scholarship has been devoted to the significance of the colours of the ribbons awarded to the Lilliputians as rewards for their antics on the high wire. For this is one of the passages Swift had in mind when, writing as Charles Ford to his publisher Motte, he complained that the first edition of the *Travels* abounded 'with many gross errors of the press'.[332] In the first edition the ribbons do not bear the colours of the three highest British orders of honour, the Garter (blue), the Bath (red), and the Thistle (green), but were instead given the more neutral–or Roman–colours, purple, gold, and silver. However, *Gulliver's Travels,* more than any other of Swift's satires, is a work in which such specific allusions are of less significance than the general impact of his satiric analysis of human nature. Indeed, it seems possible that he may deliberately have included a number of such specific 'clues' as tubs thrown out to specialist decoders and decipherers, tempting them to hunt out some esoteric political allegory in the book, and hence miss the general force of his satire altogether. In Book III Gulliver refers to the kingdom of Tribnia where

> the bulk of the people consisted wholly of discoverers, witnesses, informers, accusers . . . who can, from a close examination of a suspected person's papers, discover any number of subversive meanings in the most innocent phrases. For instance, they can decipher a close-stool to signify a privy-council, a flock of geese, a senate, a lame dog an invader, [a codshead a ——], the plague a standing army, a buzzard a prime minister, the gout a high priest, a gibbet a secretary of state, a chamber-pot a committee of grandees, a sieve a court lady . . .[333]

[331] *PW* xi. 39. For textual variants and their political significance, see David Woolley's 'Swift's Copy of *Gulliver's Travels*' in *The Art of Jonathan Swift,* ed. Probyn, 1978, and F. P. Lock. *The Politics of Gulliver's Travels,* 1980.

[332] *Corr.* iii. 194.

[333] *PW* xi. 191. For variants see notes, *PW* xi. 311-12.

In this satiric farrago Swift cunningly intermingles fact and fantasy. References to the real codes used in the Jacobite period are interspersed with characteristically anal analogies of his own invention as he ridicules the spy's and critic's desire to uncover hidden meanings, which blinds them to more obvious comic effects. In his 'Thoughts on Various Subjects' Swift observed that 'climbing is performed in the same posture with creeping'. The antics of the Lilliputian courtiers develop this epigram into a comic vignette of universal application. His harangues at Motte's cautious 'mangling' of the text may well have been an ironic extension of a joke by which Swift deliberately drew attention to these specific points, thereby implicitly pretending that the rest of the book was innocent of satire.

In formal terms, *Gulliver's Travels* neatly combines elements of a Utopian traveller's tale with a satiric fable. The first disturbing element which violates our expectations of this literary framework, is Gulliver's obsessive need to report on his bodily functions.

> I had been for some hours extremely pressed by the necessities of nature; which was no wonder, it being almost two days since I had last disburthened myself. I was under great difficulties between urgency and shame. The best expedient I could think on, was to creep into my house, which I accordingly did; and shutting the gate after me, I went as far as the length of my chain would suffer, and discharged my body of that uneasy load. But this was the only time I was ever guilty of so uncleanly an action; for which I cannot but hope the candid reader will give some allowance after he hath maturely and impartially considered my case, and the distress I was in.[334]

Throughout his time in Lilliput we are made oppressively aware of the disproportionate physical bulk of Gulliver's body, and the awkwardness of his physical needs among such a tiny and delicate people. Even dead, his body would pose them unpleasant problems.

> They considered that the stench of so large a carcass might produce a plague in the metropolis, and probably spread through the whole kingdom.[335]

[334] Ibid., p. 29. [335] Ibid., p. 32.

Gulliver's earnest and rather pained descriptions of his physical functions introduce the basic structural antithesis of the book, which is not so much the contrast between big men and little men, but the opposition between the mind and the body.

In Brobdingnag the implications of this mind-body dualism are clearly developed. Viewed through the other end of the perspective glass, humanity appears not as vain and petty, but as ugly, gross, and crude. In particular, the sight and smell of the Brobdingnagian women cause Gulliver great offence.

> Their skins appeared so coarse and uneven, so variously coloured, when I saw them near, with a mole here and there as broad as a trencher, and hairs hanging from it thicker than pack-threads; to say nothing further concerning the rest of their persons.[336]

In his *Essay on Man* Pope argued that man's senses, like his reason, were necessarily limited, in accordance with man's role and status in the great chain of all created things.

> Why has not man a microscopic eye?
> For this plain reason, man is not a fly:
> Say what the use, were finer optics giv'n?
> T'inspect a mite, not comprehend the heav'n?
> Or touch, if tremblingly alive all o'er,
> To smart, and agonize at ev'ry pore?
> Or quick effluvia darting thro' the brain,
> Die of a rose, in aromatic pain?[337]

Swift gives a powerful sense of what it might mean to smart and agonize at every pore in Gulliver's fastidious hypersensitivity to the gross physical phenomena of Brobdingnag. Gulliver's reactions to the race of giants are, understandably, dominated by terror and disgust. Yet the physical bulk of the Brobdingnagians seems to carry with it a corresponding largeness of vision. In his exchanges with the King of Brobdingnag, it is Gulliver who appears mean and petty, ludicrously boastful of his own puny powers, while the King emerges as a model of magnanimous common sense. When the King expresses his clear preference for corn rather than gunpowder, Gulliver's petulant aside, 'a strange effect of narrow principals and short views', is rendered the more

[336] Ibid., p. 119.
[337] Pope. *Essay on Man,* Epistle I, ed. Maynard Mack, Twickenham Edition, 1950, pp. 38-50.

absurd by the preposterous incongruity of the adjectives he chooses. Indeed, throughout the first two books of the *Travels* Swift deliberately plays upon the submerged metaphors implicit in such words as 'low' and 'petty', 'grand' and 'highness'. In Book I we are allowed to share Gulliver's superior viewpoint towards the Lilliputians with their absurd disputes between high heels and low heels. In Book II, Swift enables us to endorse the lofty and Olympian judgement of the King of Brobdingnag on the human race in general and the English nation in particular.

I cannot but conclude the bulk of your natives to be the most pernicious race of little odious vermin that nature ever suffered to crawl upon the surface of the earth.[338]

Thus, while breaking down one pattern of expectations–that of trusting and identifying with Gulliver as our guide and mentor in the *Travels*–Swift seems to be tempting us with another by hinting at a quasi-materialist identification of mind with matter. Gulliver's protest at the apparent plausibility of this thesis merely reinforces its imaginative status.

I one day took the freedom to tell his majesty . . . that reason did not extend itself with the bulk of the body; on the contrary we observed in our country that the tallest persons were usually least provided with it.[339]

Actually, Swift's presentation of the physico-logical relationship between the mind and the body is considerably more problematical than this teasingly simple proposition. But the importance of the voyage to Brobdingnag is that it introduces a dichotomy in our reactions to Gulliver as a narrator. While still sympathizing emotionally with his fears and vulnerabilities as he finds himself exhibited as a freak in a country fair, dismissed as a *lusus naturae* by philosophers, and manipulated as a human dildo by the maids of honour, we no longer place much intellectual reliance on the conclusions he draws from his experiences. In fact, Gulliver's exposure gives imaginative force to a frightening suggestion expressed by Locke as he contemplated the immensity of the universe.

What other simple ideas 'tis possible the creatures in other parts of the universe may have, by the assistance of senses and faculties more or

[338] *PW* xi. 132. [339] Ibid., p. 127.

perfecter than we have, or different from ours, 'tis not for us to determine. . . . The ignorance and darkness that is in us, no more hinders, nor confines the knowledge, that is in others, than the blindness of a mole is an argument against the quick-sightedness of an eagle. He that will consider the infinite power, wisdom, and goodness of the Creator of all things, will find reason to think, it was not all laid out upon so inconsiderable, mean and impotent a creature, as he will find man to be; who in all probability, is one of the lowest of all intellectual beings.[340]

However, the humiliation of finding himself dismissed as a freak of nature produces no corresponding humility in Gulliver, but merely reinforces his complacent pride and egocentricity. As he proceeds with his further travels, Swift's central preoccupation with human pride becomes increasingly clear, and Gulliver's role shifts from being the mediator of Swift's satire to becoming its main target.

Gulliver's third voyage, often criticized as being both episodic and esoteric, is like an extended trip through the various wards of Bedlam. The experiments in the Academy of Lagado parody those of the Royal Society, turning them all into alchemical dreams. Here are virtuosi with schemes for turning ice into gunpowder, for 'reducing human excrement to its original food', and for extracting sunbeams from cucumbers. As with Pope's dunces, there is an absurd and childish fantasy about many of these projects. Yet we should not forget that these are dangerous delusions, not just simple follies.

The only inconvenience is that none of these projects are yet brought to perfection, and in the meantime the whole country lies miserably waste, the houses in ruins and the people without food or clothes.[341]

The Struldbrugs, the creatures who cannot die, are the most melancholy indictment of all man's vain attempts to transcend his mortal limitations. Condemned to a state of perpetual decay, chained to their rotting bodies for all eternity, they are a grotesque parody of that ultimate Faustian dream, the desire for immortality.

Ever since the first appearance of *Gulliver's Travels,* it has been the fourth voyage that has commanded most attention and

[340] John Locke, *An Essay concerning Human Understanding,* IV. iii. 23. Edited by P. H. Nidditch, Oxford, 1975, p. 554.

[341] *PW* xi. 177.

provoked most debate. Are the Houyhnhnms the rational, hygienic ideal of a man embittered by failure and isolated among slaves? Are the Yahoos his final savage indictment of the brute unreason, apathy, and lust of the majority of mankind? Swift certainly takes every opportunity to try to convince us that these are his views. The Houyhnhnms may lack passions, but not virtues.

> Friendship and benevolence are the two principal virtues among the Houyhnhnms, and these not confined to particular objects, but universal to the whole race.[342]

Many of the Houyhnhnms' qualities are deliberately couched in terms which, we know, had a strong appeal for Swift. Thus, to take just one example, their manner of accepting bereavement with dignity and restraint was something which he frequently recommended, and makes a poignant contrast with the ghoulish death-in-life agonies of the Struldbrugs.

One of the most useful clues to understanding Swift's satiric strategy in this fourth book was provided by R. S. Crane when he reminded modern readers that the figures of the Houyhnhnms and the Yahoos have their origins in the traditional terms and formulas employed in academic disputes about the definition of man. At Trinity College Swift learnt to construct syllogisms employing the orthodox formulas 'homo est animal rationale'; 'nullus equus est rationale'. The familiarity of this opposition of horse to man is confirmed in *Hudibras:* The hero of that poem is 'in logic a great critic' who would often

> . . . undertake to prove by force
> Of argument, a man's no horse.[343]

Locke, among others, was extremely dissatisfied with this conventional and complacent notion which identified rationality as the unique, defining characteristic of humanity.

> How far men determine of the sorts of animals rather by their shape than descent, is very visible; since it has been more than once debated, whether several human foetuses should be preserved or received to baptism or no, only because of the difference of their outward configura-

[342] Ibid., p. 268.

[343] *Hudibras,* I. i. 65, 71-2. See also R. S. Crane, 'The Houyhnhnms, the Yahoos and the History of Ideas' in *Reason and the Imagination,* Studies in the History of Ideas, 1660-1800, Ed., J. A. Mazzeo, 1962, pp. 231-53.

tion from the ordinary make of children, without knowing whether they were not as capable of reason as infants cast in another mould: some whereof, though of an approved shape, are never capable of as much appearance of reason all their lives as is to be found in an ape, or an elephant, and never give any signs of being acted by a rational soul. Whereby it is evident, that the outward figure, which only was found wanting, and not the faculty of reason, which nobody could know would be found wanting in its due season, was made essential to the human species. The learned divine and lawyer must, on such occasions, renounce his sacred definition of *animal rationale,* and substitute some other essence of the human species.[344]

In his letter to Pope, outlining the strategy of the *Travels,* Swift makes clear that he is thinking in the same terms.

I have got materials towards a treatise proving the falsity of that definition *animal rationale*; and to show it should be only *rationis capax*.[345]

Reversing the terms of the conventional formulas, Swift presents us with rational horses and irrational men. His personal circumstances may have added a piquancy to this paradox. Isolated, deaf, and increasingly pessimistic about the chances of persuading the Irish to perceive and pursue their own best interests, his happiest hours were often those which he spent out riding on horseback in the countryside. As he contrasted the placid behaviour of his horses with the obstinacy of the humans he knew, it may well have strengthened his scepticism at these conventional tropes of syllogistic philosophy. As with the familiar political maxim that 'people are the riches of a nation', this complacent definition of human nature was another maxim 'controlled', that is, contradicted, by his experience of living in Ireland.

Immediately prior to being set ashore on Houyhnhnm-land, Gulliver is confined to his cabin for several weeks by a mutinous crew. This event, coming on top of the various culture shocks he has suffered on his former voyages, appears to have a disturbing effect upon him. Superficially his narrative style retains its even, matter-of-fact, travel-book tone, but gradually, as we read on, we may become aware of a more zealous note, indicating a greater degree of mental panic. Hitherto, though we may often have ques-

[344] Locke, *An Essay on Human Understanding,* III. vi. 26; ed. Nidditch, pp. 453-4.
[345] *Corr.* iii. 103.

tioned Gulliver's judgements, we have not doubted his descriptions, however fantastical the experiences he narrates. However, in Houyhnhnm-land Gulliver's descriptions themselves embody a new alienated, or visionary, consciousness. On his arrival he beholds 'several animals in a field, and one or two of the same kind sitting in a tree'. His description of these 'animals', while on the surface zoologically neutral, is actually full of an alienating horror as he concentrates on their 'thick' hair, 'frizzled' and 'lank'; their 'strong extended claws'; their 'dugs' hanging down between their 'forefeet'. At the conclusion of this description Gulliver declares that he has never beheld 'in all my travels so disagreeable an animal, or one against which I naturally conceived so strong an antipathy'.[346] When one of the animals approaches him and lifts up 'his forepaw', Gulliver instinctively sees it as an 'ugly monster' and gives it a blow with his sword. Yet, if we can reclaim the description of these 'animals' from Gulliver's emotive terminology, we must recognize them as human beings. They are humans in a wretchedly primitive state, certainly, but Robinson Crusoe would have found no difficulty in identifying them as his fellow creatures and in beginning to civilize them. It is a human being who approaches Gulliver, with his hand raised in greeting, and his face 'distorted several ways', in what a sane man might decipher as a smile.

Gulliver never once uses the words 'man' or 'human', and his complete refusal to associate himself in any way with these 'animals' is the first and most fundamental point about his experiences on this island. From now on we are in the mind of a madman. Gulliver's horror and detestation of the human body has finally led to a mental breakdown, and he no longer recognizes himself as part of the same species as the rest of humanity. He is a man; they are animals. On the other hand, he does recognize the horses in the fields as horses.

> Looking on my left hand, I saw a horse walking softly in the field, which my persecutors [the yahoos] having sooner discovered, was the cause of their flight. The horse started a little when he came near me, but soon recovering himself, looked full in my face with manifest tokens of wonder.[347]

[346] *PW* xi. 223-4. [347] Ibid., p. 224.

While recognizing these creatures as horses, it is nevertheless with them that Gulliver senses an instinctive affinity. In their 'soft' movements and gentle neighing he intuits a rational discourse, 'like persons deliberating upon some affair of weight'. Having encountered such apparently intelligent horses, Gulliver naturally assumes that the local race of *Homo sapiens* must be supermen indeed. If the horses are rational, how much more rational must the humans be.

I was amazed to see such actions and behaviour in brute beasts, and concluded with myself that if the inhabitants of this country were endued with a proportionable degree of reason, they must needs be the wisest people upon the earth.[348]

In Gulliver's mind the residual theoretical definition of man as *animal rationale,* and the physical perception of him as a dung-throwing Yahoo, have become two totally irreconcilable things. It becomes Gulliver's greatest shame that the Houyhnhnms, with whom he seeks to identify, can find no other name for him than Yahoo, assigning him, despite the distinctions of his clothes and rationality, to that violent dung-throwing species. Clothes assume a vital importance for him, being not merely 'a cover for lewdness as well as nastiness' but the outward and visible sign of his intrinsic sense of separateness. That Gulliver, without his clothes, is nothing more nor less than a Yahoo is confirmed by the behaviour of the young female Yahoo who sees him bathing.

She embraced me after a most fulsome manner; I roared as loud as I could, and the nag came galloping towards me, whereupon she quitted her grasp, with the utmost reluctancy, and leaped upon the opposite bank, where she stood gazing and howling all the time I was putting on my clothes.[349]

To convince his Houyhnhnm masters of his detestation of all things Yahoo, he describes his own countrymen, whom he now accepts to be Yahoos, in the most hostile and reductive terms. He tells them that 'a soldier is a yahoo hired to kill in cold blood as many of his own species who have never offended him as possibly he can', and that 'this whole globe of earth must be at least three times gone round, before one of our better female yahoos could get her breakfast'. Where, in Brobdingnag, Swift made his satiric

[348] Ibid., p. 225. [349] Ibid., p. 267.

attack by having Gulliver defend the vanities of English society in pompous, narrow-minded terms, here he reinforces that effect by having a 'converted' Gulliver repudiate his previous allegiance with the vehemence of an apostate.

There is no doubt that Swift shared many of the attitudes and values that he attributes to the Houyhnhnms. Thus when Gulliver's master remarks that 'although he hated the yahoos of this country, yet he no more blamed them for their odious qualities, than he did a gnnayh [a bird of prey] for its cruelty', he is reiterating the cynical tolerance that Swift had expressed in a letter to Pope.

> I am no more angry with [Walpole] than I was with the kite that last week flew away with one of my chickens, and yet I was pleased when one of my servants shot him two days after.[350]

Cleanliness, rationality, benevolence, decency, and restraint were all ideals which Swift cherished. Yet in a post-lapsarian world, the belief that such ideals might be realized was at best a paradoxical folly, at worst a criminal cheat. The inability of the Houyhnhnms to distinguish between form and content, or to understand the use of 'the thing which is not' (lying), are severe limitations which would render them helpless in the real world of humans. They would never suspect the faults that lie beneath the beau's coat, and would be unable to comprehend the ironies of Swift's own writings. The Houyhnhnm world is perfect, ideal, and they define themselves as being 'the perfection of nature'. By seeking to imitate them, Gulliver has not learnt humility. He has merely transferred his allegiance from one species, supremely complacent of its own superiority, to another. How satisfying it would be if human life could be purged of lust and war, poverty and anger, envy and desire; how wonderful if one could really get sunbeams out of cucumbers! Gulliver seeks to persuade us that such things are indeed possible. His Messianic tone confronts us with a stark dichotomy: either to acknowledge ourselves as incurable dung-throwing animals, or to retire, like him, to the sane and sanitary haven of the stables. He can conceive of no middle way between these extremes. As in all refining processes, we are left with two products: the first, a rarified spirit, the other, a condensed dross.

[350] *Corr.* iii. 118.

The complexities of human nature are simplified out in this stark physico-logical antithesis.

In the final pages of the fourth book Swift clarifies his irony with some strokes of simple comedy. Gulliver's rescuer, the Portuguese sea-captain, is deliberately portrayed as a sane and humane man, while Gulliver's reactions to him are churlish and offensive. Gulliver's supreme rationality leads him to take such an unreasonable abhorrence of humanity that he faints at the touch of his wife, and spends four hours a day conversing with his horses. Man is not in the market for cure-alls, and with this last sad picture of Gulliver in his stables, like a Bedlamite in his cell, we reach, as promised, the end 'of all such visionary schemes'.

Part Five
THE ISLE OF SLAVES

I

SWIFT ARRIVED back in Dublin, just as he had feared, to find that Stella was dying. The years since Vanessa's death had been difficult ones in their relationship, and Stella, who had previously been healthily plump, had suffered a series of illnesses, and wasted away in a most disturbing manner. It seems possible that Swift had finally suggested that they should live openly as man and wife, some time shortly after his final letter to Vanessa in August 1722. But Stella refused, on the grounds that he had grown miserly and morose, and that her health was too bad. 'It was then', she is supposed to have told a friend, 'too *late*; and therefore better that they should live on as they had done.'[1] At the time of Vanessa's death, Stella and Dingley were staying with Ford at Wood Park, while Swift journeyed 'among people where I know no creature'.[2] Ford entertained her in splendid style, and Sheridan visited her often, doing his best to revive her spirits. It was several weeks before Swift made his own brief, awkward visit to Wood Park, where he composed a poem for her, as a peace offering. His birthday poem that year had been a perfunctory effort, which pictured Stella among the domestic servants at the Deanery, Mrs Brent, Saunders the butler, Archy the footman, and Robert the valet. Ford (Don Carlos) treated her in a much more splendid manner.

> He entertained her half a year
> With generous wines and costly cheer.
> Don Carlos made her chief director
> That she might o'er the servants hector.[3]
> (ll. 3-6)

'Stella at Wood Park' is a gentle satire, which suggests that she may have been spoiled by this exposure to the good life. Swift pictures her home-coming to her humble lodgings by 'Liffey's stinking

[1] Delany, p. 56. [2] *Corr.* ii. 456. [3] *Poems,* ii. 749.

tide' as a comic descent, as she finds that both her head and her hoops have swollen too big for her surroundings.

> The coachman stopped, she looked and swore
> The rascal had mistook the door;
> At coming in you saw her stoop;
> The entry brushed against her hoop.
> (ll. 47-50)

She does what she can to imitate Wood Park for a week, but after that, gives up the pretence, and returns to normality.

> She fell into her former scene,
> Small beer, a herring, and the Dean.
> (ll. 71-2)

There is a mixture of tones here. Swift needed to have Stella back in her place; yet he wished to suggest both that she was worthy of higher things, and that she had sufficient modesty and good sense to be content with homely pleasures. The unease of the poem comes from the recognition that Stella had enjoyed being treated as mistress of a house where cheese-paring economies were not constantly practised. In her unstable mood that summer, he could not be sure that the pleasures of Wood Park might not revive fond memories of Moor Park, and regrets for the whole course of her life since leaving Temple's household. Also he was acutely aware how difficult a companion his deafness and ill-temper had made him. 'I have put Mrs Johnson into a consumption by squalling to me,' he confessed to Ford, during a serious bout of deafness in 1724.[4] So, lest Stella should think he was attempting to belittle her, he concluded his poem with one of his more successful compliments.

> . . . you quite mistake the case;
> The virtue lies not in the place;
> For though my raillery were true,
> A cottage is Wood Park with you.
> (ll. 89-92)

Most of Swift's compliments to Stella included some little condescending qualification, to suggest that only he could appreciate her peculiar qualities. In his birthday poem in 1725 he reminded

[4] *Corr.* iii. 43.

her that 'half your locks are turned to gray.'[5] In his 'Receipt to restore Stella's Youth' later that year, he went even further in the vocabulary of unflattering compliments.

> Why Stella, should you knit your brow
> If I compare you to the cow?[6]
>
> (ll. 21-2)

It was Swift's illness in 1724 which brought them together again. 'She tends me, like an humble slave,' he wrote in his poem of gratitude.

> Although 'tis easy to descry
> She wants assistance more than I;
> Yet seems to feel my pains alone,
> And is a stoic in her own.[7]
>
> (ll. 15-18)

Patiently as she bore his outbursts, she grew thinner and paler, and would often 'sigh after Wood Park like the flesh-pots of Egypt'.[8] However, the last summer which they were to spend together was passed not at Ford's hedonist haven, but at Sheridan's ramshackle health farm at Quilca. In August 1725 he told Ford that Stella was much improved for her hikes of 'three or four miles a day over bogs and mountains'.[9] But the improvement in her health was illusory and from the time that the first printed copies of 'Cadenus and Vanessa' began to circulate in Dublin the following year, she lapsed into a long final decline.

As he watched by Stella's bedside, Swift composed prayers for them both. They are bleak, penitent prayers that plead, above all, that she should die in a sound mind.

> Preserve her, O Lord, in a sound mind and understanding, during this thy visitation; keep her from both the sad extremes of presumption and despair.

He prayed also for his own sufferings as he watched her.

> Forgive the sorrow and weakness of those among us, who sink under the grief and terror of losing so dear and useful a friend.[10]

'Useful'–even before God, this is how he described Stella–a commendation not dissimilar to that applied to Alex McGee. Sor-

[5] Ibid., p. 757. [6] Ibid., p. 759. [7] Ibid., p. 754.
[8] *Corr.* iii. 16. [9] Ibid., p. 89. [10] *PW* ix. 253-7.

7 Les Voyages du Capitaine Gulliver (1728) Bk. I, vol. i, p. 9

8 Les Voyages du Capitaine Gulliver (1728) Bk. III, vol. ii, p. 7

row is something that must be forgiven; terror the result of the loss of this last useful barrier. On 30 December Stella had made her will, guided by Swift in both its form and content. She left £1,000 to endow a chaplaincy for Dr Steeven's hospital, but added the highly Swiftian stipulations that the chaplain must be unmarried, and the Church of Ireland must remain the established Church of that kingdom. She also provided for the apprenticeship of a young child 'who now lives with me and whom I keep on charity'. This child has, inevitably, been the subject of much speculation, but there is no reason to suspect any mystery here. In his last prayer for Stella, Swift reminded God that 'the naked she hath clothed, the hungry she hath fed, the sick and the fatherless . . . she hath relieved'. This child was one of the fortunate few who enjoyed her special charity. To Swift she left a strong-box containing £150 in gold, an eloquent testimony to the efficacy of his lectures on the virtue of thrift. As might be expected, legend hovers around Stella on her deathbed. According to Sheridan, a few days before her death, she addressed Swift 'in the most earnest and pathetic terms to grant her dying request'.

> That as the ceremony of marriage had passed between them, though for sundry considerations they had not cohabited in that state, in order to put it out of the power of slander to be busy with her name after death, she adjured him by their friendship to let her have the satisfaction of dying at least, though she had not lived, his acknowledged wife. Swift made no reply, but turning on his heel, walked silently out of the room, nor ever saw her afterwards during the few days she lived. This behaviour threw Mrs Johnson into unspeakable agonies, and for a time she sunk under the weight of so cruel a disappointment.[11]

As a result of this cruel disappointment, Sheridan asserted that Stella changed her will, and left her money 'to charitable uses'. But this seems quite wrong, and an unconscious echo of the story of Vanessa's final interview with Swift. Stella's bequest to Dr Steeven's bears the clear stamp of Swift's approval, and was suggested to her by him during her illness in 1726.

Stella died on Sunday, 28 January 1728. For several months Swift had been preparing himself for this event. In early December Mrs Moore, widow to an old friend, had lost a child, and Swift's letter of

[11] Sheridan, p. 361.

consolation seems an attempt to fore-arm himself against a similar loss.

> For Life is a tragedy, wherein we sit as spectators awhile and then act our own part in it. Self-love, as it is the motive to all our actions, so it is the sole cause of our grief. . . . God, in his wisdom, hath been pleased to load our declining years with many sufferings, with diseases, and decays of nature, with the death of many friends, and the ingratitude of more: Sometimes with the loss or diminution of our fortunes, when our infirmities most need them; often with contempt from the world, and always with neglect from it; with the death of our most hopeful or useful children; with a want of relish for all worldly enjoyments, with a general dislike of persons and things; And tho' all these are very natural effects of increasing years, yet they were intended by the author of our being, to wean us gradually from our fondness of life, the nearer we approach towards the end of it. And this is the use you are to make, in prudence as well as in conscience, of all the afflictions you have hitherto undergone, as well as of those which in the course of nature and providence you have reason to expect.[12]

Swift recognized that these sentiments made him 'but a sorry comforter', but they were the only comforts which he allowed himself; the grim acknowledgement of a divine wisdom that seldom left unpunished the weakness of fixing any mortal 'too deeply in one's heart'. In accordance with such a grim pronouncement on the need for self-discipline, he did not alter his routine on the night Stella died. Sunday was his night for company, and although he received the news of her death at 8 p.m., he waited impassively among his guests until the last of them departed at eleven. He then sat down to begin his account of 'the truest, most virtuous, and valuable friend that I . . . ever was blessed with'.[13]

It is a dignified but evasive account, full of omissions and inaccuracies, the authorized version of her life by the man who largely modelled it. Both in what he mentioned, and in what he ignored, Swift's motive was to control her in death, as he had in life. Thus Addison, whom she barely met, is mentioned, and his favourable opinion of her is cited with pride. But Sir William Temple's name is entirely excluded from this brief biography. Swift lists Stella's accomplishments like a schoolmaster parading his prize pupil.

[12] *Corr.* iii. 254-5. [13] *PW* v. 227-36.

She understood the Platonic and Epicurean philosophy, and judged very well of the defects of the latter. She made very judicious abstracts of the best books she had read. She understood the nature of government, and could point out all the errors of Hobbes.

He made a catalogue of her *bons mots*,[14] and did everything to prove that she was well equipped for entry into society, to which, at the same time, he blocked her access. Yet his school report manner confirms the impression of her continuing immaturity, and even her courageous act of firing at a burglar has the derring-do quality of a schoolgirl yarn.

Only the first half of this account was written on the night of Stella's death. The following day, Swift's head ached too badly for him to continue; or perhaps he was establishing the 'illness' which would excuse him from attending her funeral. This ceremony was held on Tuesday night, when he wrote that 'my sickness will not suffer me to attend'. He could not even bear to see 'the light in the church, which is just over against the window of my bed-chamber', and removed himself to another apartment, to be quite alone with his melancholy thoughts. He there completed his highly selective biography of Stella, while the remains of Esther Johnson were being laid to rest.

II

BY ONE of those ironies of fortune to which Swift often drew attention, this, his own deepest loss, coincided with John Gay's greatest triumph. On 29 January, as Swift moped alone in his room, the *Beggar's Opera* enjoyed a tumultuous opening night, leading to a record-breaking run which, as the wags put it, made Gay rich and Rich (the impresario) gay. Gay's years of patient attendance at court had recently been acknowledged by the offer of a post so lowly that he had turned it down with disdain. Swift wrote to him sympathetically; 'I am perfectly confident you have a firm enemy in the ministry,' he declared, and delivered a sweeping indictment of court life, based on thirty-six years of experience. In all courts, he wrote, one found the same 'insincerities of those who would be thought the best friends . . . the love of fawning, cringing and tale bearing', and so on.[15] Gay sent him an account of the *Beggar's*

[14] Ibid., pp. 237-8.

[15] *Corr.* iii. 250.

Opera's first triumphal fortnight, estimating that it would bring in 'between six and seven hundred pounds'.[16] Swift was delighted. 'The *Beggar's Opera* hath knocked down *Gulliver,* 'he declared. 'I hope to see Pope's Dulness knock down the *Beggar's Opera.*'[17] He asked eagerly, 'Does Walpole think you intended an affront to him in your opera? Pray God he may.'[18] He continued to insist that Gay should not forget, in his excitement, to squeeze every penny he could from his success. 'I think that rich rogue Rich should in conscience make you a present of 2 or 3 hundred guineas. I am impatient that such a dog, by sitting still, should get five times more than the author.'[19] He recommended the purchase of an annuity 'that you may laugh at courts, and bid ministers kiss . . . etc.'.

Swift's friends had been saddened by his sudden departure, particularly Pope. 'I was sorry to find you could think yourself easier in any house than in mine.'[20] Swift's last days at Twickenham had been marred by ill-temper and tetchiness as the two sick men had tried, unsuccessfully, to console each other. Finally Swift had been forced to insist that his host leave him entirely alone. 'I never complied so unwillingly in my life with any friend, as with you, in staying so entirely from you.' Swift wrote from Ireland, trying to explain his desperation in those last weeks. His 'unsociable, comfortless deafness' made him fit for nothing better than Dublin, where he had his own servants to look after him. 'You are the best and kindest friend in the world,' he protested, 'and if ever you made me angry, it was for your too much care about me.' He invited Pope to Dublin, where he could entertain him in 'a large house where we need not hear each other if we were both sick'.[21] If not in Dublin, then at least they were certain of a reunion in heaven; 'and if I were to write an Utopia for heaven, that would be one of my schemes'. By echoing one of Pope's pet fantasies, Swift hoped to ease the grief which had left Pope feeling 'like a girl'. A few months later Pope wrote, enclosing his dedication to Swift of his latest poem, now re-titled *The Dunciad.* In reply Swift repeated his invitation to Dublin, where good French wine was plentiful, and where an eighteen-penny chicken could be purchased for only seven pence. In a similar vein of pennywise

[16] Ibid., p. 266.
[17] Ibid., p. 278.
[18] Ibid., p. 267.
[19] Ibid., p. 276.
[20] *Corr.* iii. 241.
[21] Ibid., pp. 242-3.

economy he again entreated Gay to be thrifty 'and learn to value a shilling'.[22] But even when he attempted thrift, Gay was thwarted. Arbuthnot told Swift that Gay had followed his advice and purchased 'a little place in the custom house which was to bring in about a hundred a year'. But, 'when everything was concluded, the man repented, and said he would not part with his place'. Arbuthnot added, with wry humour, 'I have begged Gay not to buy an annuity upon my life. I'm sure I should not live a week.'[23]

All this while Swift's deafness continued. Motte was planning an illustrated edition of *Gulliver's Travels,* but Swift wondered pessimistically whether the world was not already 'glutted' with the book. Pope's mother was very ill, causing the poet to reiterate his contempt for the world as a place 'for ambition, false or flattering people to domineer in'. There is something increasingly formulaic about these ritualized declarations of *contemptus mundi* by the Tory satirists. Reading through such letters, Johnson had the unpleasant sense that the Scriblerians believed they had 'engrossed all the understanding and virtues of mankind' to themselves.[24]

Pope's fears of his mother's approaching death made him commiserate with Swift's own bereavement, though he was too tactful to mention Stella by name.

> God forbid you should be as destitute of the social comforts of life, as I must when I lose my mother; or that ever you should lose your more useful acquaintance so utterly, as to turn your thoughts to such a broken reed as I am.[25]

In April the *Beggar's Opera* opened in Dublin, and Carteret enjoyed the lampoon of Walpole very much. Swift spent most of his time with the 'middle kind' of companions, humble friends who were 'perfectly easy, never impertinent, complying in everything'. He advised Pope to imitate him in his choice of companions who were 'ready to do a hundred little offices' and who could always be told 'without offence, that I am otherwise engaged at present'. Pope still spoke of visiting Ireland, and Swift assured him that the weather was much milder than in England, and that 'all things for life' were better for those of 'a middling fortune'.

[22] Ibid., p. 250.
[23] Ibid., p. 253.
[24] Johnson, iii. 61.
[25] Ibid., p. 275.

The first unannotated edition of *The Dunciad* appeared in May, and Swift was loud in his praises. 'I never in my opinion saw so much good satire, or more good sense in so many lines.'[26] Knowing that Pope intended publishing a larger edition, with a commentary, Swift advised him to be as explicit as possible in his notes, 'for I have long observed that twenty miles from London nobody understands hints, initial letters, or town-facts and passages; and in a few years not even those who live in London'. Most important of all, Pope must not be afraid to name names. 'I insist you must have your asterisks filled up with some real names of real dunces.'

Stimulated by his friends' activities, Swift decided on a new satiric enterprise of his own, and on 11 May the first issue of *The Intelligencer* appeared in Dublin. Produced jointly with Sheridan, the new periodical was intended to 'inform, or divert, or correct, or vex the town', but after only a few months the strain of producing regular issues became too great, as Swift later explained to Pope.

> If we could have got some ingenious young man to have been the manager . . . it might have continued longer, for there were hints enough. But the printer here could not afford such a young man one farthing for his trouble . . .[27]

Among the most interesting issues of the *Intelligencer* are Swift's discussions of 'The Fates of Clergymen' (Nos. V and VII). A strong personal note is evident as he reviews the careers of clergymen and reflects, with bitter irony, on the indispensability of discretion.

> There is no talent so useful towards rising in the world . . . than that quality possessed by the dullest sort of people . . . called *discretion*; a species of lower prudence, by the assistance of which, people of the meanest intellectuals, without any other qualification, pass through the world in great tranquillity. . . . This sort of discretion is usually attended with a strong desire of money, and few scruples about the way of obtaining it.[28]

In Swift's own view of his career, it was for want of discretion that he had been denied promotion. His original sin was the sin of wit, a neglect of that sanctimonious appearance of solemnity without which no clergyman could be considered truly devout. He draws

[26] *Corr.* iii. 293. [27] Ibid., iv. 30. [28] *PW* xii. 38.

contrasting portraits of two clergymen. Corusodes is sober obsequious, hypocritical, and censorious.[29] His favourite pastime is darning his socks and he never visits a tavern or playhouse. Ingratiating to the rich and insincere with the poor, his career meets with constant success. By contrast Eugenio writes verses, frequents coffee-houses, and is denied a fellowship on account of a rumour that he had 'been found dancing'. As Corusodes ascends to a bishop's throne, Eugenio is forced to accept a vicarage worth £60 'in the most desert part of Lincolnshire, where, his spirit quite sunk . . . he married a farmer's widow, and is still alive, utterly undistinguished and forgotten'.

The moral of the tale is clear; yet, in autobiographical terms the implication is misleading. Sheridan might truthfully present his character as one of impulsive high spirits, but Swift's iconoclastic tendencies were balanced by equally powerful considerations of self-interest. Swift was a natural compromiser who darned his socks *and* went to playhouses. As a result he became neither a bishop nor an anonymous vicar, but the most notorious Dean in the two kingdoms.

During the summer of 1728 it was reported in Dublin that Swift had been arrested 'for joining with some great confederates to break the present ministry, and utterly destroy Sir R. W.'. The story was quite false, but clearly Walpole was a little anxious at the new Tory onslaught, and had *Polly,* Gay's sequel to the *Beggar's Opera,* banned from performance. Gay reported with glee, 'I am looked upon at present as the most obnoxious person almost in England',[30] and Arbuthnot confirmed that 'the inoffensive John Gay is now become . . . the terror of ministers'.[31] Swift found himself beset with too many 'paltry vexations' in Dublin to think of joining in any great schemes. 'You have more virtue in an hour than I in seven years,' he told Pope in June.[32] 'You despise the follies, and hate the vices of mankind, without the least effect on your temper . . . whereas with me, it is always directly contrary.' Gay hoped to induce Swift to join himself, Pope, Arbuthnot, and

[29] Corusodes = Drip nose: one who suffers from cararrh (Greek). Ehrenpreis suggests (iii. 882) that this may be meant to hint at the nasal drone that Swift associated with Dissenting preachers.

[30] *Corr.* iii. 323.

[31] Ibid., p. 326.

[32] Ibid., p. 289.

others in Bath that August, but Swift was not in the mood for another journey to England. He continued to chide Gay for his supposed fecklessness. 'I suppose Mr Gay will return from the Bath with twenty pounds more flesh, and two hundred less in money,' he wrote to Pope. 'Providence never designed him to be above two and twenty by his thoughtlessness and gullibility.'[33] When he heard that subscriptions for the banned *Polly* would bring in £800, Swift insisted that Gay 'lost as much more for want of human prudence'. As he himself grew increasingly parsimonious, it was almost inevitable that Gay should seem profligate by contrast.

Instead of visiting England, Swift made the first of his long summer visits to Market Hill, near Armagh, the home of his friends Sir Arthur and Lady Acheson. He was delighted to get away from Dublin and happy to be the guest of such an agreeable couple. Market Hill was the setting for a series of occasional poems, composed as diversions for himself and his hosts. In September he told Sheridan that his time was so taken up with 'long lampoons' for her ladyship that he had time to think of little else. 'If I do not produce one every now and then of about two hundred lines, I am chid for my idleness, and threatened with you!'[34]

The Market Hill poems comprise a distinct group, though written during three separate visits in 1728, 1729, and 1730. In one of the first, 'My Lady's Lamentation', Swift impersonates the voice of his hostess who has been teased by his puns and tired out by his long healthy hikes 'through bogs and through briars'. Before Swift arrived she had been free to lounge about at will.

> With two boney thumbs
> Could rub my own gums,
> Or scratching my nose
> And jogging my toes.[35]
> (ll. 19-22)

But the fastidious Dean allows no such unlady-like fidgeting, and rails at both her person and her pastimes. He 'loves to be bitter at/A lady illiterate'. Worst of all, having condemned Lady Acheson to tomes of Milton and Bacon, Swift himself is out chawing tobacco with the labourers and playing the teague.

[33] Ibid., p. 294. [34] Ibid., p. 297. [35] *Poems,* iii. 852.

Find out if you can,
Whose master, whose man;
Who makes the best figure,
The Dean or the digger.[36]
(ll. 167-70)

As at Quilca, Swift found relief from his mental anxieties in the hard physical exertion of hedging and ditching. In 'Lady Acheson weary of the Dean' he again pre-empts criticism by supplying his own catalogue of the complaints that Lady Acheson might legitimately make to her husband about Swift's behaviour.

The house accounts are daily rising
So much his stay does swell the bills;
My dearest life it is surprising
How much he eats, how much he swill.[37]
(ll. 29-32)

The fact that Swift could make such jokes indicates that he felt fairly confident of his welcome at Market Hill, despite the length of his visits. His sensitivity to the inconvenience he had caused at Pope's home gave rise to no such frank jesting. The Achesons were able to perform the difficult task of allowing him the independence he needed as a guest, and these poems are his way of thanking them for indulging his cantankerous ways. Never easy with compliments, his thanks are couched entirely in terms of tolerated complaints and grumblings. Yet the underlying warmth is evident in the wit and the affectionate impersonations.

In fact Swift was so contented at Market Hill that in 1729 he bought some land from Sir Arthur with the intention of building on the site, which he renamed Drapier's Hill. For a few months he was full of the scheme, but at some stage during the summer tensions began to develop between himself and Sir Arthur which could not be laughed off in self-deprecating verses. In October he remarked to Pope dismally that 'the frolic is gone off . . . I have neither years, nor spirit, nor money, nor patience for such amusements'. He was, he added, 'only a hundred pound the poorer'[38] for his romantic notion. The poem which he wrote on this occasion, 'The Dean's Reasons for not building at Drapier's

[36] Ibid., p. 856. [37] Ibid., p. 860. [38] *Corr.* iii. 355.

Hill', is a sad piece, in which all the mock-grumbles of the earlier poems finally come true. Apparently the Achesons had at last grown tired of his domineering ways and his tendency to supervise the work on the estate as if he owned it. This poem is written in Swift's own voice, and has none of the affectionate mockery of the other poems.

> He still may keep a pack of knaves
> To spoil his work, and work by halves:
> His meadows may be dug by swine,
> It shall be no concern of mine.
> For, why should I continue still
> To serve a friend against his will?[39]
>
> (ll. 109-14)

From now on Swift would have to confine his landscaping endeavours to his own domains of Naboth's vineyard and Laracor.

Appropriately enough, Swift's only permanent improvement to the estate at Market Hill was the provision of public lavatories; 'two temples of magnific size' where,

> In separate cells, the He's and She's
> Here pay their vows with bended knees.[40]
>
> (ll. 207-8)

A privy and a madhouse; Swift's benefactions to his native land were places of confinement for the worst necessities of mankind.

Meanwhile Sheridan remained proof against all Swift's good advice. Swift counselled him to 'talk a little whiggishly' to the Primate, in hopes of obtaining 'the first good vacant school' that might be available. Instead Sheridan took it into his head to start a boarding school, and took on the lease of 'a rotten house at Rathfarnam, the worst air in Ireland, for 999 years, at twelve pounds a year; the land . . . not being worth twenty shillings a year'.[41] Having spent £100 on the house he changed his mind again, and let it run to ruin. He was therefore obliged to pay £12 a year 'for a place he never sees'. His management of his estate at Cavan was scarcely better. Worth a nominal £35 a year, Swift

[39] *Poems,* iii. 902. [40] Ibid., p. 893. [41] *PW* v. 223.

estimated that Sheridan never received a penny in rent, but rather 'expended about thirty pounds *per annum*' in buildings and plantations, which were nevertheless, 'all gone to ruin'. They continued to tease each other with riddles and rhyming-games, but despite these attempts at wit, the loss of Stella weighed heavily on Swift's mind. He never allowed her name to be mentioned, but in November wrote feelingly to a fellow clergyman on the death of his wife.

> Such misfortunes seem to break the whole scheme of man's life: and although time may lessen sorrow, yet it cannot hinder a man from feeling the want of so near a companion, nor hardly supply it with another.[42]

The scheme of his own life had been broken by Stella's death. His hatred of Dublin was redoubled now, and his long sojourns at Market Hill were a vain attempt to find a new pattern for his life.

Not long after her death, his broodings led him to jot down the beginnings of his autobiography. His reminiscences may have been prompted by some unexpected letters recalling long-dead days. The first, which vexed him by having 6s. 4d. postage due on it, included the will of Betty Jones's mother. It was with the dry chuckle of an old roué that he recounted the tale to Worrall of Betty Jones, whom he called, somewhat infelicitously 'my mistress, with a pox . . . my prudent mother was afraid I should be in love with her'.[43] The letter was from Betty's daughter, who wished to settle in Ireland, and desired to borrow three guineas to facilitate the move. To Swift the notion of anyone voluntarily choosing to settle in Ireland was 'romantic' not to say foolhardy. But he was prepared to 'sacrifice' five pounds on an old acquaintance, though nostalgia did not entirely extinguish his suspicions. 'I suspect her mother's letter to be a counterfeit', he wrote, 'for I remember she spells like a kitchen maid'.[44] Evidently Betty Jones had been the first of that long line of young women to whom he had made love with the aid of a spelling bee. Two months later a distant relative of the Temple family recalled a memorable journey which he had made as a schoolboy in Swift's care, along the Thames from Sheen. The boatman

[42] *Corr.* iii. 304. [43] Ibid., p. 309. [44] Ibid., p. 310.

had been 'very drunk and insolent'[45] but Swift had put him in his place with an authority which remained in the boy's mind ever afterwards.

However, Swift's mind was not entirely taken up by nostalgia, for it was at this time that he threw himself into his most sustained campaign against the miseries and injustices suffered by Ireland though resolutely denying the title of patriot. 'What I do', he told Pope, 'is owing to perfect rage and resentment, and the mortifying sight of slavery, folly and baseness about me, among which I am forced to live'.[46] By now his programme of reforms had achieved the recognizable characteristics of a personal manifesto. He began his new campaign with *A Short View of the State of Ireland,* in which he argued that Ireland was blessed with all the natural advantages necessary for prosperity. Political constraints alone were responsible for the nation's poverty. Ireland's many ports and havens were 'of now more use to us, than a beautiful prospect to a man shut up in a dungeon', since Ireland alone, of all kingdoms ancient or modern, was 'denied the liberty of exporting their native commodities'. How could a nation flourish when 'more than half of the rents and profits of the whole kingdom' were annually exported, and when the women 'despise and abhor to wear any of their own manufactures'. An important new element in this campaign was the sense of national shame that Swift evoked.

No strangers from other countries make this a part of their travels, where they can expect to see nothing but scenes of misery and desolation.[47]

In the *Intelligencer* (VI) Sheridan wrote, 'The poor are sunk to the lowest degree of misery and poverty–their houses dunghills, their victuals the blood of their cattle, or the herbs of the field.' It was a land, wrote Swift, where only BANKERS could grow rich, a thought which stimulated some of his most vehement rhetoric.

I have often wished that a law were enacted to hang up a half a dozen bankers every year, and thereby interpose at least some short delay to the further ruin of Ireland.[48]

[45] Ibid., p. 318.
[46] Ibid., p. 289.
[47] *PW* xii. 9.
[48] Ibid., p. 11.

Worst of all–the insult added to Ireland's injuries–was to hear those in England who lived on Irish revenues, declaring that the cries of poverty were exaggerated.

I have known a hospital where all the household officers grew rich, while the poor for whose sake it was built, were almost starving for want of food and raiment.[49]

Early in 1729, Sir John Browne published *A Memorial of the poor Inhabitants . . . of Ireland,* complaining of the dearness of corn, and proposing remedies. Few of Swift's Irish tracts are sharper, or more penetrating than his *Answer* to Browne's *Memorial.* Browne had expressed wonder that the poor should starve in such a rich country. 'Are you in earnest?' demands Swift. 'Is Ireland the rich country you mean? Or are you insulting our poverty?' As later in Scotland, so in Ireland, a policy of clearances depopulated the countryside, as absentee landlords turned over their estates to pasture lands.

Thus a vast tract of land, where twenty or thirty farmers lived, together with their cottagers and labourers in their several cabins, became all desolate and easily managed by one or two herdsmen and their boys.[50]

Swift believed that this was the main reason for the crowds of beggars who thronged the Dublin streets. 'Why all this concern for the poor?' he demanded, ironically. 'We want them not, as the country is now managed; they may follow thousands of their leaders, and seek their bread abroad.' Nor was emigration the only remedy, as Swift showed in his *Modest Proposal.*

As to the younger labourers, they are now in almost as hopeful a condition. They cannot get work, and consequently pine away from want of nourishment, to a degree that if at any time they are accidently hired to common labour, they have not strength to perform it; and thus the country and themselves are in a fair way of being soon delivered from the evils to come.[51]

England's exploitation of the Irish economy had resulted in the reversal, or 'controlling' of the state maxim that 'people are the riches of a nation'. As if this were not bad enough, the use of land for grazing sheep was 'an absurdity that a wild Indian

[49] Ibid., p. 12. [50] Ibid., p. 18. [51] Ibid., p. 114.

would be ashamed of';[52] since now the whole economy was geared to producing a commodity, wool, which the country was forbidden to export. It was an irony worthy of Swift's own invention, but tragically, it was a fact. He was particularly scathing about Browne's plans for a tax on such luxury goods as wine, which, he believed, would merely lead to an increase in smuggling. Besides, as far as Swift was concerned, wine was no luxury, but a prime necessity for making life tolerable in a bleak Hibernian atmosphere. 'Good wine is 90% in living in Ireland,' he declared.

One can detect at least two motives behind his attack on Browne. Primarily he believed that the situation was now too grave for palliatives; 'there is no dallying with hunger'. But Swift also seemed to resent anyone else offering remedies when he was convinced that his own programme of reforms represented the only sure path to salvation. Pamphlets like Browne's were at best an irritation, at worst an obstacle. There is an unmistakable tone of 'I told-you-so' about some of his arguments.

> Methinks I could have a malicious pleasure, after all the warnings I have in vain given the public . . . to see the consequences and events answering in every particular.[53]

In another pamphlet that year, deploring the steady emigration of Irish families to America, he repeated his programme with growing impatience. 'The directions for Ireland are very short and plain; to encourage agriculture and home consumption, and utterly discard all importations which are not absolutely necessary for health and life.'[54] Why would not people see it? 'I laugh with contempt at those weak wise heads, who proceed upon general maxims, or advise us to follow the examples of Holland and England. Those empirics talk by rote . . .' Ireland had suffered enough from absentee theorists. Swift was soon to produce the theory to end all theories, the state maxim reduced to state madness. Amid all the wretchedness of the poor, the conspicuous indulgence of a few seemed especially distasteful. 'Are these . . . fit for us, any more than for the beggar who could not eat his veal without orange?' The double standards that the English authorities continually applied in their dealing with

[52] Ibid., p. 18. [53] Ibid., p. 22. [54] Ibid., p. 79.

Ireland, led Swift to suppose that the Irish were regarded as an inferior species, a kind of expendable yahoo. He wondered whether 'those animals which come in my way with two legs and human faces, clad and erect, be of the same species with what I have seen very like them in England'.[55] The *Modest Proposal* was the logical outcome of this process of dehumanizing the Irish poor. Swift continued to denounce the amounts wasted on the 'unnecessary finery' of 'foreign dress and luxury'. He estimated that £300,000 annually was spent on luxury imports, mainly by women. His tone became positively vitriolic when he considered this obscene self-indulgence. 'Every husband of any fortune in the kingdom is nourishing a poisonous devouring serpent in his bosom.' At times his tone verged on monomania, as he pictured the whole of Dublin as a confederacy of dunces, rogues, and knaves, all conniving at a chaos of self-destructive self-interests. In all his years in Dublin, he declared, he had never once employed a workman 'who did not cheat me at all times'.

> One universal maxim I have constantly observed among them, that they would rather gain a shilling by cheating you, than twenty in the honest way of dealing.

The anger that runs through Swift's major Irish tracts comes from his conviction that the Irish themselves were largely responsible for their own miseries. Their self-destructive ghetto instincts were the badge of their slavery. 'I do suppose nobody hates and despises this kingdom more than myself,' he told Ford, at the height of this campaign.[56]

A Modest Proposal for Preventing the Children of Poor Parents in Ireland from Being a Burden to their Parents or the Country, and for making them Beneficial to the Public[57] is the masterpiece of Swift's mature ironic style. In it, he challenges his readers to register their own humanity by supplying those humane qualities which his logical formulae deliberately leave out of account. His recommendation of organized cannibalism is such a monstrous exhibition of inhumanity that modern readers tend to decode it by converting its force into an equally impressive monument of compassion. But, reading it in the context of his other Irish tracts and sermons, we are reminded that Swift

[55] Ibid., p. 65. [56] *Corr.* iii. 322. [57] *PW* xii. 109-18.

was sufficient of an authoritarian to wish to retain many of those penal restraints upon the poor that our century has removed. The key to the *Proposal* is the voice of the proposer. His language suggests compassion: 'I have been desired to employ my thoughts what course may be taken to ease the nation of so grievous an incumbrance.' But it is the nation that must be eased, not the suffering. Like most political theorists, the Proposer assumes a distinction between the nation and those who comprise it. The word 'ease' is a typical Swiftian usage. Had the Proposer said 'rid' the nation he would have sounded considerably less modest and awakened exactly those suspicions which his mollifying tones seek to soothe.

The *Modest Proposal* begins with words which sound sympathetic.

> It is a melancholy object to those, who walk through this great town, or travel in the country; when they see the streets, the roads and cabin doors crowded with beggars . . .

Yet it quickly transpires that the 'melancholy' comes from the inconvenience, waste, and expense of so many unproductive people, not from any fellow-feeling for their sufferings. By the consistent use of such ambiguous phrases as 'helpless infants' the Proposer manages to maintain the tone of a man with everyone's good at heart, while his argument unfolds with calculated callousness. The Proposer is a complex character. Swift does not make him completely obtuse to the hideous implications of what he suggests. There is a nervous cough in his voice: 'I shall now therefore humbly propose my own thought, which I hope will not be liable to the least objection', which entreats us into a conspiracy of silence. It is a mealy-mouthed nervousness that wishes to be absolved from any unmentioned or unmentionable offensiveness in what is promised. We sense a guilty conscience in his institutional declarations of humanity. Later, when introducing refinements to his basic scheme, regarding ways of serving children at table, he includes one 'by a very worthy person and true lover of his country' for supplying the shortage of venison with specially reared children of between twelve and fourteen years, but demurs, in his greasy fashion, for 'it is not improbable that some scrupulous people might be apt to censure such a practice (although very unjustly) as a little bordering on cruelty; which I confess, hath always been with me the strongest objection against my project, how well so ever intended'. Yet this objection is only raised after three material considerations have already caused him to reject the idea; the cost of maintaining children till twelve; the toughness of their flesh; and the waste of a capital asset in slaughtering females soon before they would be ready to breed themselves. It is this obsequious strain of guilt-consciousness that gives the Proposer's ideas their real obscenity. He is not simply an absent-minded theorist so used to dealing with statistics that he cannot think of people as people. Such a figure, while yielding a sense of savage farce, would be innocent of conscious evil. But there is a shiftiness, a loathsome cast of political opportunism behind these proposals that takes the brunt of Swift's attack: 'I fortunately fell upon this proposal; which as it is wholly new, so it hath something solid and real, of no expense, and little trouble, full in our power, and whereby we can incur no danger of disobliging England.'

The modest Proposer is in the invidious position of the Jewish police in the Warsaw ghetto, mediating between the miseries of his own people, and the intransigence of an imperial power. He is not the first to apply the strict laws of economic necessity to Ireland. He merely reacts with Pavlovian reflexes to the logic imposed by the English authorities. The most repulsive element in that obsequiousness is when he begins to mingle the language of luxury with that of harsh economic logic, by offering the children as tasty dainties to tempt the appetites of English voluptuaries.

a young healthy child, well nursed, is at a year old a most delicious, nourishing and wholesome food, whether stewed, roasted, baked or boiled, and I make no doubt that it will equally well serve in a fricassee or a ragout.

There is a lip-smacking relish to these lines, which confirm, in their panderous tone, the utter corruption of the Irish nation, in the deprivations of its poor and the depravity of its leaders.

As a final proof of his probity the modest Proposer makes this declaration of disinterestedness:

I have no children by which I can propose to get a single penny; the youngest being nine years old, and my wife past child-bearing.

It is a typical Swiftian twist, ever distrustful of those who claimed to act from disinterested motives. So it is hardly surprising that an element of self-interest mingled with Swift's genuine concern for the plight of the Irish poor in these pamphlets. His desire to vex the Whig establishment was not at odds with his wish to end the ruinous exploitation of Ireland's natural resources. It was a good example of 'self-love and social' proving their interdependence. Swift's declaration that he hated 'all nations, professions and communities' and loved only individuals, is well known. It was a serious limitation of his political radicalism that he extended this orthodox conservative distrust of abstractions to include the tribe of beggars. As he showed many times in his sermons, he would single out for his charity the one beggar in a hundred with a human face, while dismissing the rest as mere statistics.

III

SWIFT ARRIVED back from Market Hill to hear that Congreve had died and that Gay was dangerously ill. Letters from England contained increasingly gruesome descriptions of fluxions, rheums, cuppings, and vomits, and ever more gloomy fears of death. Swift was heartened, however, by the cheerfully phlegmatic attitude of his friend Dr Helsham. If one of Helsham's friends died, 'it is no more than poor Tom! He gets another or takes up with the rest, and is no more moved than at the loss of his cat: he offends nobody, is easy with everybody–is not this the true happy man?'[58]

Swift wrote recommending Helsham's resilient unsentimentality to Lady Acheson, who was horrified at such callousness, and refused to drink the doctor's health. It was not an absence of feeling which led Swift to envy his friend's philosophy, however, but a need to conquer torments that he found too painful to express. 'I would give half my fortune for the same temper,' he declared. *Half* my fortune–even in his ideals, Swift retained a sense of financial prudence.

From time to time scurrilous articles appeared in Dublin reflecting on Swift's relations with Stella and Vanessa. He

[58] *Corr.* iii. 312.

attempted to shrug these off, observing that 'every man is safe from evil tongues, who can be content to be obscure', but they rankled, nevertheless. He felt sunk and depressed. 'There is not in Ireland a duller man in a gown than I.' The news from England was not all gloomy, though. The subscription edition of *Polly* was earning Gay a good deal more money and notoriety. The Duchess of Marlborough gave £100 for a single copy of the play. The Duke and Duchess of Queensbury were banished from the court for canvassing subscriptions there, and when Gay was ejected from his court lodgings in Whitehall, it was the Queensberrys who offered him accommodation at Burlington Gardens. 'I would be contented with the worst ministry in Europe to live in a country which produces such a spirit as that girl's' was Swift's cheerful reaction to the Duchess's defiant behaviour. Arbuthnot declared that 'seven or eight duchesses' were pushing forward like primitive Christians to see 'who shall suffer martyrdom' first, on Gay's account.[59] 'He is the darling of the city; if he should travel about the country, he would have hecatombs of roasted oxen sacrificed to him.'

Gay's success caused Swift to take a long sad look at himself. After a lifetime of struggle and self-denial, what had he achieved? His health was gone, his friends were in another country, if not in another world; his finances were precarious and his fame more bad than good. He had once written that 'Dignity, high station, or great riches are in some sort necessary to old men, in order to keep the younger at a distance, who are otherwise too apt to insult them on the score of their age'.[60] He now felt acutely the insults of the ignorant, ungrateful young, and confessed to Pope and Bolingbroke, with disarming frankness, 'All my endeavours from a boy to distinguish myself were only for want of a great title and fortune, that I might be used like a lord by those who have an opinion of my parts.'[61] There is more truth in this statement than many of Swift's modern admirers care to admit. Had Swift been granted the high offices that he sought, it is quite possible that he would have been remembered as the voice of high Tory authoritarianism, rather than as the protesting

[59] Ibid., p. 326.
[60] 'Thoughts on Various Subjects', *PW* iv. 246.
[61] *Corr.* iii. 330-1.

cry of an island of slaves. It was disappointment which made Swift a radical, and it was in this letter that he described that first disappointment, with the fish that got away. He was counting over his failures. 'Nothing but deafness', he told Pope, could cure his itch for 'meddling' with public affairs. He had, after all, only two ways to gain the respect which might render old age tolerable. He must either gain public fame as the terror of ministers, or private riches as a thrifty landlord. 'I want only to be rich,' he told Pope in the spring, 'for I am hard to be pleased; and for want of riches, people grow every day less solicitous to please me.'[62] It is noticeable that his instinctive thrift was becoming, at this time, an obsessive parsimony. He kept humble company to keep down the expense of hospitality. 'I give my vicar a supper, and his wife a shilling to play with me an hour at backgammon once a fortnight.' More usually he dined alone, frugally on 'half a dish of meat' and some well-watered wine. He continued to lecture Gay on money matters, insisting that he should keep the £3,000 that he had gained from his two ballad operas intact, 'and live on the interest without decreasing the principal one penny'. He was glad to hear that Pope's fortune had increased by £100 that summer. 'Those *subsidia senectuti* are extremely desirable,' he declared.[63]

Swift's increasing obsession with money can be accounted for, at least in part, as the result of a particular ill-judged investment. While haranguing Gay on the necessity for prudence, he had been prevailed upon to lend £1,000, in a venture involving young Deane Swift's estate, which now seemed poised on the brink of disaster. 'Every farthing of any temporal fortune I have is upon the balance to be lost,' he lamented. Ignorant of this predicament, Bolingbroke gently chided him on his miserliness, and Pope insisted that he should not be 'too careful of . . . worldly affairs'. Swift's maxim, which he claimed should be written in letters of diamond, was 'that a wise man ought to have money in his head, but not in his heart'.[64] In reply to his friends' criticisms, he pointed to the self-help schemes of small loans which he had initiated. 'I pretend to value money as little as you,' he assured Bolingbroke, 'and I will call 500 witnesses, (if you will

[62] Ibid., p. 314. [63] Ibid., p. 342. [64] Ibid., p. 328.

take Irish witnesses) to prove it.'[65] Yet he could not help reminding his lordship that in a matter like this, the viewpoint of a 'younger son of a younger son' was inevitably rather different from that of one 'born to a great fortune . . . I wish you could learn arithmetic, that 3 and 2 make 5, and will never make more.' He could resign himself, philosophically, to living on £50 per annum, but not *sine dignitate.* A great lord might live frugally at ease, but 'because I cannot be a great lord, I would acquire what is a kind of *subsidium*'.[66] It annoyed Swift when his plans for building at Drapier's Hill stimulated his English friends to make gentle mockery of his schemes for retrenchment. In November 1729 Pope wrote teasingly:

> You can't imagine what a vanity it is to me to have something to rebuke you for in the way of economy. I love the man that builds a house *subito ingenio,* and makes a wall for a horse; then cries, 'we wise men must think of nothing but getting ready money.[67]

Swift was not amused. He needed the friendship of the Achesons, and was prepared to spend money to secure a haven near them and away from hateful Dublin. Nor did he like to be reminded how close to ruin he had come, while building his wall around Naboth's vineyard. These places were precious sanctuaries to him, places of retirement from the hostile world. He wrote back in protest. It was true that he had built a wall, 'being tired with the knavery of grooms who foundered all my horses and hindered me from the only remedy against increasing ill health'.[68] It was true that he had lost money over his abandoned scheme of building at Drapier's Hill. But were these expenses really worthy of ridicule? For the first time he explained to Pope the true precariousness of his financial situation. 'I am in danger of losing every groat I have in the world by having put my whole fortune, no less than £1,600 upon ill hands upon the advice of a lawyer and a friend.'[69] In this situation every penny counted, and Swift became even more zealous about hauling in his tithes, which brought fresh conflicts with his wealthiest parishioner, Percival. Swift wrote to him complaining 'of the odd way of dealing among you folks of great estates'.

[65] Ibid., p. 354.
[66] Ibid., p. 355.
[67] *Corr.* iii. 363.
[68] Ibid., p. 373.
[69] Ibid., p. 374.

It is strange that clergymen have more trouble with one or two squires and meet with more injustice from them than with fifty farmers. If your tenants paid your rents as you pay your tithes, you would have cause to complain terribly.[70]

It was appropriate that Swift's public pronouncements on the evils of landlords should have this local application. Just as when Stella was dying he found it almost a comfort to suffer a sympathetic illness, so at this time of wretchedness and starvation, it seemed almost a guarantee of his integrity that his own finances should be at the mercy of cunning lawyers and ignorant, avaricious squires. In May he described to Chetwode how he was forced to borrow money to pay his servants and common expenses. 'I have within these ten days borrowed the very poor money lodged in my hands to buy clothes for my servants.'

These pinches are not peculiar to me, but to all men in this kingdom, who live upon tithes or high rents, for as we have been on the high road to ruin these dozen years, so we have now got almost to our journey's end.[71]

In a short while Swift expected to sell or pawn his plate 'for subsistence'. Arbuthnot was 'sensibly touched'[72] by Swift's melancholy accounts of the poverty and famine in Ireland. Yet, on the whole, Swift could not help feeling that his London friends had little real conception of the true state of affairs in Ireland. In February 1730, Bathurst wrote a facetious letter in praise and imitation of the *Modest Proposal.* A father of nine, he wrote that it was 'reasonable the youngest should raise fortunes for the eldest' or, that in the case of twins, 'the selling of one might provide for the other.'[73] Several more jokes of this kind indicate that Bathurst had relished the wit, but completely missed the point of Swift's pamphlet.

Even the recognition that Swift received was soured by detractors. When in February 1730 the Dublin Corporation finally presented him with the freedom of the City in a gold box, he was immediately denounced by Lord Allen as a Jacobite libeller. The occasion of Allen's surprise outburst was Swift's recently published poem, 'A Libel on D.D.', which attacked an ingratiating

[70] Ibid., p. 366.
[71] Ibid., p. 333.
[72] Ibid., p. 337.
[73] Ibid., p. 372.

verse 'Epistle' that Delany had addressed to Carteret. Delany's gauche flattery of the Lord Lieutenant as 'thou wise and learned ruler of our isle' had stimulated Swift to deliver an attack on the deviousness and false promises of politicians who encouraged men of wit to prostitute their talents in this way. He expressed contempt for those hangers-on at court who repeated the witticisms of the great as a means of showing how well 'in' they were, while actually only demonstrating their fatal misunderstanding of the rules of the power game.

> Deluded mortals, whom the great
> Choose for companions *tète à tête*
> Who at their dinners, *en famille*
> Get leave to sit whene'er you will;
> Then boasting tell us where you dined,
> And how his lordship was so kind.[74]
>
> (ll. 1-6)

Although never mentioned by name, it is Oxford, who had manipulated Swift in just this way, who is the main target of this attack.

> Suppose my lord and you alone;
> Hint the least int'rest of your own;
> His visage drops, he knits his brow,
> He cannot talk of bus'ness now
>
> (ll. 13-16)

The lines inevitably recall those painful episodes, related at length in the *Journal to Stella,* when Swift had tried to pin down Oxford's promises, only to be put off again, and again. He looked back bitterly on those years when he had lived the life of a spider, admitted to the dinner-table, but not the council-rooms of the great.

> For, as their appetites to quench
> Lords keep a pimp to bring a wench;
> So, men of wit are but a kind
> Of pander to a vicious mind,
> Who proper objects must provide
> To gratify their lust of pride,

[74] *Poems,* ii. 480.

When weary'd with intrigues of state
They find an idle hour to prate.
Then, should you dare to ask a place,
You forfeit all your patron's grace,
And disappoint the sole design
For which he summoned you to dine.
(ll. 21-32)

He had deceived himself into believing that he was Oxford's trusted confidant and adviser, but now saw that he had merely been a court jester, and hack. In October 1729 he had told Bolingbroke that 'you were my hero, but the other [Oxford] never was'.[75] This was untrue, and from this point on Swift's memory and judgement became increasingly unreliable. A lonely, deaf, autocratic man of sixty-two, whose thoughts, he declared, had been pre-occupied with death for the last fifteen years, he was a prey to obsessions and contradictory impulses. This abrupt statement to Bolingbroke is a further example of that disconcerting facility for rearranging the past which made Swift a fine satirist, but a highly unreliable historian. He now regarded his period of dining *en famille* with Oxford in the same light as he looked back on his time at Moor Park. Both Oxford and Temple had patronized him, treating him like a clever schoolboy, tantalizing him with the prospects of positions of which they assured him he was worthy. Yet both had betrayed him, and left him to rot in Ireland, dying there, day by day, 'in a rage, like a poisoned rat in a hole'.

Swift was genuinely contrite at the mortification Delany had suffered as a by-product of the *Libel on D.D.* 'He is a man of the easiest and best conversation I ever met with in this island,' he told Pope, 'a very good listener, a right reasoner.' Such high praise was probably not unrelated to the fact that Swift had gained the use of Delany's pleasant country house, a mile outside Dublin, as a lodging for Pope, if he should visit Ireland. And in June Delany's flattery in his *Epistle* paid off, when Carteret disproved Swift's cynicism by appointing Delany Chancellor of St Patrick's. Swift was highly pleased by this move, and regretted the pain he had caused his friend, 'for I would see all the little

[75] *Corr.* iii. 353.

rascals of Ireland hanged, rather than give them any pleasure at the expense of disgusting one judicious friend'. Yet he could not entirely cease writing his libels and lampoons, 'bad prose or worse verses, either of rage or raillery', though he burned most of what he produced. Only a few escaped the flames, he told Bolingbroke, 'to give offence, or mirth'.[76] Increasingly he made himself the main subject of these ironic sallies, composing two mock-autobiographies in verse, the 'Life and Genuine Character of Dr Swift' and 'Verses on the Death of Dr Swift'.

The 'Life and Genuine Character' which appeared in London with the date of April 1733 was denounced as spurious by Swift, even though he had already told Gay that he had written 'near five hundred lines' of a self-portrait, based on one of La Rochefoucauld's maxims. He asserted that there was not 'a single line, or bit of a line, or thought, any way resembling the genuine copy' in the poem.[77] At the very least, this is a considerable overstatement, since the disowned 'Life and Genuine Character' bears many strong resemblances to the acknowledged 'Verses on the Death', both in form and content. Pope was clearly suspicious of Swift's disclaimer when he made this tentatively ironic comment.

> The man who drew your character and printed it here was not much in the wrong in many things he said of you; yet he was a very impertinent fellow, for saying them in words quite different from those you had yourself employed before on the same subject; for surely, to alter your words is to prejudice them and I have been told, that a man himself can hardly say the same thing twice over with equal happiness.[78]

The first of April was a favourite Swiftian date for hoaxes, and Pope obviously suspected that the vehemence of Swift's repudiation was all part of the joke. Certainly this ironic description of Swift in the poem has the spiky quality of the genuine article.

> 'He was an honest man, I'll swear.'
> 'Why sir, I differ from you there,
> For I have heard another story,
> He was a most confounded Tory!'[79]
>
> (ll. 80-3)

[76] Ibid., p. 382.
[77] Ibid., iv. 151-2.
[78] Ibid., p. 194.
[79] *Poems,* ii. 547.

Swift's authorship of the 'Verses on the Death of Dr Swift' is not in doubt, but the problem of identifying the precise tone of this, his most famous poem, remains. Swift undertakes to prove one of La Rochefoucauld's most cynical maxims: 'Dans l'adversité de nos meilleurs amis nous trouvons quelque chose, qui ne nous déplait pas.[80] He seemed determined to show that the misanthropy he professed was more than a pose. In November 1730 he told Gay that hatred was the sauce he used to give relish to his meals, and he assured Pope that avarice and hardness of heart were 'the two happiest qualities a man can acquire who is late in his life'.[81] Most of such remarks amounted to no more than his usual shock tactics for gaining attention, but his English friends clearly suspected that he was beginning to grow into his own misanthropic mask, just as in his *Project for the Advancement of Religion* he had suggested that the pretender to virtue might at last become virtuous in fact.

The couplets in which Swift admits to envying his friends' satiric achievements are typical of the poem's ambiguities. It was all very well to confess,

> In Pope I cannot read a line
> But with a sigh I wish it mine.
> (ll. 47-8)

But when he writes of Gay

> Why must I be outdone by Gay
> In my own humorous biting way.[82]
> (ll. 53-4)

one senses a tinge of genuine jealousy. Gay, the 'terror of ministers', had recently boasted that he was determined to write to Swift, 'though those dirty fellows of the post office do read my letter'.[83] One is left to wonder whether the letter was more likely to be intercepted because it was *from* Gay or *to* Swift? Which man now posed the greater threat? Swift, safely isolated in Dublin, or Gay, who had precipitated a little court revolution

[80] 'In the adversity of our best friends, we find something that doth not displease us.'

[81] *Corr.* iii. 421, 435.

[82] *Poems,* ii. 555.

[83] *Corr.* iii. 403.

with his banned play? Of course Swift, who had first suggested the idea of a Newgate work, liked to think of Gay as his protégé. It came as something of a shock therefore when Gay, typically maladroit, flattered him by denying any such influence. 'You and I are alike in one particular,' he wrote in July 1731, (I wish to be so in many), I mean that we hate to write upon other folks' hints. I like to have my own scheme and to treat it in my own way.'[84] It is impossible to believe that Gay intended any snub by this, but Swift was hurt by the remark, and replied that while it was 'past doubt' that a writer could best 'find hints for himself . . . sometimes a friend may give you a lucky one just suited to your own imagination'.[85] This was just a nudge to remind Gay not to become too conceited.

Swift's presentation of his friends' reactions to his death is deliberately flippant.

> Poor Pope will grieve a month: and Gay
> A week; and Arbuthnot a day.[86]
> (ll. 206-7)

We know that Swift approved of such brevity in mourning, and commended the example of his phlegmatic friend Dr Helsham. The problem presented by this poem is that of distinguishing between a cheerful and rational restraint, and simple insensitivity. The reactions of Swift's 'female friends' exhibit a callousness that translates self-interest into selfishness. Two females in particular, Mrs Howard, now Lady Suffolk, and Princess–now Queen–Caroline, demonstrate an indifference which is condemned, not commended.

> Kind Lady Suffolk in the spleen
> Runs laughing up to tell the Queen.
> The Queen, so gracious, mild and good
> Cries, 'Is he gone? Tis time he should.
> He's dead you say; why let him rot . . .[87]
> (ll. 179-83)

For several years now Swift's criticisms of Lady Suffolk had grown into a fierce antipathy, though Pope and Gay still did their best to blunt the edge of his anger.

[84] Ibid., p. 478.
[85] Ibid., p. 495.
[86] *Poems*, ii. 561.
[87] Ibid., p. 559.

The ambiguities of the poem are at their most acute in the final five hundred lines, where an apparently impartial speaker offers a posthumous summary of Swift's qualities. Much of what Swift claims for himself here is both vain and false. How can one take seriously such a line as,

> He lashed the vice, but spared the name,
> (l. 460)

when this poem itself contains the names of at least a dozen victims. Likewise the claim,

> To steal a hint was never known,
> But what he writ was all his own,
> (ll. 317-18)

is a direct self-contradiction, being itself stolen from Denham's Elegy on Cowley:

> To him no author was unknown
> Yet what he wrote was all his own.[88]

Pope was so embarrassed by the exaggerations of this section of the poem that he edited out several of the more contentious or self-glorifying couplets before forwarding the poem to the press. The resulting *Verses* were a great success with the public. Several thousand copies were sold, and Pope hoped that Swift would not 'dislike the liberties'[89] he had taken with the text. But Swift did dislike them very much, and insisted on publishing his own unexpurgated Dublin edition. Pope and his colleague William King were convinced that they had acted in Swift's best interest, believing 'that the latter part of the poem might be thought by the public a little vain, if so much were said by himself of himself'.[90] King added, somewhat hesitantly, that the claim to have spared names was not entirely true. Swift must have smiled when he read these well-intentioned objections, that so completely missed the irony of his poem. For a great deal of the poem is comprised of examples of 'the thing which is not'.

> Yet, malice never was his aim
> He lashed the vice but spared the name.

[88] 'On Mr Abraham Cowley', *The Poetical Works of Sir John Denham*, ed. T. H. Banks, 1969, p. 150.
[89] *Corr.* v. 133.
[90] Ibid., p. 139.

At the very time that he wrote this, he was declaring to Bathurst that he preached nothing but 'revenge, malice, envy and hatred and all uncharitableness'. Obviously, both statements cannot be true, but understanding Swift's art is a matter of identifying a local context and persona. It should come as little surprise that Swift could offer these two antithetical versions of himself. Writing to Bathurst he took a malicious yahoo view of the world. Writing as the impartial obituarist in the poem he asserts a Houyhnhnm-like idealism.

> Fair Liberty was all his cry.

Both statements are partial, unreliable views, and the latter declaration needs to be read in the context of a work whose avowed purpose is to prove La Rochefoucauld's contention that self-love conquers all. The most interesting irony of this poem is that the 'Swift' who is memorialized here emerges as the only altruist in a society of narcissists. His money is left to 'public uses . . . to do strangers good'; his safety is hazarded in the cause of liberty.

> For her he stood prepared to die
> For her he boldly stood alone.

Unconcerned for wealth and power, he

> Kept the tenor of his mind
> To merit well of human kind.

It is a picture of selflessness, dedication, and high principles, in apparent contradiction of La Rochefoucauld. Yet who is saying all this? Why, Swift himself; hence the panegyric on altruism becomes an exercise in self-love, confirming the poem's thesis. The only one who will really speak well of the dead is the dead person himself. And yet, is not the poem ironic? Is Swift really trying to prove La Rochefoucauld right, or merely pretending to do so, as a means of chastening our pride? We are in a *galerie des glaces,* where self-images ricochet in an infinite perspective. Swift, who well knew the power of the printed word to create its own reality, deliberately left posterity to ponder on these eulogistic couplets. The final irony of the poem is that the man who privately declared himself to have been motivated by simple malice and the desire to be 'used like a lord' should invite us to see him as a patriot and a philanthropist. It is as though he was determined to have the last laugh, whatever judgement we should pass on him. Many people have expressed a desire to read their own obituaries. Swift went one better and wrote his. He also wrote his own epitaph.

IV

IT IS impossible to ignore Swift's growing ill temper at this period. After Percival and Lady Suffolk, Chetwode was the next victim of his spleen. For several years Swift had regarded Chetwode's financial and marital problems as irritations which he had brought upon himself, and turned aside Chetwode's frequent appeals for advice with condescending snubs. During the winter of 1729/30 Chetwode spent three months in Dublin, but was never once invited to the Deanery. As he was leaving, after pursuing an unsuccessful amour, he wrote to protest that he was returning to Woodbrooke where he could enjoy 'the sun and fresh air without paying a fruitless attendance upon his eminence of St Patrick's; my fruit will bloom, my herbs be fragrant, my flowers smile though the Dean frowns, and looks gloomy'.[91] Chetwode was convinced that Swift had warned the 'half a dozen females' who had been the objects of his attention, to be on their guard against him, and was understandably piqued. It was some months before Swift could bring himself to complete a reply, but when he did so it was an abusive catalogue of Chetwode's blunders, from his Jacobite adventure, to his wooing of a sixty-year-old widow. 'Your whole scheme of thinking, conversing and living differs in every point from mine',[92] he declared loftily, renouncing any further interest in Chetwode's existence.

Certain themes recur with regularity in Swift's letters to friends in England. Most notably there are his complaints of deafness and giddiness, and his descriptions of the strenuous remedies of walking and riding that he undertook to combat them. Often he paints a forlorn picture of his isolation in Dublin, with only 'half a dozen middling clergymen, and one or two middling laymen' for company. He read only sparingly, fearing to

[91] *Corr.* iii. 442-3. [92] Ibid., pp. 461-3.

strain his eyes, and found little inclination to write. Meals were frugal affairs, taken along with 'Sir Robert' alias Mrs Brent, prime minister of his domain. 'Sometimes,' he wrote, 'I hardly think it worth my time to rise, and would certainly lie all day a bed if decency and dread of sickness did not drive me thence.' Often his letters include a little travel fantasy, with Swift and his English correspondents urging each other to cross the Irish sea. These invitations soon became ritual exchanges, with little likelihood of being fulfilled. Swift had several reasons for not risking another visit to England. The first was financial. Throughout 1730 and 1731 he was in constant expectation that his lawsuit would be quickly settled, but each time some new twist delayed a resolution. In June 1731 he complained, 'I am at sea again for almost all I am worth.' Ill health was another strong disincentive, as were his genuine fears that he had grown too dull and morose for civilized society. 'Time and the miseries I see about me have made me almost as stupid as the people I am among, and altogether disqualified me from living with better.' Nevertheless his friends in England all claimed to see through his curmudgeonly bluster. In November 1731 he was offered his pick of the stately homes of England as accommodation. Lady Betty told him not to tantalize her further, but 'choose your residence, summer or winter, St James's Square or Drayton'.[93] Meanwhile Gay declared, 'You make your own conditions at Amesbury, where I am at present; you may do the same at Dawley, and Twickenham . . .' If none of these would serve, then Gay went on, 'I will purchase the house you and I used to dispute about over against Ham walks, on purpose to entertain you.'[94]

Swift was flattered by all these entreaties, which were part of a game in which his role was to refuse ever more lavish and extravagant offers. It was a game which enhanced the asceticism of his isolation, making it seem a choice rather than a punishment. Meanwhile Bolingbroke continued to seek for a permanent position in England for Swift, and assured him in August that he had 'two or three projects on foot for making such an establishment here as might tempt you to quit Ireland'.[95] Swift still found that idea attractive, though he pretended to

[93] *Corr.* iii. 505. [94] Ibid., p. 503. [95] Ibid., p. 485.

dismiss it. 'What should I do in England?' he demanded of Lord Oxford. At least in Dublin 'I have the rabble on my side.'[96] The height of his ambition, he now claimed, was to be a vicar in Wales, 'to keep my nag, myself, and a glass of ale'.

He wrote little, 'My invention & judgement are perpetually at fisty cuffs,' he told Gay. Even his 'zeal for liberty' found no effective outlet, but, as he told Oxford, had rather 'eaten me up, and I have nothing left but ill thoughts, ill words and ill wishes . . . of all which I am not sparing, and like roaring in the gout, they give me some imaginary ease'.[97] The relief that he obtained from these 'roarings' reminded him of a monkey he had once seen 'overthrow all the dishes and plates in a kitchen, merely for the pleasure of seeing them tumble and hearing the clatter they made in their fall'.[98] Only the most malicious of political ironies brought a smile to his lips. In June 1730 the Dean of Ferns was indicted for rape, but was acquitted according to Swift, on the grounds that he was drunk at the time. 'I name him to your lordship,' Swift wrote to Oxford, 'because I am confident you will hear of his being a bishop.'[99] He composed a sardonic little ballad in which he used the incident as a perfect metaphor for England's guiltless rape of Ireland.

Gay had recently expressed an interest in a Mrs Drelincourt, and written to Swift for his advice. But Swift was impatient with his friend's hesitant gallantries and circumlocutions. 'If you like Mrs D why do you not command her to take you; if she does not, she is not worth pursuing.'[100] Free from emotional entanglements himself now, Swift adopted a crusty, peremptory style which was quite at odds with his own previous behaviour. Pope and Bolingbroke were closeted together in profound philosophical speculations which were to lead, eventually, to the 'Essay on Man'. In January 1731 Bolingbroke announced that Pope was writing a work of extraordinary significance, that would attempt 'a real reformation' of the nation's vices.[101] In March he wrote again, with the born-again idealism of a reclaimed *débauché*, urging Swift to repair his fences against the evils of bitterness and envy. Uninspired, Swift replied that

[96] Ibid., p. 405. [97] Ibid., p. 405. [98] Ibid., p. 383.
[99] Ibid., p. 405. [100] Ibid., p. 471. [101] Ibid., p. 438.

life was in general a farce, but old age was 'a ridiculous tragedy, which is the worst kind of composition'.[102]

Pope was still very ill, and Arbuthnot prescribed for him a diet of asses' milk. Bathurst had a similar idea for curing Swift of the spleen. He recommended that he should dismiss all his servants, except for one 'sound wholesome wench', and that he should 'contrive some way or other that she should have milk'. For he assured him that the best physicians all agreed 'that women's milk is the wholesomest food in the world'.[103] With this modest proposal, Bathurst hoped to solve both Swift's financial and his emotional problems; but Swift was too committed to his stance of lonely intellectual warfare to re-enter the human family in a second babyhood at the breast.

'I am glad you resolve to meddle no more with the low concerns and interests of parties,' wrote Pope in December 1731.[104] He hoped that Swift's retirement from politics would free his mind for loftier speculations. But he was wrong. Swift was still fascinated with low concerns as he revealed in the group of poems known commonly as his 'scatological poems', which have done more to blacken his reputation with later generations than anything else he wrote. The four most notorious poems are 'The Lady's Dressing Room' (1730), 'A Beautiful Young Nymph going to Bed', 'Strephon and Chloe', and 'Cassinus and Peter' (all 1731), and although he protested to Pope that they had been published without his consent from 'a stolen copy' we can discount that as a typical piece of prevarication. 'A Beautiful Young Nymph going to Bed' has certain obvious similarities with his earlier poem 'The Progress of Beauty' (1719). The 'heroines' of both poems are ageing whores who use every cosmetic aid to conceal the ravages of time and the pox. But the satire in 'The Progress of Beauty' is literary rather than moral; poets, not prostitutes are the subject of his attack. 'A Beautiful Young Nymph' and 'The Lady's Dressing Room'[105] are both anatomizing poems, in which Swift takes a forensic delight in lifting the silk petticoats to expose what lies beneath. The motivation

[102] Ibid., p. 456.
[103] Ibid., p. 454.
[104] Ibid., p. 510.
[105] *Poems,* ii. 524-30, 580-3.

for this examination is partly explained in the 'Digression on Madness', where the carcass of a beau is stripped and 'we were all amazed to find so many unsuspected faults under one suit of clothes'. The two poems represent two stages of such an operation. In 'The Lady's Dressing Room' a high-class lady's secrets are detected 'as physicians discover the state of the whole body, by consulting only what comes from behind'. In 'A Beautiful Young Nymph', the whore's living body is itself dismembered to reveal how beauty and disease intermingle like rose and bindweed in the deepest recesses of her body. 'The Lady's Dressing Room' makes a before-and-after diptych with Belinda's toilette in 'The Rape of the Lock'. Pope describes how awful beauty puts on all its charms by using the treasures of a plundered world.

> This casket India's glowing gems unlocks,
> And all Arabia breathes from yonder box.[106]

Swift's poem leaves us with the debris; the throne-room once the goddess has departed.

> The various combs for various uses
> Filled up with dirt so closely fixed
> No brush could force a way betwixt.
> A paste of composition rare,
> Sweat, dandriff, powder, lead and hair.[107]
> (ll. 20-43)

However, an important new element in this poem is the presence of the male character, Strephon, to mediate between Swift and his subject. It is Strephon, not Swift, who rummages through Celia's soiled underwear, reading the runes of her laundry-basket for damning evidence of human animality.

> No object Strephon's eye escapes,
> Here petticoats in frowzy heaps;
> Nor be the handkerchief forgot
> All varnished o'er with snuff and snot.
> (ll. 47-50)

Strephon is ridiculed for idealizing his goddess, and for taking her at her own made-up face value. But what is the tone of this

[106] *The Rape of the Lock,* Twickenham Edition, 1962, p. 156.
[107] *Poems,* ii. 526.

ridicule? What is the controlling attitude of Swift, who declared that the majority of women were 'bêtes en jupes', to the activities of his clown/fetishist who follows a trail of clues in smocks and stockings back to the original sin of the chamber pot?

> Thus finishing his grand survey,
> Disgusted Strephon stole away,
> Repeating in his amorous fits,
> Oh! *Celia, Celia, Celia shits!*
> (ll. 115-18)

In some ways the poem conforms to a recognizable Swiftian pattern. Strephon is punished for his naïvety and for his presumptuous curiosity by being condemned, like Gulliver, to a fixed association of ideas forever afterwards.

> And if unsavoury odours fly
> Conceives a lady standing by.
> (ll. 123-4)

In other words he lurches from one extreme to the other; from a universal idealization of women, to total execration. His blindness to any middle course earns the author's pity, but in terms that are disquietingly ambiguous.

> Should I the Queen of love refuse
> Because she rose from stinking ooze?
> (ll. 131-2)

Swift did refuse the queen of love, and in his repeated insistence upon cleanliness and masculinity in the women that he praised, one cannot escape a suspicion that his refusal was associated with his own conviction that sex and 'stinking ooze' were inseparable.

There is an interesting blend of tones in the poem 'A Beautiful Young Nymph going to Bed'. As she removes her tawdry streetwalker's charms, it is like watching an accelerated film of the process of ageing. There is a gruesome humour in the nightly dismemberment of this wretchedly maimed debauchee, but it is inseparable from a certain pity. Her pretensions are pathetic, but they are all that she has; shorn of her adornments, she is a skeleton whose dreams are nightmares of the lash. Here there is no prying clown to display her props to us, but we observe her dismemberment with a voyeuristic horror, as though watching a patient on the table or a prisoner on the rack.

> Then seated on a three-legged chair,
> Takes off her artificial hair:
> Now, picking out a crystal eye,
> She wipes it clean, and lays it by.
> Her eyebrows from a mouse's hide
> Stuck on with art on either side,
> Pulls off with care, and first displays 'em,
> Then in a play-book smoothly lays 'em.
> Now dextrously her plumpers draws
> That serve to fill her hollow jaws.
> Untwists a wire, and from her gums
> A set of teeth completely comes.
> Pulls out the rags contrived to prop
> Her flabby dugs, and down they drop.[108]
> (ll. 1-22)

When she awakes from her dreams of torment, she finds that her moth-eaten props have suffered yet further devastation:

> A wicked rat her plaster stole
> Half-eat, and dragged it to his hole.
> The crystal eye, alas, was missed
> And puss had on her plumpers pissed.
> (ll. 59-62)

To repair the ravages of such a night requires a dedication more akin to metaphysics than mere cosmetics. There is more of the pathos of a Struldbrug about this piece of human jetsam, than the sexual aggression of the Yahoo.

'Strephon and Chloe' is the longest and most puzzling of this little group of poems. Chloe is no painted strumpet or diseased coquette, but a virgin of exemplary habits and cleanliness.

> Such cleanliness from head to heel
> No humours gross, or frowzy steams
> No noisome whiffs, or sweaty streams
> Before, behind, above, below–[109]
> (ll. 10-13)

[108] *Poems,* ii. 581-2. [109] *Poems,* ii, 584.

The adverbial echoes of Donne's 'American' mistress are deliberate; this is a goddess of some brave new hygienic world where women's bodies are quite without odorous secretions.

> Her dearest comrades never caught her
> Squat on her hams, to make maid's water.
> You'd swear, that so divine a creature
> Felt no necessities of nature.
>
> (ll. 17-20)

As the Houyhnhnms are without passions and opinions, as Wharton is without a moral sense, so Chloe is without sweat, immaculately dry. Strephon is correspondingly cleanly. Apprehensive of imposing his sullied flesh on such a divine creature, he washes as scrupulously as even Swift could have wished.

> His hands, his neck, his mouth, and feet
> Were duly washed to keep 'em sweet;
> (With other parts that shall be nameless).
>
> (ll. 77-9)

And yet, when this well-scrubbed pair come to bed–disaster; the result of too much tea and beans.

> The nymph, oppressed before, behind,
> As ships are tossed by waves and wind,
> Steals out her hand by nature led,
> And brings a vessel into bed:
> Fair utensil, as smooth and white
> As Chloe's skin almost as bright.
> Strephon who heard the fuming rill
> As from a mossy cliff distill;
> Cried out, ye Gods, what sound is this?
> Can Chloe, heavenly Chloe ——?
> But, when he smelt a noysom steam
> Which oft attends that luke-warm stream . . .
> He found her, while the scent increased
> As mortal as himself, at least.
>
> (ll. 169-80, 185-6)

Once again the coy adverbs maintain the erotic mystery, but the mock-heroics of this couple's decorous courtship soon collapse into fabliau slapstick. The utensil, 'smooth and white as Chloe's skin', seems to preserve her alabaster sanctity; even the 'fuming rill' reminds the idealistic Strephon at first of a picturesque waterfall. But the descent into bathos, following the shock of his disillusionment, is sudden and total. From being afraid to touch his bride, he is soon happily matching her fart for fart. This time there has been no cosmetic deception–merely the conventional naïvety of young love. But Swift seems to argue–for there is no mediator in this poem–at the first smell of urine, the reverence departs.

> Adieu to ravishing delights,
> High raptures, and romantic flights . . .
> How great a change! how quickly made!
> They learnt to call a spade a spade.
> They soon from all constraints are freed;
> Can see each other *do their need*
> On box of cedar sits the wife,
> And makes it warm for *dearest life,*
> And by the beastly way of thinking
> Find great society in stinking.
>
> (ll. 197-8, 203-10)

The pair, no longer idealistically bashful, can watch each other on the privy, and Chloe happily warms the toilet seat for her hubby. Such pleasant mutuality, however, strikes Swift as a degrading denial of decency. In this poem we have the pattern as before; a sudden transformation from sublime idealism to humiliating bathos. Yet this time there is no obvious comic target for the satire–only human nature itself. This time, instead of denouncing cosmetics, Swift seems to see them as necessary for maintaining some decent cover over the baser necessities of our nature. He adds a homily to this effect, which has his personal endorsement.

> Authorities both old and recent
> Direct that women must be decent
> And from the spouse each blemish hide
> More than from all the world beside.
>
> (ll. 251-4)

Decency is the concept that must be added to cleanliness. But decency here seems to imply a certain discreet deception. However, it is a deception whose essence is self-respect. The real target of Swift's satire in these poems is the deception whose only aim is superficial attraction; the display that is the corollary of an inner corruption. So here, it is not simply that married women should preserve a certain distance and discretion. They should also treat intellectual pursuits as more than mere accomplishments to help them snare a mate. Women, in short, should be prudent, decent, sensible, clean; the roll-call of virtues has a familiar and conventional ring. It is when one reads a passage like the following that one has qualms about Swift's route to such an orthodox conclusion.

> O Strephon, e'er that fatal day
> When Chloe stole your heart away
> Had you but through a cranny spy'd
> On house of ease your future bride,
> In all the postures of her face,
> Which nature gives in such a case;
> Distortions, groanings, strainings, heavings;
> 'Twere better you had licked her leavings,
> Than from experience find too late
> Your goddess grown a filthy mate.
> Your fancy then had always dwelt
> On what you saw, and what you smelt . . .
> And, spight of Chloe's charms divine,
> Your heart had been as whole as mine.
> (ll. 235-46, 249-50)

'Death before defecation' is Swift's motto, claims Nora Crow Jaffe,[110] and we can see what she means. Swift's excremental obsession has seemed to many critics to confirm Swift's own description of himself as a misanthropist, and more particularly a misogynist. Middleton Murry spoke for traditional critical opinion when he described these poems as 'so perverse, so unnatural, so mentally diseased, so humanly *wrong*'.[111] More recently scholarly defenders have discerned traditional Christian themes in Swift's satires, while psychoanalytical critics, led by Norman O. Brown, have found prefigurements of Freud in his works. Too many critics, argues Brown, have explained this 'excremental vision' as a personal aberration, as 'an attempt to justify his genital failure (with Varina, Vanessa and Stella) by indicting the filthiness of the female sex'.[112] Certainly there is

[110] N. C. Jaffe, *The Poet Swift,* University Press of New England, 1977, p. 109.
[111] Murry, p. 440.
[112] 'The Excremental Vision' in *Life against Death* (1959), Sphere Books, 1968, pp. 163-81.

ample evidence in both Swift's life and works to accuse him of an anal fixation. In 'Cassinus and Peter' the distraught undergraduate Cassinus declares, in Macbeth's words, 'you cannot say 'twas I'. Yet the vision he confronts is not a ghost, but a turd in a chamber-pot. Like Lady Macbeth, Swift seems to have believed that no amount of washing could purge the stain of this original guilt. Quite apart from the excremental images which flow through his works from *A Tale of a Tub* to *Directions to Servants,* there are corroborating indications of such a fixation in his personal biography. His 'loss' of both parents in infancy; his defensiveness with women; his childlike fascination with words and sounds, treating them as objects of play; his obsessive retentiveness with money; these all amount to a fairly classic case-study. Yet Swift was far more than his own Strephon. While struggling to convince humankind of their own animality, he pleaded guilty to the common vices of the species and included himself within the scope of his satires. The last of these four poems ends with Swift's most notorious couplet, as Cassinus complains to Peter:

> Nor wonder how I lost my wits;
> Oh! Celia, Celia, Celia shits!
>
> (ll. 117-18)

The effect is not horror, but bathos. Worse than death, disease, or dishonour in Cassinus's eyes, is this. Love, no more than science or religion, can raise man (or woman) above this common denominator, and Cassinus's fastidious horror convicts him of the same absurd pride as Gulliver, who sought to renounce his own human form. Swift gives the old homily 'Regard thy end' a new application. On the privy, all men are equal.

V

In the opening months of 1732 the chorus of invitations from friends in England grew to a new crescendo. Pope dreamed of re-establishing something like the old Scriblerian fellowship. 'I fancy if we three were together but for three years,' he wrote, thinking of himself, Bolingbroke, and Swift, 'some good might be done even upon this age.'[113] All thoughts of travel were

[113] *Corr.* iv. 8.

dashed in February, however, when Swift strained a leg in a fall downstairs. The pain and isolation of this injury plunged him into deeper gloom. Gay tried to cheer him up, assuring him that the Queensberrys' estate at Amesbury Downs was 'so smooth that neither horse nor man can hardly make a wrong step'.[114] But Swift, whose lameness persisted for several months, was in no mood to be cheered up, even by Gay. 'I am not in a condition to make a true step, even on Amesbury Downs . . . to talk of riding and walking is insulting to me, for I can as soon fly as do either.'[115] His enforced immobility made him the captive host for any passing well-wisher. He complained bitterly that 'to increase my misery, the knaves are sure to find me at home, and make huge void spaces in my cellar'.[116] However, he was even more plaintive a few weeks later, when he was forced to recognize that even the prospect of a drink was insufficient inducement for most people to bear with his ill temper. 'Even my wine will not purchase company, and I begin to think the lame are forsaken as much as the poor and blind.'[117]

In his solitude he became more 'talkative upon paper', and his letters for this period are longer and more numerous than before. 'My poetical fountain is drained,' he told Pope in June, 'even prose speculations tire me.'[118] He showed a growing obsession with his own status, while naturally denying any such vain concern. 'I am ten times more out of favour than you,' he told Gay in October, in a boast disguised as a regret.[119] He was entering upon his second childhood, he told Pope, having lost the capacity to amuse himself with thinking. Nothing happened in Dublin that was worth notice. 'A murder now and then is all we have' for conversation, he observed with customary grim humour. In June he wrote to 'welcome' a newcomer to Ireland with this devastatingly bleak account of the country.

Tipperary . . . is like the rest of the whole kingdom, a bare face of nature, without houses or plantations; filthy cabins, miserable tattered, half-starved creatures, scarce in human shape; one insolent ignorant oppressive squire to be found in twenty miles riding; a parish church to be found only in a summer-day's journey, in comparison of which, an English farmer's barn is a cathedral; a bog of fifteen miles round; every

[114] Ibid., p. 9.
[115] Ibid., p. 14.
[116] Ibid., p. 15.
[117] Ibid., p. 35.
[118] Ibid., p. 31.
[119] Ibid., p. 72.

meadow a slough, and every hill a mixture of rock, heath and marsh; and every male and female, from the farmer, inclusive to the day labourer, infallibly a thief, and consequently a beggar, which in this island are terms convertible.[120]

By midsummer he was sufficiently recovered to be able to ride a little with the help of a contraption called 'gambadoes', a sort of wooden leggings that saved any strain on his Achilles tendon. Then, in July, Bolingbroke wrote to inform him that he had finally found an independent living, at Burghfield, near Reading, where Swift could have a permanent residence close to London and his friends. The living was worth £400 a year 'over and above a curate paid'. The present incumbent was 'desirous to settle in Ireland', and to Bolingbroke it seemed the perfect opportunity for a transfer. In fact he had already set the wheels in motion for such a swap by contacting the Lord Lieutenant and Archbishop Hoadly of Dublin.

I suppose it will not be hard to persuade them that it is better for them you should be a private parish priest in an English county, than a Dean in the metropolis of Ireland, where they know, because they have felt, your authority and influence.[121]

However, it was this very argument, that the Whig grandees in Church and State would be pleased at Swift's move, and would smile at his 'foolish bargain' of 'quitting power for ease and a greater for a less revenue' that gave Swift pause. For some years, Pope and Bolingbroke had been attempting to turn his attention away from politics and towards philosophy and metaphysics. They evidently hoped that, as a private clergyman at Burghfield, Swift would finally be persuaded to give his mind to such higher things. But they ignored, or sought to minimize, the very real if masochistic pleasure that Swift derived from his involvement in grubby political squabbles. 'I have not the courage . . . to be such a satirist as you,' confessed Pope, 'but I would be as much, or more, a philosopher.'[122] In moving to Burghfield, Swift would be regaining the company of friends, but would be relinquishing a position of authority which had made him a notorious, even a legendary figure. In August he told Gay that if he was to come to

[120] Ibid., p. 34. [121] *Corr.* iv. 44. [122] Ibid., p. 147.

England 'I want to be minister of Amesbury, Dawley, Twickenham, Riskins and prebendary of Westminster, else I will not stir a step.'[123] At the very least he would require an extra £300 a year to make him consider the move. His income from tithes had 'sunk almost to nothing' and his lawsuit dragged on, placing his fortune in 'the utmost confusion'. Even in these straitened circumstances, however, Burghfield 'would not answer . . . The dignity of my present station damps the pertness of inferior puppies and squires, which without plenty and ease on your side the channel, would break my heart in a month.' He concluded, 'I would rather be a freeman among slaves, than a slave among freemen.' It was an important admission, and signified the end of his long-lived fantasy of returning to England. He had been offered an English post, modest, but by no means disgraceful. And he had thrown in his lot with Ireland. Immediately, in an instinctive reflex, his vituperation against his native land burst out with fresh venom. In letters to ladies in England he pictured himself 'banished to a country of slaves and beggars, my blood soured, my spirit sunk'. Yet even in his grim isolation, he could present himself as a lone warrior, 'fighting with beasts, like St Paul, not at Ephesus, but in Ireland'.[124] The fight would continue. He would not retire to become an English vicar, but would remain the embattled Dean till death.

At the end of November another link with England was broken with the death of Gay. Swift affected the sang-froid of Helsham when he heard the news. 'I endeavour to comfort myself upon the loss of friends as I do upon the loss of money,' he told Pope, 'by turning to my account-book, and seeing whether I have enough left for my support.'[125] Pope was shocked at this reaction, and the Duchess of Queensberry wrote to protest that friends could never be equated with money. She could live penniless but happy, as long as she had good friends, etc. She renewed her invitation to Swift to visit Amesbury more emphatically than before, as if in tribute to their dead friend. In response to her courteous reproof, Swift allowed himself to be teased into a more conventional pose. 'Sure I was never capable of comparing the loss of a friend with the loss of money,' he lied

[123] Ibid., p. 58. [124] Ibid., p. 79. [125] *Corr.* iv. 103.

in his best courtly manner.[126] Pope too renewed his invitations with a morbid insistence.

I . . . vehemently wish, you and I might walk into the grave together, by as slow steps as you please, but contentedly and cheerfully.[127]

The image of these two satirists shuffling cheerfully towards death like Darby and Joan is both cloying and macabre, yet it recalls the similar sentimentality of Pope's description of the *Miscellanies* in which they walked 'hand in hand to posterity'. Actually Swift was far from happy about the figure that they cut in the *Miscellanies*, as he told Motte. 'I am not at all satisfied with the last miscellany . . . My part . . . is very incorrect.'[128] Increasingly anxious to prove that 'men of wit . . . may be the most moral of mankind', Pope did all he could to suppress the 'unguarded and trifling *jeux d'esprit*' that they had written years before. He sought to make a clear distinction between 'our studies and our idleness, our works and our weaknesses'.[129] To Swift no such division seemed possible, or appropriate. 'I think all men of wit should employ it in satire,' he told Ford. 'If my talent that way were equal to the sourness of my temper, I would write nothing else.'[130] He was also angry that, since there was no copyright law in Ireland he had gained 'no advantage by any one of the four volumes' of *Miscellanies*. This anger grew to fury the following May when George Faulkner proposed publishing a subscription edition of Swift's collected works, from which he would still not gain a penny. He explained the situation to Pope.

A printer came to me to desire he might print my works (as he called them) in 4 volumes by subscription. I said I would give him no leave, & should be sorry to see them printed here. He said they could not be printed in London. I answered they could if the partners agreed. He said he would be glad of my permission, but as he could print them without it, and was advised that it could do me no harm, & having been assured of numerous subscriptions, he hoped I would not be angry at his pursuing his own interest . . . I determine not to intermeddle, though it be much to my discontent, and I wish it could be done in England, rather than here . . . neither will it be one farthing in my pocket, for among us, money for copy is a thing unheard of.[131]

[126] Ibid., p. 126. [127] Ibid., p. 147. [128] Ibid., p. 89.
[129] Ibid., p. 115. [130] Ibid., p. 138. [131] Ibid., p. 154.

In a futile defence against such legal piracy, Swift repudiated the authorship of many works, and in a letter to Ford he denied all knowledge of at least five of his pamphlets written in England during Oxford's ministry.

By the beginning of 1733 his leg was no longer giving pain, and he promised Pope that he would come over in August and spend the winter in England. Apart from any other consideration, he felt increasingly out of touch with literary developments in England. As Pope and Bolingbroke teased him with hints about the forthcoming *Essay on Man* he cried out in protest, 'How can I judge of your schemes from this distance?' What he could detect, however, was a new excitement resulting from Pope's most recent satires. 'Pope . . . has made the town too hot to hold him,' wrote Bathurst with mischievous glee.[132]

The political temperature too was rising, following Walpole's decision to introduce a comprehensive extension of Excise duties. These duties, first imposed during the Civil War, were easily portrayed as a symbol of tyranny, and Swift gave a sardonic smile when the first of them, on tobacco, was approved by the Commons in March 1733. 'I give you joy of your first large stride to slavery, the Excise,' he wrote to Ford.[133] Perhaps now, at last, the English would understand something of the restrictions that Ireland had suffered for so long. 'Go into what company you will' declared Bathurst, 'you can hear of nothing else.'[134] An old friend of Swift's, John Barber, now Lord Mayor of London, played an important role in the opposition which quickly formed to resist and finally defeat the Excise Bill. Swift wrote to congratulate him on the City of London's defiant address to the Commons, and, when news of the defeat of the Bill reached Dublin, he caused bonfires to be lit on the steeple of St Patrick's and before the Deanery.

The death of Pope's mother in June gave Swift new hope that his friend might venture a journey to Dublin. Finally released from his filial obligation to 'rock the cradle of reposing age', Pope might welcome an opportunity to recover from his grief in fresh surroundings. It happened that some friends were actually travelling to Ireland, and offered to accompany him. But Pope did not go. He made up a facetious excuse; he would be 'killed

[132] Ibid., p. 132. [133] *Corr.* iv. 138. [134] Ibid., p. 131.

by kindness', forced to eat on a vast Hibernian scale. Swift was deeply hurt by his refusal. In an obstinate and yet endearing way he subjected Pope's 'excuse' to detailed refutation, assuring him that 'there is nothing in excess in either eating or drinking'.[135] With a mixture of sorrow and recrimination he enumerated the many conveniences that Pope might have enjoyed; a warm apartment, a garden 'as large as your green plot that fronts the Thames', small dinners 'of what you liked', and good wine. Swift wanted so much to show off his domain to his friend, as an explanation of his reluctance to visit London. 'I hate the very thought of London, where I am not rich enough to live otherwise than by shifting.' He could not forbear boasting of his pre-eminence in Dublin.

> I walk the streets in peace, without being justled, nor ever without a thousand blessings from my friends, the vulgar. I am lord Mayor of 120 houses, I am absolute lord of the greatest cathedral in the kingdom; am at peace with the neighbouring princes, the Lord Mayor of the City, and the Abp of Dublin.[136]

He might be a big fish in a small pond, but in England he would be a minnow, totally unregarded.

It was two months before Pope replied. When he did so, it was a brief, wretched, hopeless letter. He felt Swift's unspoken reproaches and was bitterly sorry for the rift that had opened between them. Yet he could not have brought himself to travel to Ireland.

> I write more to show you I am tired of this life, than to tell you anything relating to it. I live as I did, I think as I did, I love you as I did; but all these are to no purpose: the world will not live, think, or love as I do. I am troubled for, and vexed at, all my friends by turns. . . . There is a great gulph between! In earnest, I would go a thousand miles by land to see you, but the sea I dread. My ailments are such, that I really believe a sea-sickness, (considering the oppression of cholical pains, and the great weakness of my breast) would kill me.[137]

The gulf between them was far wider than the Irish sea, for they had also grown apart in ideas. Swift was content to pay lip-service to Pope's brave new fatalistic doctrine that 'whatever is, is right' as 'the thought of Socrates and Plato'. Privately, his

[135] Ibid., p. 170. [136] Ibid., p. 171. [137] Ibid., p. 194.

opinion of such pious optimism can be gauged from his remarks to a fellow clergyman a year before.

> Whoever really believes that things are well, is many ways happy. He is pleased with the world (as I was formerly) and the world with him; his merit is allowed, and favours will certainly follow . . . in what appears to my eyes a very dirty road.[138]

Like all sublime philosophers, Pope had lifted his eyes above the dirty roadways of the world, but Swift could never avert his attention from the mud, ruts, and filth of human failure.

Despite Swift's frequent complaints of loneliness, this was actually a period when he made a number of new friendships. Among the most notable of his new intimates was young Laetitia Pilkington, only seventeen, and newly married to 'little Matthew', her clergyman husband. Swift took an immediate liking to her, and she became the latest in a long line of young female admirers, vivacious, imaginative, and flattering. In her chatty, anecdotal *Memoirs* she has left us many vivid glimpses of Swift in his declining years. 'The Dean always prefaced a compliment with an affront,' she noted. She also observed that the Deanery had 'I know not how many pair of back stairs in it',[139] and that whenever ill health or foul weather denied Swift his outdoor recreations, he would march up and down the stairs for exercise, with a servant at the bottom to count his steps. On one occasion he insisted on pulling off both her boots convinced that she must have 'either broken stockings or foul toes'. It was she who asserted that she never saw Swift laugh. 'When any pleasantry passed which might have excited it, he used to suck in his cheeks, as folks do when they have a plug of tobacco in their mouths, to avoid risibility.' She failed to comprehend Swift's dead-pan style, and her stories merely confirm that, even at this late stage, he loved to hoax and mislead his friends. Sometimes, she admits, he would pinch her black and blue for her mistakes.

It is noticeable that Swift continued to attract the friendship of young women. In addition to Laetitia Pilkington, the poetess Mary Barber was a woman for whom he expressed an admiration bordering on infatuation. She was, he insisted, 'the best

[138] Ibid., p. 29.
[139] Laetitia Pilkington. *Memoirs . . . with Anecdotes of Dean Swift,* 1748.

poetess in both kingdoms', and he urged all his friends to subscribe to her volume of *Poems on Several Occasions* (1734). He also began a flirtatious correspondence with three admiring young ladies, Miss Kelly, Miss Donnellan, and Mrs Pendarves. In her first letter to him, Frances Kelly, whose beauty and good humour had, according to Mrs Pendarves, 'gained an entire conquest' over Swift, talks of having danced the previous night away.[140] Unfortunately she was far less healthy than this suggests. She spent most of the summer at Bristol Hot Wells in a forlorn effort to recover from an attack of pleurisy, and wrote from thence to Swift, declaring how much she missed his conversation. Mrs Pendarves, an attractive widow, was acquainted with many of Swift's circle, and was later to become the second Mrs Delany. She wrote Swift entertaining letters, and he replied with agreeable banter, and spelling corrections. But even the correspondence of such vivacious young women did not lift his spirits, since Miss Kelly's health continued to decline, and on the last day of October she died 'in the flower of her youth and beauty'. 'For God's sake try to keep up your spirits,' wrote Ford. 'Divert yourself with Mrs Worral at backgammon; find some new country to travel in; anything to amuse.[141] Other lady correspondents included Mrs Caesar, wife to the former treasurer of the navy, and Lady Worsley, a great-grandmother who wrote flirtatious letters in which she claimed to deserve more of his attention than these 'girls you coquet with'. The Duchess of Queensberry sought his friendship and advice, now that Gay, to whom she 'could sometimes lay open all my rambling thoughts', was dead. Lady Betty, the most loyal of all his female friends, preserved her place in his affections, despite her spirited defence of Lady Suffolk from his bitter denunciations. Indeed, it is clear that these charming and intelligent women could still find ways to lift Swift's mask of misanthropy, and bring back a smile to his lips.

Two Bills which were currently being promoted in Parliament by the Irish bishops brought Swift into fierce conflict with his predecessor at St Patrick's, John Stearne. Swift described the Bills of Residence and Division as 'two abominable bills for enslaving and beggaring the clergy'. For many years Swift and

[140] *Corr.* iv. 108. [141] Ibid., p. 203.

Stearne had eaten, drunk, played cards, and joked together, but Swift had never forgotten or forgiven the broken promise over St Nicholas Without. Now, suddenly, his long-nursed anger exploded against Stearne with a tremendous charge of self-righteousness.

> I call God to witness, that I . . . shall forever firmly believe, that every bishop who gave his vote for either of these bills, did it with no other view (bating further promotion) than a pre-meditated design, from the spirit of ambition and love of arbitrary power, to make the whole body of the clergy their slaves and vassals until the Day of Judgement, under the load of poverty and contempt.[142]

Such a curse was the price of crossing Swift. It was a year before Stearne replied to this withering attack, and when he did so, he sent £50 to augment Swift's charitable fund.

In July 1732 Delany married his first rich widow, a Mrs Tenison, worth, Swift estimated, about £1,600 a year. Swift wondered if his Chancellor could find another like her for himself, worth 'not less than two thousand'. He was pleased to note that marriage had no ill effect on Delany's habits, who continued to entertain his old friends 'very commendably . . . at an elegant, plentiful table'. It was at this time that Swift met John Boyle, Earl of Orrery, a young man of wit and learning who accompanied Swift and Sheridan on a holiday outing to Tallaght hill. Using a two-gallon bucket, Swift undertook a practical experiment to estimate the flow of the stream that rose there. His conclusion was that within three years the hill would burst asunder, and Dublin would be flooded. These dire prognostications were dutifully noted in the English and Irish press, and Miss Kelly advised him to pack his bags immediately and join her in the safety of England. Swift also resumed his old habit of April Fool's Day jokes in 1732, solemnly informing Lady Acheson that a woman in Wicklow had given birth to half a child: 'It had one eye, half a nose, a mouth, one leg and so from top to bottom.'[143] His old antagonist Tisdall continued to sting him with occasional lampoons. Swift had been assured that Tisdall 'had produced a dozen of his libels wholly against me' in a single evening, 'desiring I might be told of it'.[144] Obviously Swift was

[142] Ibid., p. 183. [143] Ibid., pp. 11-12. [144] Ibid., p. 28.

not the only one who could nurse a grievance for many years. With Sheridan, Swift kept up a more congenial satiric exchange. When Sheridan published a poem entitled 'A New Simile for the Ladies', in which he compared women to clouds, Swift composed an 'Answer', in which a cloud protests at this scandalous analogy.

> Tis true; a woman on her mettle
> Will often piss upon a nettle;
> But, though we own she makes it wetter,
> The nettle never thrives the better;
> While we, by soft prolific showers
> Can every spring produce fresh flowers.[145]
>
> (ll. 147-52)

Swift also composed some rather more distinguished verses at this time, and in 1733 Mary Barber carried two major poems, 'Epistle to a Lady' and 'On Poetry, A Rhapsody', across to London to be published. 'On Poetry' has its origins in the common stock of Scriblerian hints accumulated long since, which Pope was also exploiting in his satires. Swift had evidently read Pope's portrait of Addison/Atticus presiding over his little senate of admirers at Button's, and offers his variant on the same theme by describing Battus/Dryden pontificating at Will's.

> At Will's you hear a poem read
> Where Battus from the table-head
> Reclining on his elbow-chair
> Gives judgement with decisive air.
> To whom the tribe of circling wits
> As to an oracle, submits.[146]
>
> (ll. 63-8)

Yet Swift's couplets lack the ironic compression of Pope's. Where Pope's Atticus is a generalized heightening of a specific and detailed portrait, Swift's Battus is a conventional image of the dictator of taste. The 'Epistle to a Lady', addressed to Lady Acheson, begins as another Market Hill 'libel', full of teasing criticisms of his long-suffering hostess. In the middle, however, Swift suddenly switches from domestic to public concerns, and delivers an attack on Walpole and George II.

[145] *Poems,* ii. 628 (for MS variants see also p. 621).
[146] Ibid., p. 649.

Should a monkey wear a crown,
Must I tremble at his frown?
Could I not, through all his ermine
Spy the strutting, chatt'ring vermin?[147]
(ll. 149-52)

Lines such as these led to the arrest of the publisher, Gilliver. Under examination, he named Mrs Barber, Motte, and Pilkington as his associates in publishing the poem. All four were detained and Walpole was apparently determined to have Swift arrested, but was dissuaded from such an action by the warning that it would take an army of ten thousand men to remove Swift from the Deanery.

Swift continued to supply a trickle of pamphlets to the press on such familiar political topics as the woollen industry, the coinage, and the need to resist corruption in the disposal of public offices. In 1733 he supported the 'patriotic' candidate Eaton Stannard for the post of Recorder of the City of Dublin, and after Stannard was elected, Swift chose him as an executor for his will. Commenting on the election Carteret remarked; 'I know by experience how much that city thinks itself under your protection, and how strictly they used to obey all orders fulminated from the sovereignty of St Patrick's.'[148] That summer Swift also interested himself in a parliamentary election, publishing his *Advice to the Freemen of the City of Dublin* in which he argued that there were only two parties in Ireland.

I do not mean Popish and Protestant, High and Low Church, Episcopal and Sectarians, Whig and Tory; but of these *English* who happen to be born in this kingdom . . . and the gentlemen sent from the other side to possess most of the chief employments here.[149]

So vehement were Swift's arguments that when the Lord Lieutenant returned in September, it was rumoured that the pamphlet was about to be prosecuted. Whereupon the publisher immediately rushed out plentiful new supplies. As expected, Swift's favoured candidate was successful in the election and was 'carried amid the acclamations of many thousand of people to the parliament house'.

[147] Ibid., p. 634. [148] *Corr. iv. 128.* [149] *PW* xiii. 80.

VI

IN FEBRUARY 1734 a rumour spread that Swift was dying. There was general rejoicing in Dublin when it proved to be false. This report in the *Dublin Journal* confirms his remarkable influence over the inhabitants of the city.

Several weavers having lately assembled in great bodies to search for foreign manufactures, accidentally met with that worthy patriot, the Rev. Dr Swift D.S.P.D., who exhorted them to be quiet, and not to do things in a rash manner, but to make application in a peaceable way, and he did not make the least doubt, but proper means would be found to make them all easy &c. Whereupon they immediately dispersed to their several homes, crying out, long live Dean Swift, and Prosperity to the Drapier, and returned him thanks for his good advice, which they said they would follow.[150]

Nor was his fame confined to Dublin alone, for Sheridan reported the existence in Cavan of a 'Drapier's Club . . . of about thirty good fighting fellows; from whence I remark you have the heart of all Ireland'.[151]

Not everyone admired him, though. Towards the end of 1733, a Bill was introduced to encourage the growing of flax by commuting the tithes on that crop. Swift deplored any reduction in the value of tithes, and his opposition to this Bill brought him into conflict with Richard Bettesworth, sergeant-at-law. He included a short lampoon of Bettesworth in his brief verse satire 'On the words Brother Protestants' which objected to the way in which opponents of the Test would apply those words to Anglicans and Dissenters alike.

A ball of new-dropp'd horse's dung
Mingling with apples in a throng
Said to the pippin, plump and prim,
See brother, how we apples swim.[152]
(ll. 11-14)

[150] Ibid., p. xxix.
[151] *Corr.* iv. 282.
[152] *Poems,* iii. 811. Ehrenpreis points out (iii. 769) that the imagery in this poem is derived from a political fable of the time. See 'Of the Apple and the Horse-Turd', Fable XX in *Poems on Affairs of State,* II (1703), 85-6.

In a sardonic rhyme, Swift suggested that half a crown for Bettesworth's legal labours was 'all his sweat's worth'. Highly insulted, Bettesworth swore to cut off Swift's ears. He called at the Deanery, and finding Swift out, followed him to the home of John Worral. There followed a heated argument, which Swift subsequently laughed off in a facetious note to the Lord Lieutenant, Dorset. 'He repeated the lines that concerned him with great emphasis; said I was mistaken in one thing, for he assured me he was no booby, but owned himself to be a coxcomb. . . .'[153] Swift assured Dorset that Bettesworth's 'peevish' threats were mere bluster, but some of his friends were less confident of that. Thirty-one inhabitants of the Liberty of St Patrick's signed a declaration of their resolution to defend 'the life and limbs' of the Dean against any 'ruffians or murderers' who might threaten him.

Swift's prolonged illness that year led to a mood of depression and self-pity in which he again complained of neglect and isolation. Letters from England seemed fewer than before, and he complained particularly of Pope's silence. 'I fear he hath quite forsaken me,' he told Oxford in August, 'for I have not heard from him in many months'.[154] 'It is a very cold scent,' he wrote to Mrs Pendarves, 'to continue a correspondence with one whom we never expect to see. . . . Mr Pope and my Lord Bolingbroke themselves begin to fail me.'[155] At least he excluded Mrs Pendarves herself from his criticism. She had suddenly sent him a wonderful letter which began, 'I find your correspondence is like the singing of the nightingale.'[156] Such artless flattery won him over completely, and he wrote back in his most gallant manner to assure her that he had never once found anything amiss in her disposition, 'although I watched you narrowly'. He also played the courtly old gentleman to Miss Hoadly, only daughter of the Archbishop of Dublin. His *rapprochement* with this young lady and her father is particularly surprising since at Hoadly's enthronement in St Patrick's in January 1730 'sharp words' had been exchanged between the Tory Dean and Whig Archbishop, and Swift had refused to dine with him after the ceremony. However, the 1734 Swift was happy to receive gifts of a pig and butter from Miss Hoadley, and returned his thanks and best wishes to her and to His Grace.

[153] *Corr.* iv. 220.
[154] Ibid., p. 249.
[155] Ibid., pp. 258-9.
[156] Ibid., p. 251.

His English friends were not slow to repudiate the charge of neglect. 'You have no reason to put me amongst the rest of your forgetful friends,' protested Arbuthnot,[157] but his letter was far from cheerful. 'I am afraid my dear friend, we shall never see one another more in this world.' A month later Lady Betty joined the protest. 'Don't accuse me of forsaking you, indeed, tis not in the least in my thoughts.' But Pope made no such excuses, recognizing the justice of Swift's reproaches. Instead he offered this melancholy explanation.

I assure you it proceeded wholly from the tender kindness I bear you. When the heart is full, it is angry at all words that cannot come up to it; and you are now the man in all the world I am most troubled to write to; for you are the friend I have left whom I am most grieved about. Death has not done worse to me in separating poor Gay, or any other, than disease and absence in dividing us.[158]

He begged only that Swift would not 'laugh at my gravity' but would view his increasingly moral writings with a sympathetic eye. Giddy, deaf, and disconsolate, Swift wrote back to deny that he had ever thought Pope 'inconstant' in his friendship, but the denial lacks conviction. There is an overwhelming sense of weariness and resignation in the letter.

God be thanked I have done with everything . . . except now and then a letter, or, like a true old man, scribbling trifles only fit for children, or schoolboys of the lowest class.[159]

He had finally recalled the money that Gay had held for him in England, 'which I had set apart to maintain among you'. It was a clear sign that his hopes of ever visiting England again were at an end. His most serious composition during this bleak period was his sad little poem 'On his own Deafness'.

Deaf, giddy, helpless, left alone
To all my friends a burden grown
No more I hear my church's bell
Than if it rang out for my knell.[160]
(ll. 1-4)

Still, he endeavoured to keep up his pun-ic exchanges with Sheridan, who wrote, 'Eye am sore eye two here ewer health is knot bet

[157] Ibid., pp. 255-6.
[158] Ibid., p. 253.
[159] Ibid., p. 262.
[160] *Poems,* ii. 673-4.

her' on Christmas Day.[161] Swift had hoped to spend Christmas at Wicklow with the Reverend Blachford's family, and had written a fortnight beforehand to invite himself on fixed terms, allowing 'one shilling and sixpence English for his commons, ale and small beer included'. He set out from Dublin on the 16th, got as far as Howth Castle, but was there seized with such a severe fit of giddiness that he was forced to lie down. It was the kind of humiliation that he loathed, and turned back, as soon as he was able to ride. 'I dare not accept' Blachford's invitation, 'for fear of another attack'.[162] It was another lonely Christmas at the Deanery for himself and Mrs Brent, with only Sheridan's letter to cheer him. He sent two bottles of wine to Rebecca Dingley, together with her usual Christmas box. There was a letter from Pope in which, by the magical process of time, those two awkward summers of 1726 and 1727 that Swift had spent at Twickenham now appeared to him, he claimed, 'like a vision which gave me a glimpse of a better life'. 'I wish to God we could once meet again,' he adds but is still certain that whatever is, is best–'he who made us, not for ours, but his purposes, knows only whether it be for the better or the worse'; in other words he has accepted the separation as God's will; though it affects him 'like a limb lost,[163] he bears it stoically.

In his depression, Swift doubted whether he would ever complete the three 'Treatises', *Directions to Servants, Polite Conversation,* and *History of the Last Four Years* that had 'lain by me several years, & want nothing but correction'. Yet, as during the year, Faulkner's plans for an edition of his collected works took shape, his interest in completing his *œuvre* was reawakened.

In December 1734 he offered his opinions on the recent spate of student riots at Trinity College, which had culminated in the death of a Fellow who had been attempting to maintain order. 'I had the honour to be for some years a student at Oxford,'[164] he wrote, declaring that such disturbances were more effectively controlled by the Oxonian authorities. It is difficult to know how to treat such a preposterous assertion. His actual residence at Oxford, while taking his MA, had lasted for less than a month. We know that he was becoming increasingly forgetful, but there is

[161] *Corr.* iv. 280-1. 'I am sorry to hear your health is not better.'
[162] Ibid., p. 277.
[163] Ibid., p. 279.
[164] Ibid., p. 274.

also an implied boast of his Englishness, in this old man's harmless pretence of being an Oxford man.

After considerable delays, the first three volumes of the *Works* were published to subscribers in November, with the fourth promised for January. Swift continued to insist that the edition had been prepared 'utterly against my will', and that he would receive 'not a farthing' from it. He had always hoped that, after his death, a prestigious memorial edition would have been produced in London by Motte and Gilliver. It was another typical disappointment that the first edition of his works should appear in 'so obscure a place' as Dublin. 'I would as willingly have it done in Scotland,' he complained. But his claims that he 'never looked into' the edition were untrue. Since Faulkner could not be dissuaded from publication, Swift made a virtue of necessity and sought to ensure that the *Works* which bore his name should be as accurate–or at least as approved–as he could make them. All this while Motte had been under strict surveillance in London following his arrest for publishing the *'Epistle to a Lady'*. He dared not risk sending a letter to Swift, but eventually in July 1735 found means to communicate a letter by hand, in which he complained a great deal of Pilkington's 'unnecessary' behaviour under arrest, and also of Faulkner's edition of the *Works*. He was not at all pleased at Faulkner's attempt to scoop him. 'I was advised that it was in my power to have given him and his agents sufficent vexation by applying to the law; but that I could not sue him without bringing your name into a court of justice, which absolutely determined me to be passive.'[165] However, in recognition of this forbearance from litigation, Motte asked that Swift would 'lay your commands upon him . . . to forbear sending them over here'–that is, to England. Even in this, however, he conceded that 'if you think this request to be reasonable, I know you will comply with it; if not, I submit.' Swift's reply was evasive. It is true that, in preparing this edition, Faulkner 'did what I much dislike', but Faulkner had 'always behaved himself so decently to me that I cannot treat him otherwise than as a well-meaning man, although my desire was that those works should have been printed in London'. By now Faulkner's edition was already appearing in London, and Motte was seriously considering legal action. 'Upon the whole,' Swift

[165] *Corr.* iv. 372.

wrote, 'I think you had better suspend your suit . . . for books are not yet among prohibited goods.'[166]

Swift's relationship with Motte was complicated by the publisher's residual function as his agent and banker in England. Motte was in the habit of paying a regular annuity of £15 to Swift's sister Fenton. Reminded of this, tactfully, by Motte in October 1735, Swift replied with exasperation that he did not 'employ one thought upon her except to her disadvantage'.[167] Nevertheless, since Motte had been performing this duty for him, he demanded to know 'how far I am got in your debt, and I will discharge it as fast as I can get any money in, which is almost as impossible to find here as honesty'. The news of this debt did not improve his attitude to Motte, and when, the following year, Motte filed a bill in Chancery to prohibit Faulkner from exporting his edition to England, Swift moved to adopt Faulkners' side of the dispute. He wrote to Motte in May 1736 complaining of his 'ill treatment' of Faulkner. Motte's lawsuit now smacked rather too much of all the other legal restrictions that English traders forced upon their Irish rivals. 'The cruel oppressions of this kingdom by England are not to be borne,' he declared, adding, 'If I were a bookseller in this town, I would use all the safe means to reprint London books, and run them to any town in England that I could, because, whoever neither offends the laws of God, or the country he liveth in, commiteth no sin.'[168] This is just one further indication of Swift's gradual and grudging identification with the city of his birth, rather than with the city of false hopes and court promises. Nevertheless, he still told Pope in October that 'my flesh and bones are to be carried to Holyhead, for I will not lie in a country of slaves'.[169]

Swift's financial affairs continued to be complicated. For several years, in accordance with his own political priorities, he had sought to become a landowner, rather than merely a 'moneyed man', and early in 1735 he endeavoured to buy some land near Laracor. Declaring himself, somewhat disingenuously, as 'the most helpless man alive in such affairs,' he employed a friend to negotiate on his behalf with the vendor, whom he described as one of the 'cunning of mankind'. He turned out to be too cunning, however, and in the end, Swift withdrew from the deal, complaining of the

[166] Ibid., p. 414.
[167] Ibid., p. 411.
[168] Ibid., p. 493.
[169] Ibid., p. 408.

man's 'ill character'. He remained a moneyed man, with his funds invested in mortgages and loans. £1,500 which would have gone towards the purchase of land was lent out as a mortgage to Helsham's stepson. All this while Swift continued to plead poverty, however, since he insisted on living solely upon his income from rents and tithes, without touching his accumulated capital. 'My fortune is so sunk,' he complained in March, 'that I cannot afford half the necessaries or conveniences that I can still make a shift to provide myself with here.' Yet this parsimony, which led him to subsist on crusts and inferior wine, did not make him subscribe to the traditional vices of the money-lender. On the contrary, he continued to champion the tenant against the landlord, and in 1736 sent a stinging rebuke to Sir John Stanley on behalf of one of the baronet's tenants. 'You neither must nor shall act as an Irish racking squire,'[170] he thundered, with all the authority of Drapier and Dean. He could argue, of course, that his determination to preserve his capital was less a personal than a public concern, since he had already earmarked the money for charitable uses. His darling scheme, for a madhouse, first mentioned in the 'Verses on the Death of Dr Swift', took on a more definite shape at the beginning of the year. In the Dublin paper *Pue's Occurrences* for 18 January we read:

> Yesterday the City of Dublin made a grant of a piece of ground, viz., part of Oxmanton Green to the Rev. Dean Swift, whereon the Dean intends to build a convenient house at his own expense for the reception of lunatics.

In April Swift approached Eaton Stannard to be his 'director in the methods I ought to take for rendering my design effectual'. In July he told Orrery that he had made out his will, 'wherein I have settled my whole fortune on the city, in trust for building and maintaining an hospital for idiots and lunatics'.[171] Not only did he use his own wealth in such acts of public philanthropy, he also encouraged others to follow his example. Theophilus Bolton had once declared that he would 'zealously promote the good' of his country if ever it lay in his power. Now promoted to the archbishopric of Cashel, Swift reminded him of his promise. Bolton's reply sounds almost like Swift himself. Accepting the reproach, he protests:

[170] *Corr.* iv. 537. [171] Ibid., p. 367.

> . . . to tell you the truth I have for these four or five years past met with so much falsehood, treachery, baseness & ingratitude among mankind, that I can hardly think it incumbent upon any man to endeavour to do good to so perverse a generation.[172]

Despite Swift's constant advice and threats, Tom Sheridan remained as thriftless and improvident as ever. In 1735 he got into a mix-up over exchanging his living at Dunboyn for a schoolmastership at Cavan, and parted with his house before his appointment was confirmed. Swift was forced to intercede with the Lord Lieutenant to prevent Sheridan from becoming homeless. He also evidently lent Sheridan some money, and thereafter Sheridan's letters had a desperate and embarrassed quality whenever the matter of money arose. In July 1735 he wrote:

> The moment I can raise the devil among the tenants, I will secure your poor money. At present I have not a sous but a guinea and an half, till some bird of passage brings me some. You must know that I have lately been be-Sheridan'd. A damnable rogue, one William Sheridan, cousin to counsellor Sheridan, has run away three score and six pounds in my debt.[173]

Eight months later Sheridan was still chasing up his namesake, while hiding from duns himself. 'I pray God you may never feel a dun to the end of your life; for it is too shocking to an honest heart,' he wrote.[174] Throughout the rest of the year Sheridan's movements were furtive as he sought to evade his creditors. Swift regularly received begging letters, and treated each according to its merits. An actor, Tom Griffiths, threw himself on Swift's mercy to save himself from the 'determined cruelty' of his creditors in early 1736. Later the same year Mrs Barber, now severely affected with gout, begged permission to publish Swift's *Polite Conversation,* together with some of his uncollected poems, since 'everybody would gladly subscribe for anything Dr Swift wrote and indeed sir, I believe, in my conscience, it would be the making of me'.[175]

In April Sheridan wrote to Swift in sheer desperation at his wife's 'diabolical proceedings' in attempting to marry their daughter to a 'finical thorough fop' named Sheen. 'For heaven's

[172] Ibid., p. 330.
[173] Ibid., p. 357.
[174] Ibid., p. 463.
[175] Ibid., pp. 540–1.

sake . . . talk to the monster,' he pleaded, giving up any pretence of defending his spouse from his friend's censure. 'I have been linked to the devil for twenty-four years, with a coal in my heart, which was kindled in the first week I married her.'[176] He earnestly pleaded for 'this last friendship' but, despite the combined efforts of schoolmaster and Dean, Mrs Sheridan had her way, and the girl was married to John Sheen. Sheridan never forgave her. Swift, infuriated at Sheridan's impotence, insisted in a long 'ling' letter that he should treat his wife 'like an under*ling,* and stop her rail*ing,* rat*ling,* rang*ling,* behaviour. I would cure her ram*ling,* and rum*ling*; but you are spoi*ling* all, by wrigg*ling* into her favour, and are afraid of ruff*ling* her.[177]

This letter is typical of the frequent exchange of ingeniously facetious missives between them. Puns in English, 'Hibernian', or their own special 'anglo-latin' vocabulary kept alive the spirit of *la bagatelle* in their darkest moments. Sheridan would address Swift as De armis ter de an (Dear Mister Dean) or sometimes as DRDN (Dear Dean) and sign off 'Eye am ewers' (I am yours). Letters were headed from 'mice cool' (my school) and were written throughout in this strange schoolboyish language. Both men took delight in capping each other's efforts in these harmless sallies of wit. Orrery tried to join in the game, and in July he sent Swift a letter written entirely backwards. Throughout the summer Sheridan urged Swift to come and join him at Cavan, which, he assured him was an Arcadia compared with Quilca. 'You will wonder and be delighted when you see it.' Visiting Dublin briefly in August, he planned to bring Swift back with him, 'without ifs, ands, and ors'. Swift declined the invitation, but promised to come when the 'club'–his name for the Irish Parliament–met again in Dublin. 'For I am not able to live within the air of such animals.' Parliament met at the beginning of October, and Sheridan sent him careful instructions for travelling, reminding him to bring 'a cheese-toaster to do a mutton-chop now and then'.[178]

Swift left Dublin on 3 November, and arrived at Cavan three days later. There he found that Sheridan had considerably exaggerated the charms of the place. It was, he wrote back to Mrs Whiteway, 'the dirtiest town and . . . the dirtiest people I ever saw, particularly the mistress, daughter and servants of this house".

[176] *Corr.* iv. 314-15. [177] Ibid., pp. 346-8. [178] Ibid., p. 403.

However, in his half of their joint letter, Sheridan protested that Swift was exaggerating as usual.' I am tired of him, for I can never get him out of the dirt.'[179] These antiphonal epistles, with Swift grumbling and Sheridan apologizing became part of the ritual of Swift's two-month stay at Cavan. Mrs Whiteway entered into the spirit of the game, sending replies which addressed them both by turns. She was a cousin of Swift's, recently widowed, who had come to spend much of her time supervising the affairs of the Deanery. Increasingly, as Swift's memory and other faculties began to decline, it was Mrs Whiteway who cared for him. She had her own house in Abbey Street, north of the Liffey, but looked in at the Deanery most days to put Swift's mind at rest that anarchy had not broken out in his absence.

Unfortunately, on his way to Cavan Swift had barked his shin, which prevented him from enjoying his two favourite forms of exercise, walking or riding. Instead he was forced to sit indoors, with his leg propped up, a prisoner of 'the dirtiest people I ever saw'. His irritation at this helplessness made him more sour than ever. 'Our kitchen is a hundred yards from the house,' he complained. Food was always either raw or burnt; rooms were either intolerably stuffy or miserably damp. 'You know that he talks ironically,' Sheridan assured Mrs Whiteway, rather nervously.[180] She was extremely concerned for the Dean's health, and recommended a poultice of egg yolk and turpentine. After a fortnight of Swift's grumbles, however, the strain began to show, and Sheridan discovered, like Pope and Lady Acheson before him, what an awkward and demanding house-guest Swift could be. 'Pray write to the Dean to behave himself better to me,' Sheridan pleaded with Mrs Whiteway in a secret note towards the end of November. 'I want you to stand by me.'[181] The truth was that the injury to Swift's leg had spoiled his whole visit, 'for I intended to have been a constant rider'. Mrs Whiteway was sad that he should be away from Dublin on his birthday, which she and her family celebrated with a dinner of 'wild duck, plover, turkey and pullet', washed down with 'an ocean of punch'. Candles were lit, and bells rung in Swift's honour, while he was away in Cavan attempting to divert himself with the 'sublime amusement' of having one of the maids 'lugged' for her habit of leaving doors open. However, Swift's visit had not

[179] Ibid., pp. 426-7. [180] Ibid., p. 427. [181] Ibid., p. 432.

been entirely without festivities. Sheridan had done his best to honour his distinguished guest, and on one occasion eight local dignitaries had made a special journey to visit him. In return, Swift invited all the principal men in the town to sup with him at the 'best inn'. There were sixteen of them, he told Mrs Whiteway, 'and I came off rarely for about thirty shillings.'[182] The note of exultation at having spent so little is a characteristic gesture which Thomas Sheridan the younger – a schoolboy at the time – found very disillusioning. He cites this incident as an example that by this time Swift's avarice 'had taken possession of him'.

> He gave them a very shabby dinner at the inn, and called for the bill before the guests had got half enough wine. He disputed several articles, said there were two bottles of wine more charged than were used, flew into a violent passion, and abused his servants grossly for not keeping better account.[183]

By the beginning of December Swift was well enough to ride again, but there was now a light covering of frost on the ground which made riding 'like a life at court, very slippery'.[184] Sheridan set out for Dublin on the 11th to beg subscriptions for a schoolhouse. Swift promised to stay at Cavan until his host returned, but privately he confided to Mrs Whiteway that he intended to return 'two or three days after him'. It had not been a happy visit, and he was pleased to return to the Deanery. Sheridan took his friend's removal without rancour. 'I received your letter of reproaches with pleasure,' he remarked, buoyantly,[185] and promptly invited Swift for a return visit the following summer. He promised that his house was now transformed into a haven of domestic felicity and bountiful produce. The roads were good, the house clean and 'I have at present forty chickens, all fat, twenty sheep of my own, and sixteen lambs . . .'. But Swift was not persuaded, and contented himself with sending back a parody of Sheridan's catalogue of the joys of Cavan. A second visit would represent the triumph of hope over experience, something which it was not in Swift's character to venture.

Swift's contact with Delany was rather more limited now,

182 Ibid., p. 430.
183 Sheridan, p. 387.
184 *Corr.* iv. 446.
185 Ibid., p. 454.

since Delany's wife insisted on living almost entirely at Delville, some miles north of Dublin. Swift was unwilling to spend the 'two thirteens' (that is, two Irish shillings) that it cost for coach hire to visit him, 'by which I should be out of pocket nine pence when I dine with him'. He became increasingly convinced that Walpole's apparent stranglehold on the political constitution in England, represented the end of liberty in that country. It appeared to Swift that the slavery of Ireland and the tyranny of France were the models for the rest of Europe. He communicated his bleak vision of the final collapse of that careful balance of powers that he had described in his *Discourse of the Contests and Dissensions in Athens and Rome,* in a letter to Pulteney.

We see the Gothic system of limited monarchy is extinguished in all the nations of Europe. It is utterly extirpated in this wretched kingdom, and yours must be the next. Such hath ever been human nature, that a single man, without any superior advantages either of body or mind, but usually the direct contrary, is able to attack twenty millions and drag them voluntary at his chariot-wheels.[186]

To Pope he wrote that all things were 'tending towards absolute power, in both nations, (it is here in perfection already), although I shall not live to see it established'. 'It is very natural', he wrote again to Pulteney, 'for every king to desire unlimited power . . . but what puzzles me is, to know how a man of birth, title and fortune, can find his account in making himself and his posterity slaves.' Swift's vision of politics now was apocalyptic rather than satirical. 'I am as sick of the world as I am of age and disease, the last of which I am never wholly without. I live in a nation of slaves who sell themselves for nothing.' Letters between himself and Pope were an exquisite torture, despairing, bitter, but, thankfully, infrequent.

I never am a day without frequent terrors of a fit of giddiness; my head is never well, and I cannot walk after nightfall. My memory is going fast; my spirits are sunk nine parts in ten. You will find in this letter probably fifty blunders, mistakes not only literal and verbal, but half sentences either omitted or doubled.[187]

The deaths of friends were wounds that added to his loneliness and despair. He continued with a habit, grown deeper than

[186] Ibid., p. 303.

[187] Ibid., p. 333.

affectation, to look upon dead friends as capital investments that had sunk. In February 1735 Arbuthnot died. 'The death of Mr Gay & the doctor hath been terrible wounds near my heart. Their living would have been a great comfort to me, although I should never have seen them, like a sum of money in a bank, from which I should receive at least annual interest.'[188] Lady Masham and Lord Peterborough also died during the year, and, nearer home, Mrs Whiteway's son Theophilus. That loss affected Swift as deeply as any. 'I was born to a million of disappointments,' he wrote in an attempt to console her. 'I had set my heart very much upon that young man, but I find he has no business in so corrupt a world.'[189] Three months later he guaranteed to pay the £100 fee for her second son, John, to be entered as an apprentice surgeon.

Swift had few hopes or requests to make of his surviving English friends. He demanded only that Pulteney should deserve the title *Ultimus Britannorum* by preserving his defiant spirit of liberty while all about him were selling theirs. Of Pope, his final and frequently repeated wish, was more simple and specific– *Orna me.* He was delighted that the *Dunciad* was inscribed to him, but what he really wanted was a separate 'Epistle to Dean Swift'. Several times he asked Pope to write such a poem as a final tribute to their shared ideals and friendship. Pope's 'Epistle to Dr Arbuthnot' had appeared only weeks before the doctor's death, and one senses a kind of death-wish about Swift's repeated requests for such a memorial poem. 'I expect you shall perform your promise,' he wrote; 'it need not be very long . . .' Such an epistle might be the blessed break that would release him from the servitude of mortality. Pope did indeed plan such an epistle, but doubted his ability to perform the task with sufficient strength and dignity.

Both men shared a desire to supervise their own appearances before posterity. This meant not only producing their own official versions of their works, but also preventing any of their early indiscretions, squibs, libels, or letters, from falling into piratical hands. Pope, who carefully edited, revised, and reassigned his early correspondence with as few qualms as he revised his poems, was obsessed with this matter, and wrote

[188] Ibid., pp. 333-4.

[189] Ibid., p. 460.

several times to Swift for the return of his early letters. Swift assured him that he need have no fears, since there were 'strict orders in my will to burn every letter left behind me'. But that was not what Pope wanted. He wished to keep a record, but an adjusted record. 'Believe me,' he wrote to Swift in August 1736, 'great geniuses must and do esteem one another, and I question if any others can esteem or comprehend uncommon merit.' The 'great geniuses' he had in mind were himself, Swift, and Bolingbroke. 'You and Lord Bolingbroke are the only men to whom I write. . . . You are indeed almost the only men I know, who either can write in this age, or whose writing will reach the next: others are mere mortals.[190] Having engrossed a monopoly of genius and merit to this post-Scriblerian triumvirate, it is hardly surprising that Pope should have felt the need to edit and annotate their correspondence, lest mere mortals in later ages should have been blinded by the light.

Anything that happened to Swift now became public news, in both England and Ireland. On 9 September the *London Daily Post* reported that he had been riding along the strand in Dublin one day when Butler, a schoolmaster, shooting at some larks, frightened Swift's horse, almost causing him to be thrown. Bathurst wrote immediately with facetious plans for hounding the schoolmaster through the world for this 'brutal attempt upon the Drapier'. Swift's riding was more limited these days anyway. 'I cannot venture to be half a day's journey from Dublin, because there is no sufficient medium of flesh between my skin and bone.' His birthday in 1736 was marked by public rejoicing, the firing of guns, and the publication of congratulatory verses. Otherwise there were few diversions in Dublin. Sheridan reported a 'merry' hanging at which the condemned man entertained the crowd with a long and witty speech before making his exit from the world, and Orrery was pelted with meal and flour at a mayoral ceremony in Cork.

It was in 1736 that Swift published his last major poem, 'The Legion Club',[191] an attack upon the Irish Parliament. It was written in response to a move by the landowners in Parliament to abolish the tithe of pasturage. In April, Swift spoke of printing his poem in a threepenny book, but publication was delayed and the

[190] Ibid., pp. 526-7.

[191] *Poems,* iii. 827-39.

poem circulated in manuscript, gaining many additions and interpolations. In May Swift complained to Sheridan that other pens had now added several extra names to the thirteen members he had originally selected for attack. 'The Legion Club' is a howl of anger and anguish at the men who had betrayed the liberty and independence of Ireland.

> Let them, when they once get in
> Sell the nation for a pin;
> While they sit a picking straws
> Let them rave of making laws;
> While they never hold their tongue,
> Let them dabble in their dung . . .
> We may, while they strain their throats,
> Wipe our a——s with their votes.
> (ll. 47-52, 61-2)

In this poem, old enemies, such as Tighe and Bettesworth, wallow in the ordure with new, like Bingham, Dilke, and Clements, all of them 'toasting old Glorious in [their] piss'. Yet, if the poem seems crude, it is worth remembering that this body, effectually disabled by English placemen and absentee land-owners, did nothing to prevent the slide into constitutional impotence and economic devastation of the country it professed to represent. In the famines of 1726 and 1729 thousands of Irish families perished, or emigrated. In the great famine of 1739-41 it is generally reckoned that another 400,000 people, or a fifth of the entire population, died of starvation. As one reads through Swift's letters and pamphlets on the miseries of Ireland, one sometimes has the sense of an old man's monotonous self-righteousness and gloom. But the facts of the situation were, if anything, even worse than the pictures he gave. The appalling unconcern of the English Government and the Irish Parliament at these wretched conditions is part of the long legacy of hatred and mistrust that has poisoned Anglo-Irish relations. It is no exaggeration to say that Swift, as the Drapier, had done more for Irish self-respect than the Irish Parliament had done over several decades. 'The Legion Club' has the savagery of justified anger.

As he was completing this poem, there seemed a possibility that Swift might even take up his pen as the Drapier once more, to combat a new English monetary scheme. With the continuing shortage of small change in Ireland, silver coins were at a premium, and had begun to encroach upon the value of gold. In 1687 the value of an Irish guinea had been fixed, by proclamation, at twenty-four shillings, but by now its value had slipped to twenty-three. Archbishop Boulter sought to regularize the exchange rate by devaluing the guinea to £1 2*s* 9*d*. Swift was entirely opposed to this proposal, believing that the absentee placemen who administered the country would be the principal beneficiaries. 'Can there be a greater folly than to pave a bridge of gold . . . to support them in their luxury and vanity abroad, while hundreds of thousands are starving at home for want of employment?'[192]

He delivered his opinion to an assembly of merchants at the Guildhall in April, who had gathered to draw up a petition to the Lord Lieutenant against devaluation. 'He made a long speech for which he will be reckoned a Jacobite,' wrote Mrs Whiteway to Sheridan. 'God send hanging does not go round.' Despite Swift's opposition, the devaluation was proclaimed in September the following year. On the day of the proclamation, Swift met Boulter at the Lord Mayor's banquet, where he told him that 'had it not been for him, he [Boulter] would have been torn to pieces by the mob; and that if he had held up his finger he could make them do it that instant.' In a short verse that he composed on the occasion, he presented himself advising Boulter to warn Walpole:

> Go tell your friend Bob and the other great folk,
> That sinking the coin is a dangerous joke.
> The Irish dear joys have enough common sense
> To treat gold reduced like Wood's copper pence.[193]
> (ll. 13-16)

Swift caused a black flag to be hoisted from the steeple of St Patrick's and bade the bells toll a funeral peal all day to mark the proclamation. Boulter was so fearful of an attack by Swift's army of teagues that he had a military guard surround his house. The image of Swift the demagogue is not entirely a pleasant one, particularly when one recalls that his pose as a man of the people went hand-in-hand with an avowed contempt for the

[192] *PW* xiii. 119-20.
[193] *Poems,* iii. 843. See also pp. 841-2.

majority of mankind. He confessed to Pulteney at just this time that he hated the Irish people 'worse than toads'.

At this time too, he finalized proposals, which he had first jotted down a decade earlier, for a scheme to identify all the bona fide beggars of Dublin with badges.[194] Badges should be worn 'well sewn upon one of their shoulders, always visible, on pain of being whipped and turned out of town'. His argument was not that 'foreign beggars' should be left to starve, but that they should be compelled to remain in their own country parish where, he believed, there might well be more of the necessities of life than in the overcrowded city. It is not so much Swift's argument, but his tone of righteous anger, that a modern reader finds disquieting. His language is full of prejudice and contempt. 'To say the truth, there is not a more undeserving, vicious race of human kind than the bulk of those who are reduced to beggary, even in this beggarly country.' A beggar family is described as 'he and his trull and litter of brats'. Swift presents himself, on the one hand, as a man 'personally acquainted with a great number of street beggars' while on the other, he is roused to fury by their residual pretensions to dignity.

> They are too lazy to work; they are not afraid to steal, nor ashamed to beg, and yet are too proud to be seen with a badge.

His own reaction to this 'absurd insolence' and yahoo pride is both petty and petulant.

> I must confess, this absurd insolence hath so affected me, that for several years past, I have not disposed of one single farthing to a street beggar, nor intend to do so, until I see a better regulation. . . . For, if beggary be not able to beat out pride, it cannot deserve charity.[195]

It is a chilly and unforgiving note on which to end his political career.

VII

Swift's *Polite Conversation*, finally published in 1738, was the outcome of his lifelong habit of observing and collecting examples of linguistic abuse. Several of the items in the collection may well have been gathered while he was preparing *A Tale*

[194] *PW* xiii. 131-40. [195] Ibid., pp. 134-5.

of a Tub; many more were added as part of the Scriblerian programme for ridiculing false learning. In his *Letter to a Young Gentleman* (1720) he declared,

I have been curious enough to take a list of several hundred words in a sermon of a new beginner, which not one of his hearers among a hundred could possibly understand.[196]

However, it was the fashionable clichés of courtly small talk that particularly fascinated him. The intention of his *Polite Conversation*[197] was to banish such affectations, and to insist that simplicity, 'without which no human performance can arrive to any great perfection', is nowhere more essential than in conversation. To make his point, he dignifies these meaningless vacuities with the mock-order of an artistic form, converting his list of clichés into a little comedy of manners. However, most readers soon tire of inanity depicted at such remorseless length. An exchange such as this must rate as one of the sharper pieces of repartee in the work.

Mr Neverout: Here's a poor Miss, has not a word to throw at a dog. Come, a penny for your thought.
Miss Notable: It is not worth a farthing. I was thinking of you.

Directions to Servants,[198] the other 'treatise' with which Swift amused himself in his declining years, demonstrates the same mischievous delight in dignifying anarchy with its own Aristotelian system. Throughout Swift's life, the occasional glimpses that we have of his servants enrich our picture of his domestic life. His most famous servant was Patrick, who played a minor role in the *Journal to Stella* as a comic character, an Irish joke. Swift was forever threatening to 'turn him off', but actually put up with him for over two years. In one early reference, he noted, 'Patrick is drunk about three times a week, and I bear it; . . . but one of these days I will positively turn him off.'[199] Eighteen months later he repeated that Patrick was 'drunk every day, and I design to turn him off'. Patrick's hand sometimes shook so much from drinking that Swift was afraid to let him shave him. At Christmas 1711 he gave Patrick a half-crown 'on condition he

[196] *PW* ix. 65.
[197] Ibid., iv. 97-202.
[198] Ibid., xiii. 1-66.
[199] *Journal*, i. 28.

would be good'.[200] Inevitably Patrick came home roaring drunk. 'I never design to give him a groat more' was Swift's solemn memorandum. The blunders of Patrick in London, like the blunders of Wat at Holyhead, offer a slapstick counterpoint to his more serious preoccupations. On one occasion Patrick broke the master key to all six locks of Swift's chest of drawers. Another time he simply disappeared with the key to Swift's lodgings. When he finally returned, drunk, several hours later, Swift 'gave him two or three swinging cuffs on the ear, and I have strained the thumb of my left hand with pulling him'. However, despite all his faults, Patrick had one valuable skill–that of lying. Swift was soon boasting that Patrick was almost as good at denying his master to unwelcome callers as Harley's notorious porter Read. Indeed when Read fell ill in 1711, Swift not only wished for all his half-crown tips back, but also had a mind to recommend Patrick as a replacement. It was Patrick who warned Swift that the Mohocks were out to get him in March 1712.[201] Swift found the below-stairs world of Patrick, Wat, and Frances Harris a continual source of sympathetic amusement. Throughout his life, he was acutely conscious of status, and his own insecure position at the fringe of the court gave him an uncomfortable understanding of a dependant's position. A childhood of neglect, an early career spent in a dependent position at Moor Park, a life of suspense and disillusionment at court, all gave him a considerable insight into the psychology of the servant who seeks, by minor depredations, to wreak revenge upon his masters. Just as he found greater happiness with 'hedge-companions' whom he could patronize, than with courtiers who patronized him, so Swift's writings show a greater understanding of the mind of a footman, than of that of a chief minister. His most successful political writings are not his histories, in which he tries to unravel the skein of state events with the authority of an insider, but his satires, which demonstrate that ministers of state are merely footmen on stilts.

Directions to Servants is a handbook for domestic guerrilla warfare. No union rule-book could improve on its elaborate system of demarcations and perquisites. Servants in general are cautioned:

[200] Ibid., ii. 445.

[201] Ibid., p. 511.

When your master or lady call a servant by name, if that servant be not in the way, none of you are to answer, for then there will be no end of your drudgery: And masters themselves allow, that if a servant comes when he is called, it is sufficient.[202]

Among the advice offered to particular ranks of servants is this, to the cook:

Never send up a leg of a fowl at supper, while there is a cat or dog in the house, that can be accused for running away with it: But, if there happen to be neither, you must lay it upon the rats, or a strange greyhound.[203]

No work better illustrates the conflicting impulses between authoritarianism and libertarianism than this one. Swift's instinctive gift for creating anarchic scenes and his sense of the simple comedy of humanity combine with his equally enduring conviction that self-love rules everywhere. For the order which he seeks to subvert is his own. *Directions to Servants* is an anarchists' handbook compiled by the chief of police, and often, in the obsessive, fastidious details of its recommendations, one detects a darker tone.

To save time and trouble, cut your apples and onions with the same knife, for well-bred gentry love the taste of an onion in everything they eat.

Leave a pail of dirty water with the mop in it, a coal-box, a bottle, a broom, a chamber-pot, and such other unsightly things, either in a blind entry, or upon the darkest part of the back-stairs, that they may not be seen; and, if people break their shins by trampling on them, it is their own fault.

There is a note of despair behind the comedy, as Swift, who felt himself increasingly helpless and besieged, in the power of these incompetent and careless yahoos, indicates the bitter acuity with which he observed their crude attempts to cheat and deceive him. Whatever tricks they tried, he had foreseen them all. The *Directions* are ostensibly written by one who was formerly a footman himself, but the voice often has the sardonic tone of a long-suffering master. All Swift's familiar obsessions, with cleanliness, thrift, and the operations of the bowels, are in evidence and chamber-pots appear on almost every page.

[202] *PW* viii. 7. [203] Ibid., p. 28.

I am very much offended with those ladies, who are so proud and lazy, that they . . . keep an odious implement sometimes in the bed-chamber itself, or at least in a dark closet adjoining, which they make use of to ease their worst necessities . . . which maketh not only the chamber, but even their clothes offensive to all who come near. Now, to cure them of this odious practice, let me advise you, on whom this office lies, to convey away this utensil, that you will do it openly, down the great stairs, and in the presence of the footmen; and, if anybody knocks, to open the street-door, while you have the vessel filled in your hands; This, if anything can, will make your lady take the pains of evacuating her person in the proper place.[204]

Not for the first time, Swift is on both sides; he is the ingenious, revengeful servant and he is the fastidious, despairing master fighting a rearguard action against the yahoo hordes. It is worth noting that in the Deanery itself no such anarchy, organized or accidental, was tolerated. Swift's 'Laws' for his own servants, promulgated on 7 December 1733, are precise and punitive. They include such provisions as these:

If either of the two men-servants be drunk, he shall pay an English crown out of his wages for the said offence.

When the Dean is abroad, no servant, except the woman, shall presume to leave the house for above one half-hour; after which, for every half-hour's absence, he shall forfeit sixpence . . .

Whatever servant shall be taken in a manifest lie, shall forfeit one shilling out of his or her board-wages.[205]

He was a stern task-master, but Alex McGee and Mrs Ridgeway at least found the terms acceptable, and Sheridan tells of another servant, Blakely, who was in tears when Swift threatened to dismiss him. *Directions to Servants*, which Swift left unfinished at his death, is a fantasy version of the dream of order, a domestic dystopia, bearing a strong affinity to those many other works in which he counterfeited the voices of his antagonists, the moderns, the virtuosi, the free-thinkers, projectors, and fanatics.

Swift still wielded great power and influence in Dublin. His sovereign authority over the Cathedral and the liberty of St Patrick's was a symbol of the political independence for which

[204] Ibid., pp. 60-1. [205] Ibid., p. 161.

he had always striven, and allowed him to harry and pester higher authorities with virtual impunity. He boasted to Pope that, although the courtiers and gentry might despise him, the common people loved him. 'I walk the streets, and so do my lower friends, from whom, and from whom alone, I have a thousand hats and blessings.' In 1737 he remonstrated with the Irish Society for their scheme of 'improving' Coleraine, which involved quadrupling rents and driving some thousands of tenants from their land. His pressure eventually yielded results, and he was assured by Alderman Barber that the Society had 'resolved to relieve their tenants in Coleraine from their hard bargains'.[206] His political initiatives were not always so successful, and failures brought on depression and ill temper. When he was unable to obtain a small preferment for John Jackson, a cousin to the Grattans, it led to an angry final break with his oldest woman friend, Lady Betty, who had withstood the vagaries of his temper over many years. The letter which he sent her had been revised and corrected several times, and its ferocity came not from haste, but from deliberate ill humour.

> I now dismiss you, Madam, for ever from being a go-between upon any affair I might have with his Grace. I will never more trouble him either with my visits or applications. His business in this kingdom is to make himself easy; his lessons are all prescribed him from court, and he is sure at a very cheap rate to have a majority of most corrupt slaves and idiots at his devotion.[207]

Pope wrote once more, with some urgency, for the return of his letters 'to secure me against that rascal printer'. He began to fear that Swift's memory might be unequal to the task of finding and gathering them all together, but in fact Swift had already collected them into bundles, to be carried across to England by Orrery, who was also taking the manuscripts of Swift's *Polite Conversation* and his *History of the Four Last Years of the Queen.* As soon as the young Lord Oxford heard of Swift's plan to publish this *History* he wrote, with tender regard to his father's reputation, to insist that 'this history be not printed and published till I have had an opportunity of seeing it'.[208] But Swift was tired of deferring to the Harley family, and he replied,

[206] *Corr.* v. 50. [207] Ibid., p. 3. [208] Ibid., p. 27.

describing at some length the rift between Oxford (senior) and Bolingbroke. There was some bitterness in his tone when he observed:

> I had no obligation to [your father] on the score of preferment, having been driven to this wretched kingdom (to which I was almost a stranger) by his want of power to keep me in what I ought to call my own country; though I happened to be dropped here.[209]

Erasmus Lewis, under the guise of renewing an old friendship, also tried to dissuade him from publishing. 'Won't you let me see it before you send it to the press. . . . I may possibly contribute a mite.'[210] In July, Oxford wrote again, with more urgent pleas. 'At least defer this printing until you have had the advice of friends.' Swift, annoyed at all this pressure, delayed replying until he was sure that Orrery had safely delivered the manuscript to King in England. He then told Lewis that Orrery was under strict instructions not to let the manuscript out of his hands, although he might be prevailed on 'to let it be read to either of you, if it could be done without letting it out of his hands'. However, in June King told Mrs Whiteway that he had still not received the History, adding a little ironic postscript to the gentlemen of the post-office: 'When you have sufficiently perused this letter, I beg the favour of you to send it to the lady to whom it is directed.'[211] In August Lewis wrote again, and this time he was more candid in his objections. 'I am told you have treated some people's characters with a severity which the present times will not bear,' he wrote, adding, 'It is now too late to publish a pamphlet, and too early to publish a history.'[212] Swift was unmoved by these appeals, and when persuasion failed, other means were sought to suppress the work. For months he heard nothing from King, and in February, suspecting treachery, he asked Orrery to find out why. King wrote back immediately, expressing surprise that Swift had received no letters, since he had sent two, both explaining the difficulties he was under with the *History.* Evidently the gentlemen of the post office had taken their revenge on King for his little witticism. The fact was, as King explained to Deane Swift, that publication of the *History*

[209] Ibid., p. 46.
[210] Ibid., p. 56.
[211] Ibid., p. 53.
[212] Ibid., pp. 65-6.

was 'by no means agreeable to some of our great men, nor indeed to some of the Dean's particular friends in London'.[213] Somehow Oxford and Lewis had finally managed to see Swift's manuscript, and in April Lewis wrote to inform him 'that it was the unanimous opinion of the company, a great deal of the first part should be retrenched, and many things altered'. If these changes were not made, he feared that 'nothing could save the author's printer and publishers from some grievous punishment'.[214] This conspiracy among his 'friends' to thwart his wishes annoyed Swift greatly, but he was too old and tired to fight them any further. He endorsed Lewis's letter with a sardonic note, from 'mon ami prudent', and reluctantly gave up the struggle.

Meanwhile he heard that he had been offered the freedom of the City of Cork. He was not flattered. The Freedom was to be presented in a silver box. 'I will certainly sell it,' he complained, 'for not being gold.' When he finally received the box, he was even more dismayed. Not only was it not gold there was 'not so much as my name upon it'.[215] He returned it in disgust. The Mayor of Cork immediately wrote to apologize, agreeing that 'so great and deserving a patriot merits all distinctions that can be made', and he arranged to have the box inscribed. Swift was still not particularly thrilled at the honour. In his will he bequeathed the box to John Grattan as a receptacle for 'the tobacco he usually chewed, called pigtail'.

At Christmas 1737, an old blind actor, Michael Clancy, asked Helsham and John Grattan to show Swift the manuscript of his comedy *The Sharper*, in hopes that he might help to have it performed. Neither man cared to brave Swift with such a request, but Robert Grattan left the play on Swift's table one day when the Dean was out of the room. Swift read the play, liked it, and sent Clancy five pounds, but regretted that he had 'no interest with the people of the playhouse'.[216] The Pilkingtons escaped less lightly. He had been suspicious of their conduct ever since the affair of the 'Epistle to a Lady'. Now that Laetitia Pilkington had found her way into the divorce court, charged with adultery, he denounced the pair of them as 'the falsest

[213] *Corr.* v. 100.
[214] Ibid., p. 105.
[215] Ibid., pp.67-8.
[216] *Corr.* v. 81.

rogue, and the most profligate whore' in either kingdom.[217] But however much his heart was cankered against false friends and the great ones of the land, he could still show sympathy for poor suffering individuals. One morning, when he was out riding near Belcamp, he encountered a lame boy following a plough. Stopping to question him, he found that the boy's leg had been bitten by a dog three months before and had still not healed. He took pity on the boy, and gave him a note of admission to Dr Steevens's hospital.

In July he sought to increase the value of his endowment for an asylum by calling in his mortgage money and using it to buy 'a convenient estate'. His account books show that he had at least £7,500 out on loan at rates of interest that ranged between 5 and 6 per cent. Such a sum testifies to his thrift, when one recalls the precarious state of his finances only ten years earlier. However, he was unable to find anyone willing to sell him the kind of land that he sought, at the kind of price he had in mind, and his dream of dying a landed man remained unfulfilled. There is no doubt that he could have purchased land had he really wanted to, and his failure to do so probably indicates a residual reluctance to identify entirely with this island of slaves. Among his new friends at this time was Katharine Richardson of Summerseat, who took pleasure in sewing shirts for him. Miss Richardson has the distinction of being the last in the long line of admiring young women in Swift's life. He endorsed her gift of half a dozen shirts with a note, 'I accept her as my mistress because she did her duty in making the first advances.'[218] On his seventy-first birthday he sent her a diamond ring with a lock of his own hair. After which he got down on his knees, and read over the third chapter of Job.

Increasingly, Mrs Whiteway took over the management of Swift's affairs. In September 1738 he regretted that the poor woman 'is almost got into a consumption by bawling in my ears'.[219] She was the only one of his relations that he excepted from his general curse at the time. Deane Swift made him a present of 'gimcracks', a novel kind of shaving tackle. Not surprisingly, Swift was unimpressed. 'I cut my face once or twice,

[217] Ibid., p. 95. [218] Ibid., p. 85. [219] Ibid., p. 122.

was just twice as long in the performance and left twice as much hair behind.'[220]

In October 1738 Swift's last remaining true friend in Ireland, Tom Sheridan, died. Sadly there is evidence of a final rift between them in the last months of Sheridan's life. For years he had been a regular guest at the Deanery, where a room was set aside for his use. However, as his financial problems increased, his visits became more prolonged. Unfortunately one such visit coincided with one of Swift's severe bouts of giddiness and ill temper. After a while he let Sheridan know, through Mrs Whiteway, that his presence was no longer welcome. According to his son, Sheridan was 'quite thunderstruck . . . He immediately left the house . . . nor did he ever enter it again.'[221] Within a few weeks he was dead. In the 'Character' of Sheridan that Swift wrote when he heard of the schoolmaster's death, there is none of the gaiety or high spirits that had characterized their friendship over twenty years. Impartial to the last, Swift wrote like a stranger, and produced an obituary as cold and impersonal as any funeral stone.[222]

In 1739 William King again incurred Swift's wrath for tampering with the text of the 'Verses on the Death of Dr Swift'. King protested to Mrs Whiteway that all Swift's English friends agreed that the contentious, vainglorious, and ungrammatical couplets which he had removed from the poem, could only harm Swift's reputation. But Swift was infuriated at this new conspiracy to thwart him, and commissioned Faulkner to produce a Dublin edition of the whole poem. He had already made his own arrangements to control his reputation by burning, some years earlier, all his unpublished writings, except for a few loose papers which, Mrs Whiteway told Pope, 'shall never be made public without your approbation'.[223] There is a general air of suppression among all those who surrounded Swift in his final years. In October 1740 Mrs Whiteway and Orrery did all in their power to prohibit Faulkner from publishing a volume of *Letters to and from Dr Swift*, even though, as they suspected, Swift was probably responsible for leaking the letters to the publisher himself.

[220] Ibid., p. 125.
[221] Sheridan, p. 392.
[222] *PW* v. 216-18.
[223] *Corr.* v. 188.

At the beginning of 1740 we have the first mention of Dr Wilson in a note from Swift to Mrs Whiteway. The Reverend Francis Wilson plays an unenviable role in the story of Swift's declining years. A prebendary of the cathedral, he came to live in the Deanery, where his behaviour was soon the subject of criticism among Swift's friends. The lesser crimes alleged against Wilson were that he stole Swift's books, defrauded Swift of money, and used Deanery funds to make a personal display of charity. The gravest charge against him was that he threatened and assaulted Swift. According to one servant in the Deanery, Wilson forced Swift to drink a very great deal of wine on one occasion,

> which in a short time, did so intoxicate him, that he was not able to walk to the coach without being supported: and after all this, Wilson called at an ale house on his way to Dublin, and forced the poor Dean to swallow a dram of brandy. It was not long after when Wilson began to grow very noisy, and to curse and swear, and to abuse the Dean most horribly. . . . Whether he struck the Dean or not is uncertain, but one of the Dean's arms was observed, next morning to be black and blue. The noise of this bustle in the street, sudden as it was, drew a small handful of the common people together, who have since declared, that if they had known it was the Dean whom Wilson had abused, they would have torn the wretch to pieces.[224]

The evidence for this assault is slight, but the incident fits well into the atmosphere of Grand Guignol that surrounds Swift's final years as, little by little, he came to resemble one of his Struldbrugs. The text of Swift's last surviving note to Mrs Whiteway, dated 26 July 1740, is bleak in the extreme.

> I have been very miserable all night, and today extremely deaf and full of pain. I am so stupid and confounded, that I cannot express the mortification I am under both in body and mind. All I can say is, that I am not in torture; but I daily and hourly expect it. . . . I hardly understand one word I write. I am sure my days will be very few; few and miserable they must be.[225]

In August 1742, apparently at Wilson's request, a Commission of Lunacy was appointed to investigate Swift's condition. The commissioners decided that Swift

[224] Ibid., pp. 210–11.

[225] Ibid., p. 192.

hath for these nine months past been gradually failing in his memory and understanding, and of such unsound mind and memory that he is incapable of transacting any business, or managing, conducting, or taking care either of his estate or person.[226]

Three months later, Mrs Whiteway gave this harrowing description of Swift's melancholy condition.

He walked ten hours a day, would not eat or drink if his servant stayed in the room. His meat was served up ready cut, and sometimes it would lie an hour on the table before he would touch it, and then eat it walking. About six weeks ago, in one night's time, his left eye swelled as large as an egg ... and many large boils appeared under his arms and body. The torture he was in, is not to be described. Five persons could scarce hold him for a week, from tearing out his own eyes.[227]

Swift still had lucid intervals, when he could recognize Mrs Whiteway and speak to her. But they were becoming few and far between. She described one day when

he knew me perfectly well, took me by the hand, called me by my name, and showed the same pleasure as usual in seeing me. I asked him, if he would give me dinner? He said, to be sure, my old friend ...[228]

But in a day or two this respite was over, and his mind lapsed back into darkness. He lived on in this terrifying manner for a further three years. They were years of agony to himself, torture to his friends, and macabre fascination to visitors. In 1744 Deane Swift called on Swift and found him rocking himself back and forth in his chair, repeating over and over again, 'I am what I am, I am what I am', or more dismally still, 'I am a fool.' Among the many stories that have accumulated around Swift in these final humiliating years is one that his servants took money for admitting strangers to view him–the man of reason reduced to gibbering inanity. 'And Swift expires, a driveller and a show,' wrote Johnson, who shared with Swift a terror of such a final indignity. Most modern authorities cast doubt on this legend, however, which seems too neatly ironic an example of the teagues and yahoos taking their revenge on the man who

[226] See *The Correspondence of Jonathan Swift,* ed. F. Elrington Ball, 6 vols., 1910-14, vi. 182, 184.

[227] *Corr.* v. 207.

[228] Ibid., p. 207.

scourged them. It is surely ironic enough that Swift, whose last 'satiric touch' was to bequeath a lunatic asylum to Dublin, should have spent the last three years of his life under the care of guardians appointed by a Commission of Lunacy.

Swift died on 19 October 1745. According to the precise instructions which he had given in his revised will, three days elapsed before his burial. During that time crowds of people came to see him where he lay in an open coffin in his hall. He was bare-headed, with his hair 'long and thick behind, very white, and was like flax on the pillow'. But when his attendant was out of the room, his head was entirely stripped of hair by gruesome souvenir hunters. He was buried, according to his strict instructions, 'as privately as possible' at midnight, in the great aisle of the cathedral, on the south side, next to the monument to Narcissus Marsh. Seven feet above the ground he had decreed that a black marble tablet should be erected, into which the following epitaph, in Latin, should be 'deeply cut and strongly gilded'.

> Here lies the body of Jonathan Swift, Doctor of Divinity and Dean of this Cathedral Church, where savage indignation can no more lacerate his heart. Go, traveller, and imitate if you can one who strove with all his might to champion liberty.[229]

It is a proud epitaph, the last of his long line of self-assessments. It stresses a Roman courage rather than a Christian humility. Daring rather than hope, suffering rather than faith combine to challenge the rest of mankind to match his courage. He is, as always, silent on the prospects of another world.

Most characteristically, Swift's major satires confront us with the image of a monster in a maze. He leads us, with the insinuating charms of his plausible rhetoric, through a maze of conflicting definitions, only to abandon us before some terrifying image of ourselves as baby-eating politicians, or dung-throwing Yahoos. But the most enduring monster that he

[229] Swift's Will, in which he finally decided to forsake his dream of burial at Holyhead, was made on the third of May, 1740. It can be found in *PW* xiii, 149-58. The Latin text of the Epitaph reads: HIC DEPOSITUM EST CORPUS JONATHAN SWIFT, S.T.D. HUJUS ECCLESIAE CATHEDRALIS DECANI, UBI SAEVA INDIGNATIO ULTERIUS COR LACERARE NEQUIT, ABI VIATOR, ET IMITARE, SI POTERIS, STRENUUM PRO VIRILI LIBERTATIS VINDICATOREM.

created to provoke and vex us was not Celia or Strephon, or a Struldbrug or a Yahoo, but himself. He left us with the carefully cultivated image of a lonely misanthrope, chiselling his savage indignation on his tombstone, and leaving, as his benefactions to mankind, a privy and a madhouse. Only by acknowledging the ineradicable self-interest that makes each human being his own tempter, tormentor, and judge, can we face up to the challenge of Swift's ironies, and recognize the essential honesty and humanity that made him prefer to seem a monster rather than a hypocrite.

INDEX

In general, works are listed under the name of the author. Peers are listed under their best-known title.